Foulards

A Picture Book of Prints for Men's Wear

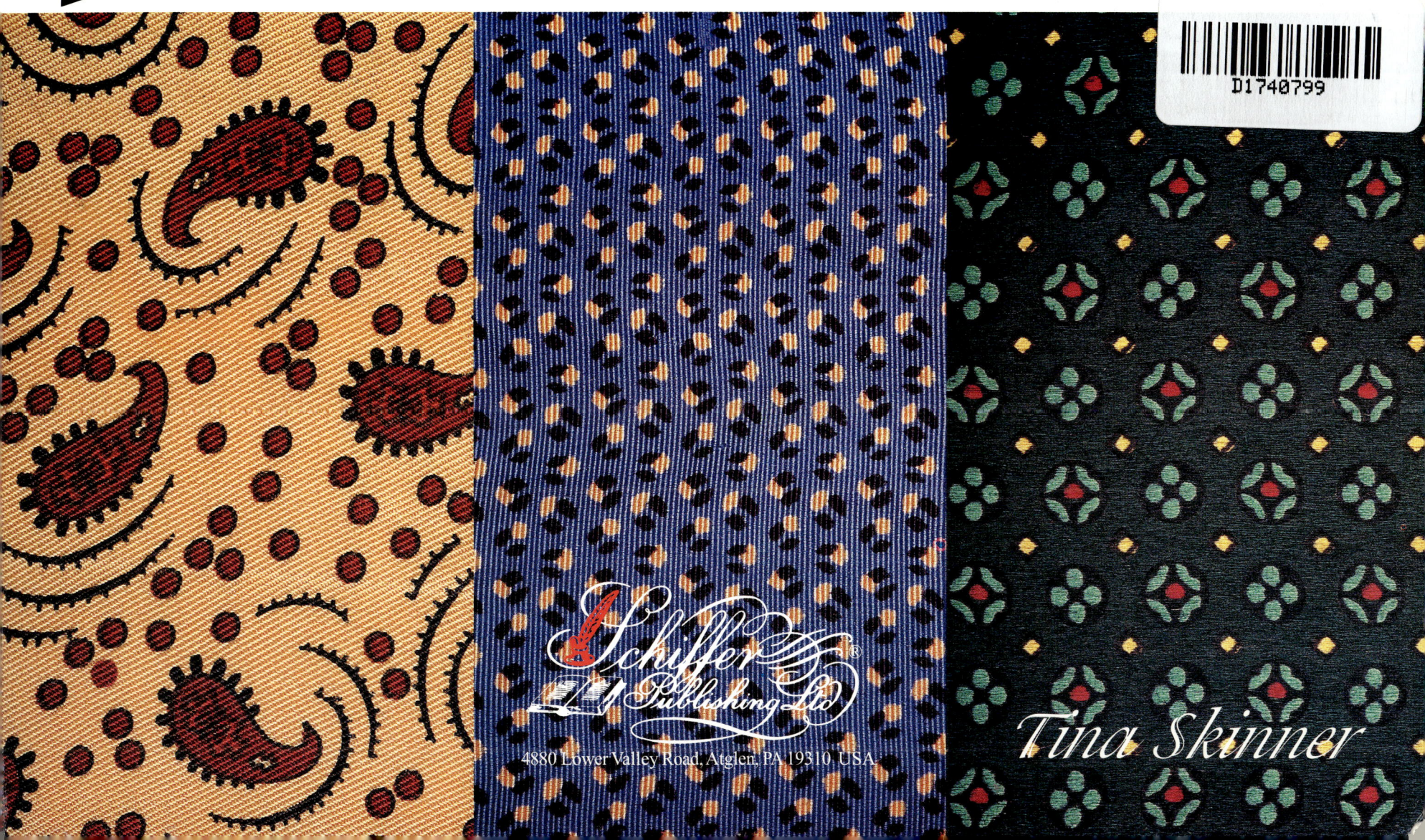

Schiffer Publishing Ltd

4880 Lower Valley Road, Atglen, PA 19310 USA

Tina Skinner

D1740799

Acknowledgments

To Tammy Ward, once again, for invaluable assistance in the photo studio.

Copyright © 2001 by Schiffer Publishing, Ltd.
Library of Congress Card Number: 00-105399

All rights reserved. No part of this work may be reproduced or used in any form or by any means—graphic, electronic, or mechanical, including photocopying or information storage and retrieval systems—without written permission from the copyright holder.

"Schiffer," "Schiffer Publishing Ltd. & Design," and the "Design of pen and ink well" are registered trademarks of Schiffer Publishing Ltd.

Designed by Anne Davidsen
Type set in Isadora /Zurich

ISBN: 0-7643-1256-1
Printed in China

Published by Schiffer Publishing Ltd.
4880 Lower Valley Road
Atglen, PA 19310
Phone: (610) 593-1777; Fax: (610) 593-2002
E-mail: Schifferbk@aol.com
Please visit our web site catalog at

www.schifferbooks.com

We are always looking for people to write books on new and related subjects. If you have an idea for a book, please contact us at the above address.

This book may be purchased from the publisher.
Include $3.95 for shipping.
Please try your bookstore first.
You may write for a free catalog.

In Europe, Schiffer books are distributed by
Bushwood Books
6 Marksbury Ave.
Kew Gardens
Surrey TW9 4JF England
Phone: 44 (0) 20 8392-8585; Fax: 44 (0) 20 8392-9876
E-mail: Bushwd@aol.com
Free postage in the UK. Europe: air mail at cost.

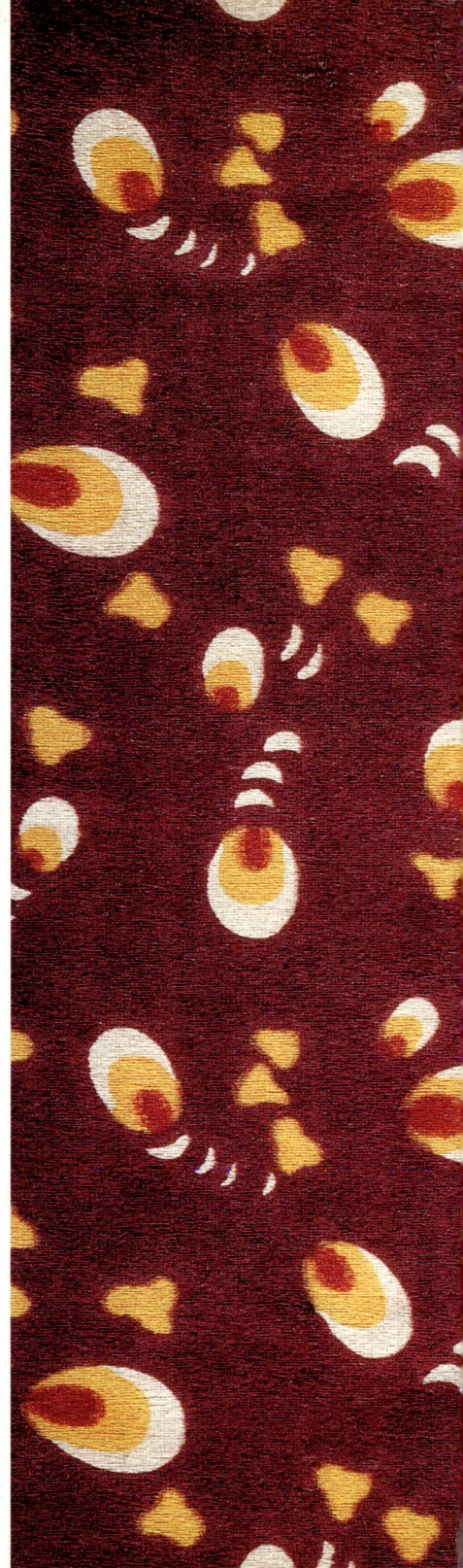

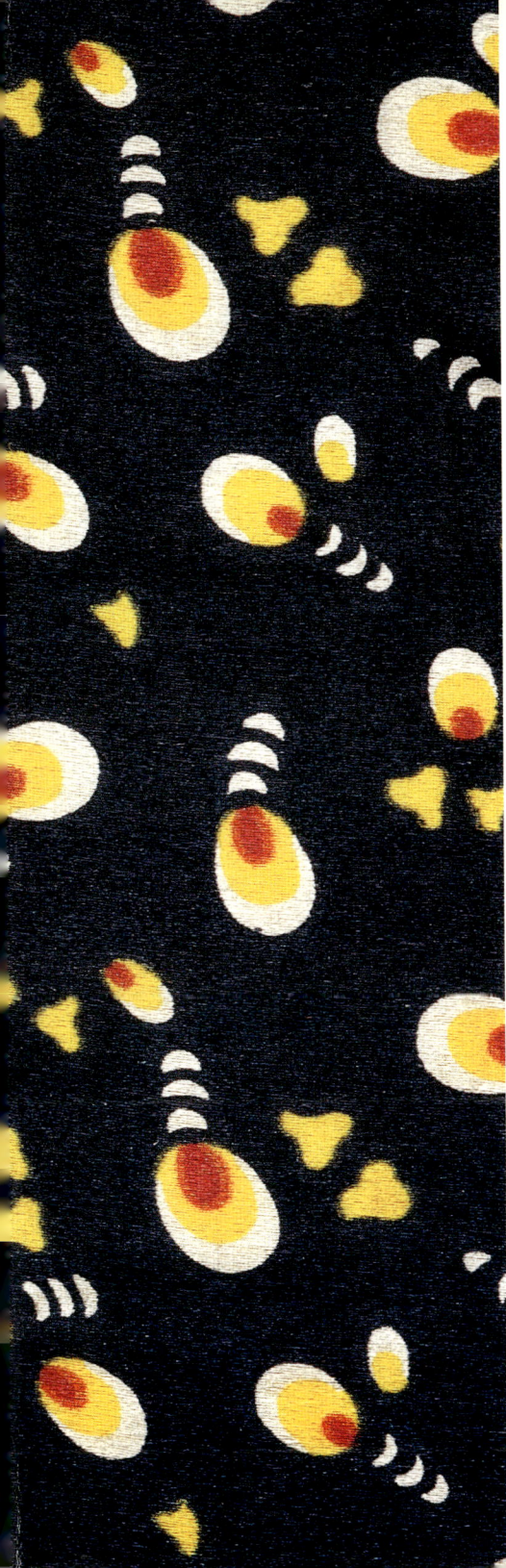

Contents

Introduction

"Foulard" is a term that first defined a fabric, later it defined a print style.

Simply, "foulard" originally referred to lustrous, fine silk fabrics that were primarily used for delicate handkerchiefs. These were usually block-printed with small, repeating patterns.

Through time, the fabric itself evolved from being inclusive of expensive silks to include more affordable imitations of acetate, rayon, and other manufactured fibers. Foulards are primarily a twill weave that is dyed or printed.

Early foulard patterns were printed in dark reds, blues, and greens, though the colors have fluctuated with changing fashions—as have the size and subject of the small geometrically-spaced print motifs, or medallions, and the width of the ties they adorn.

Foulards, as they are generically called, became the textiles of choice for men's neckties, dressing gowns, and prints: the preferred pattern when men weren't choosing conservative solids.

The foulard examples pictured here were drawn from sample books published in the 1960s and '70s. This is an excellent era in which to study foulards, as it was a period of history when people took pride in donning brighter colors and when designers were stretching creative boundaries.

The fabric swatches shown vary in size from roughly 2 x 3 inches to 3 x 4 inches. Though most of the patterns came with upward of eight color variations, in most cases I have shown only one color variation for each pattern. These variations offer a valuable lesson in how flexible such tiny patterns can be.

The foulards presented here were organized on several grounds. First, I wanted to show the impact of placement, be it on a straight, gridiron pattern, randomly scattered, or in alternating sequences. The final chapters are arranged by design characteristics of the medallions.

In each chapter, I have tried to present a design most exemplary of traditional foulard patterns, though this is impossible in such chapters as Space Age or Conversational. Yet it is the exceptions that prove most interesting. I'm sure you will enjoy them.

The book finishes up with paisleys, the traditional exception to the traditional rule that foulard patterns adhere to fairly strict geometric patterns. For more on paisley, I recommend my book *Paisley: A Visual Survey of Pattern and Color Variations*.

It is hoped that the foulards pictured here will prove fodder for further creation. I hope you enjoy the colors and the creativity that went into making these prints.

Gridlock

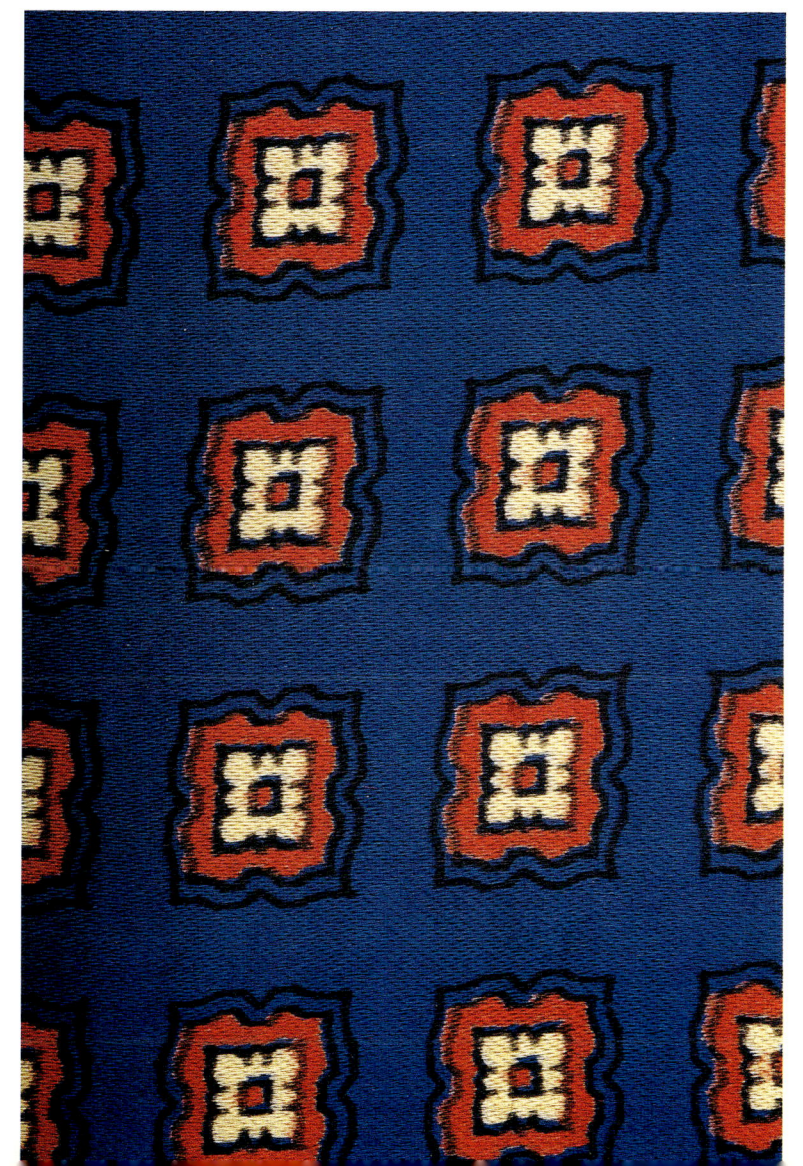

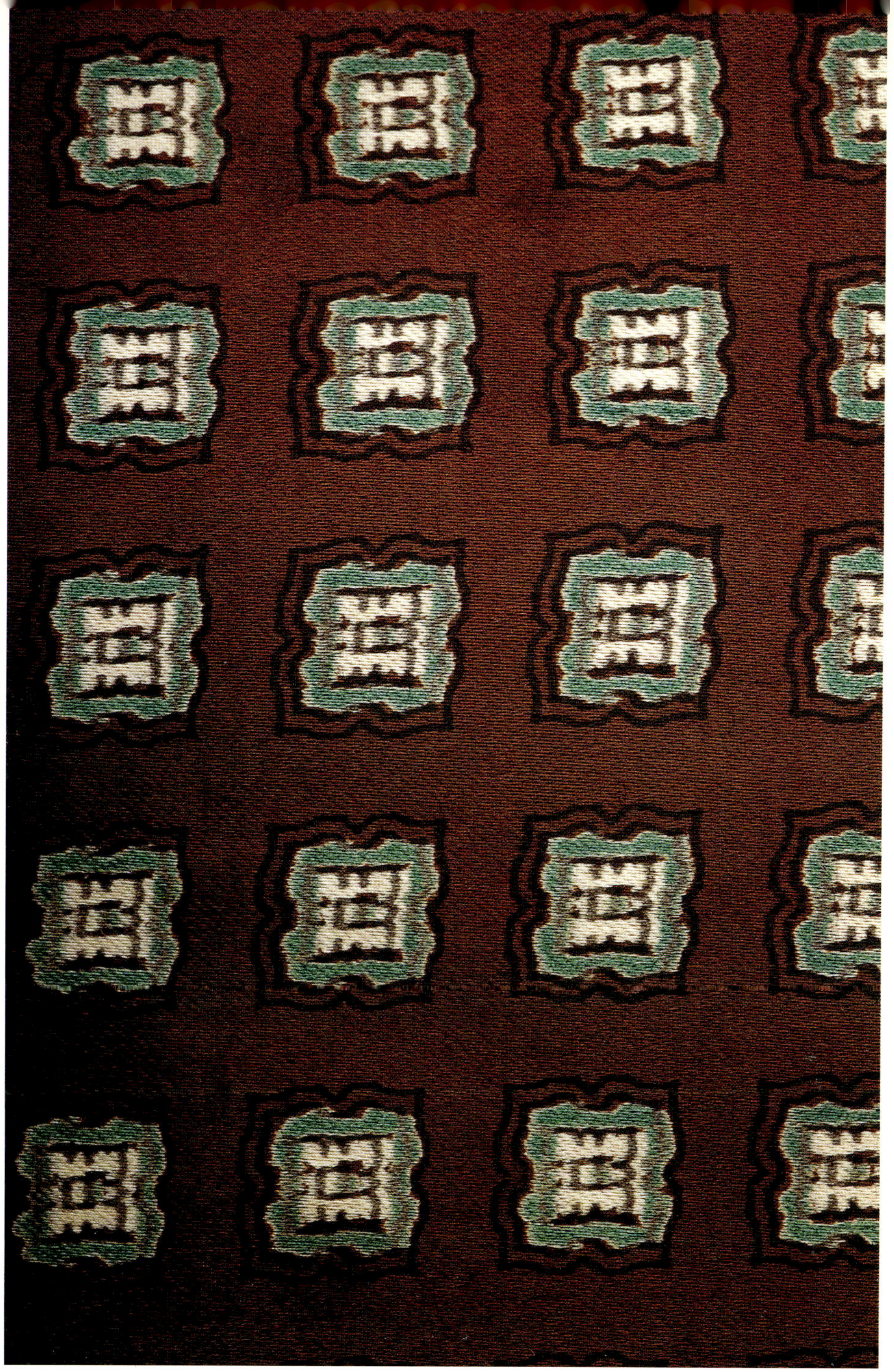

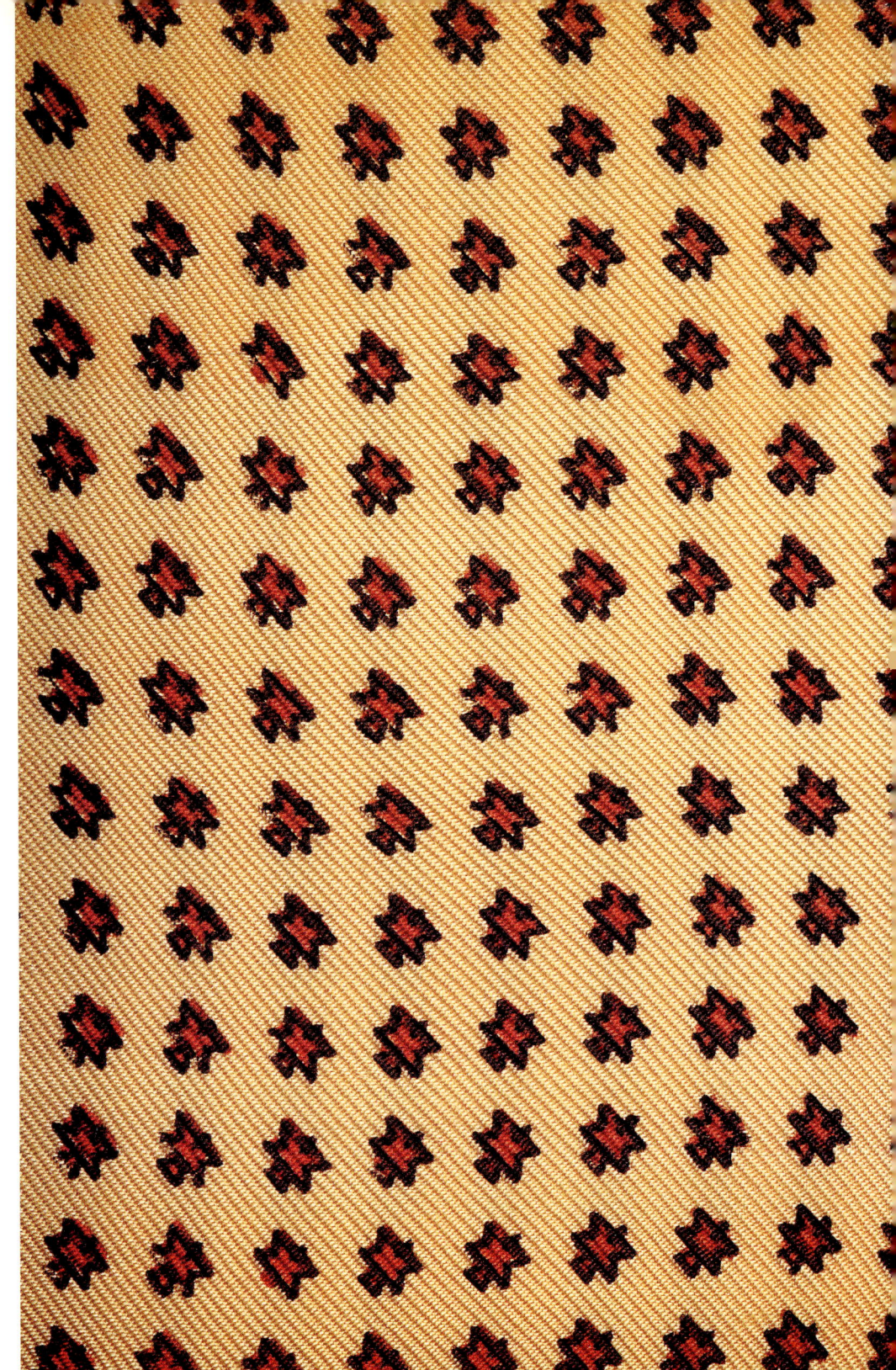

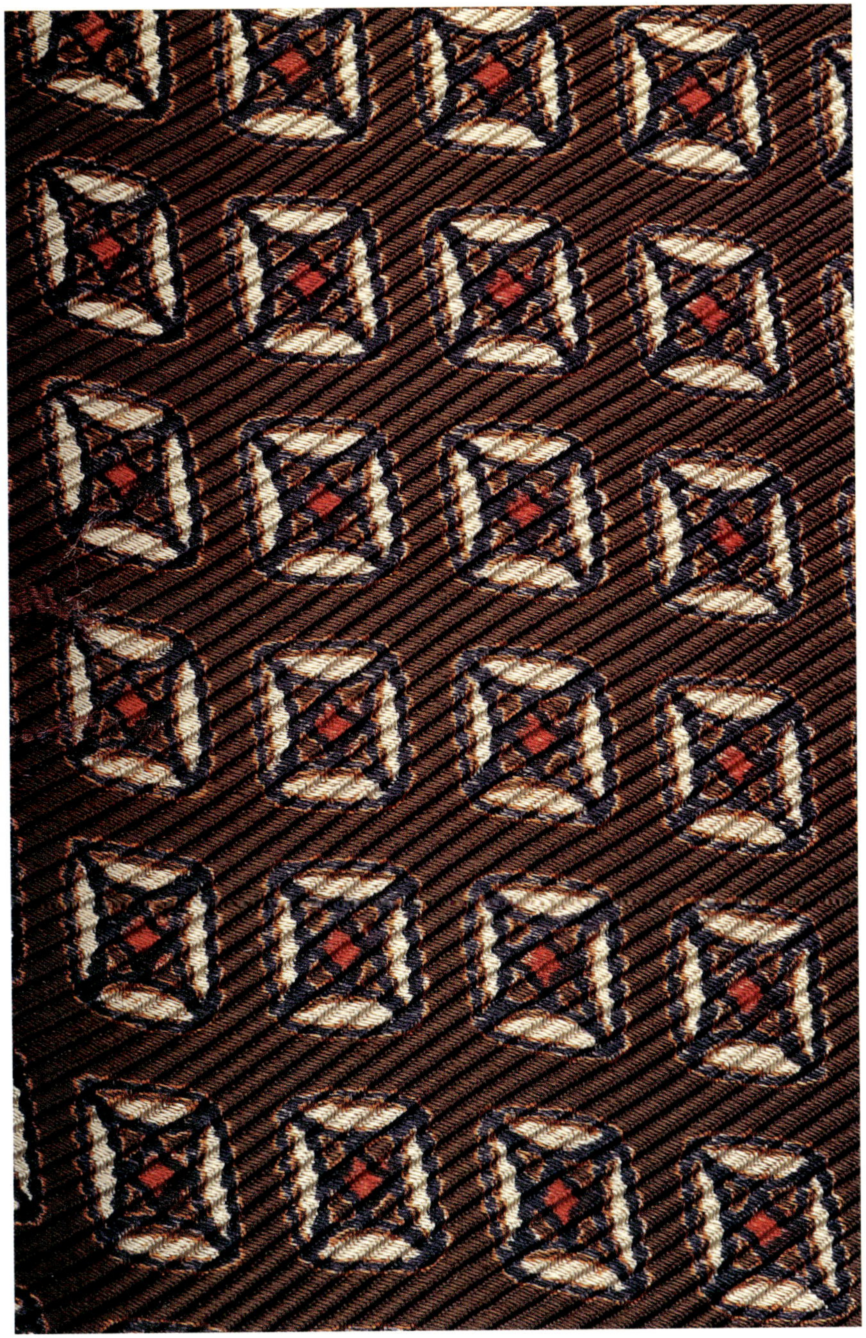

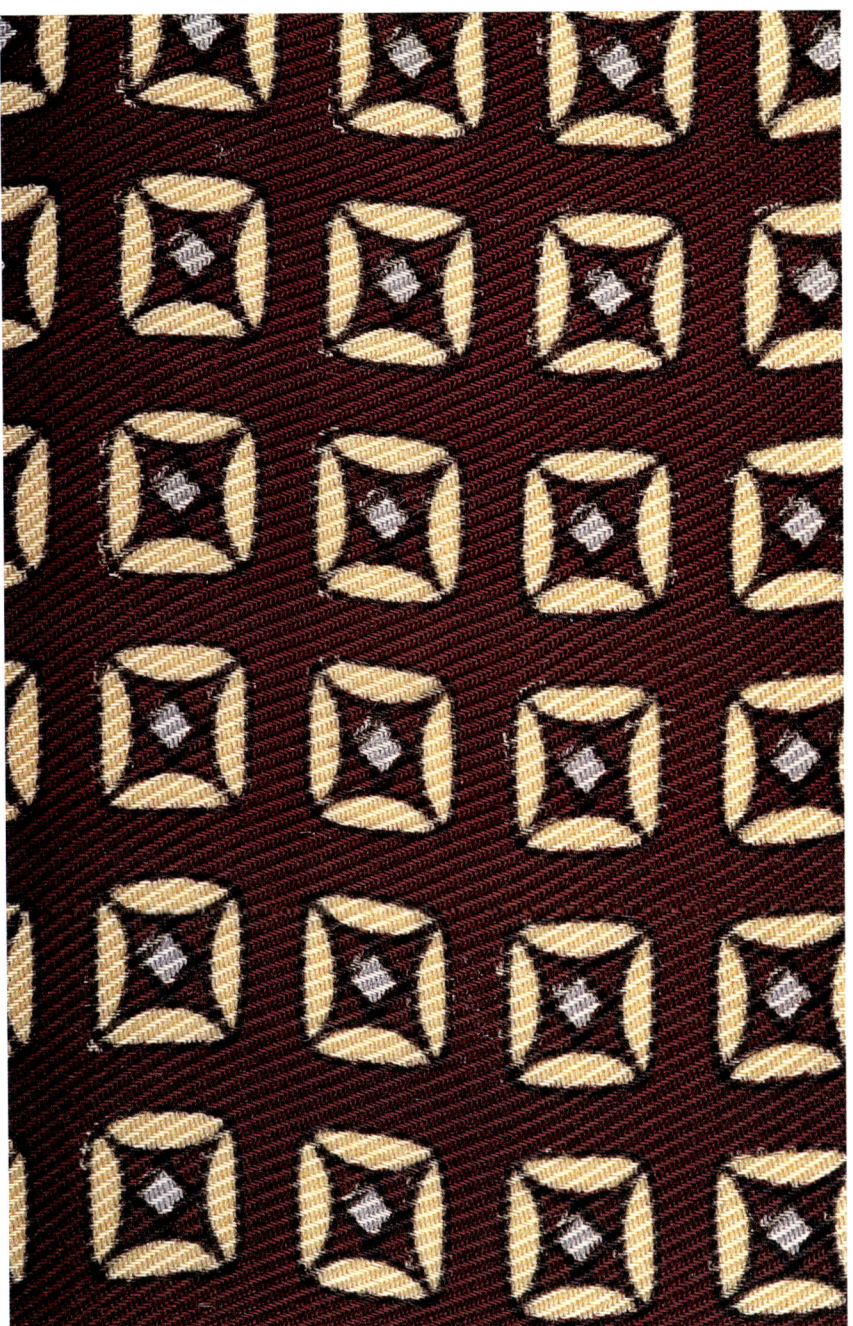

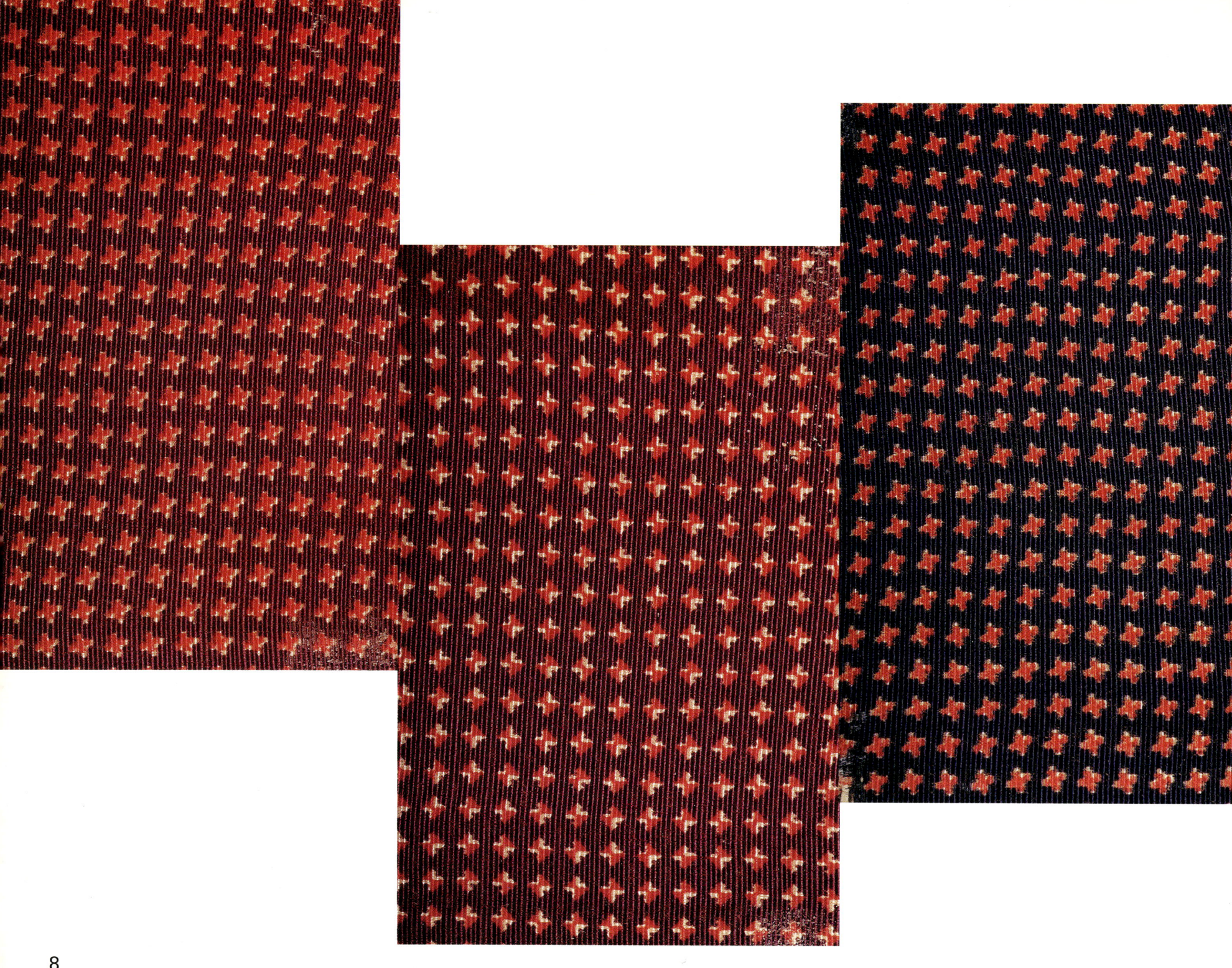

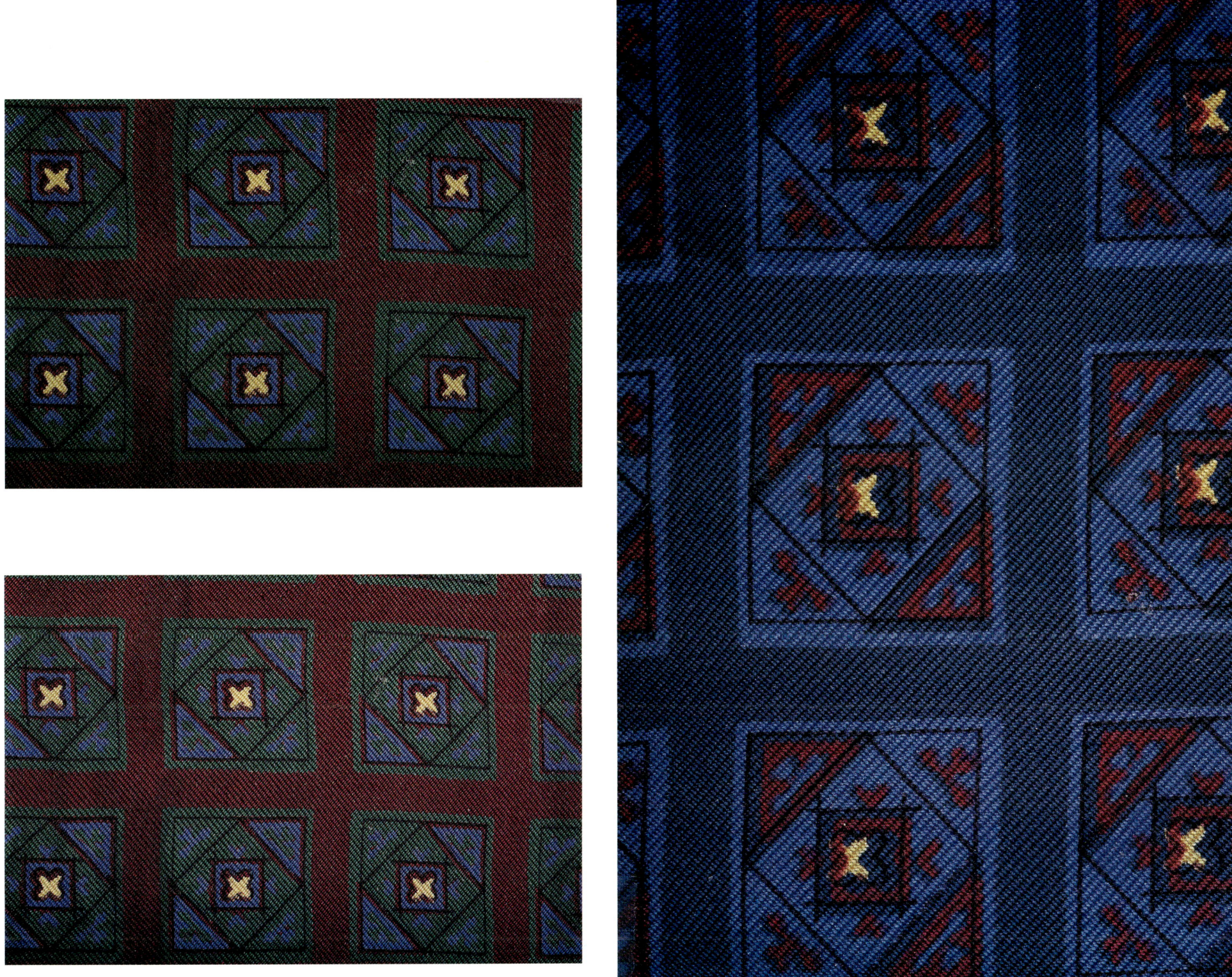

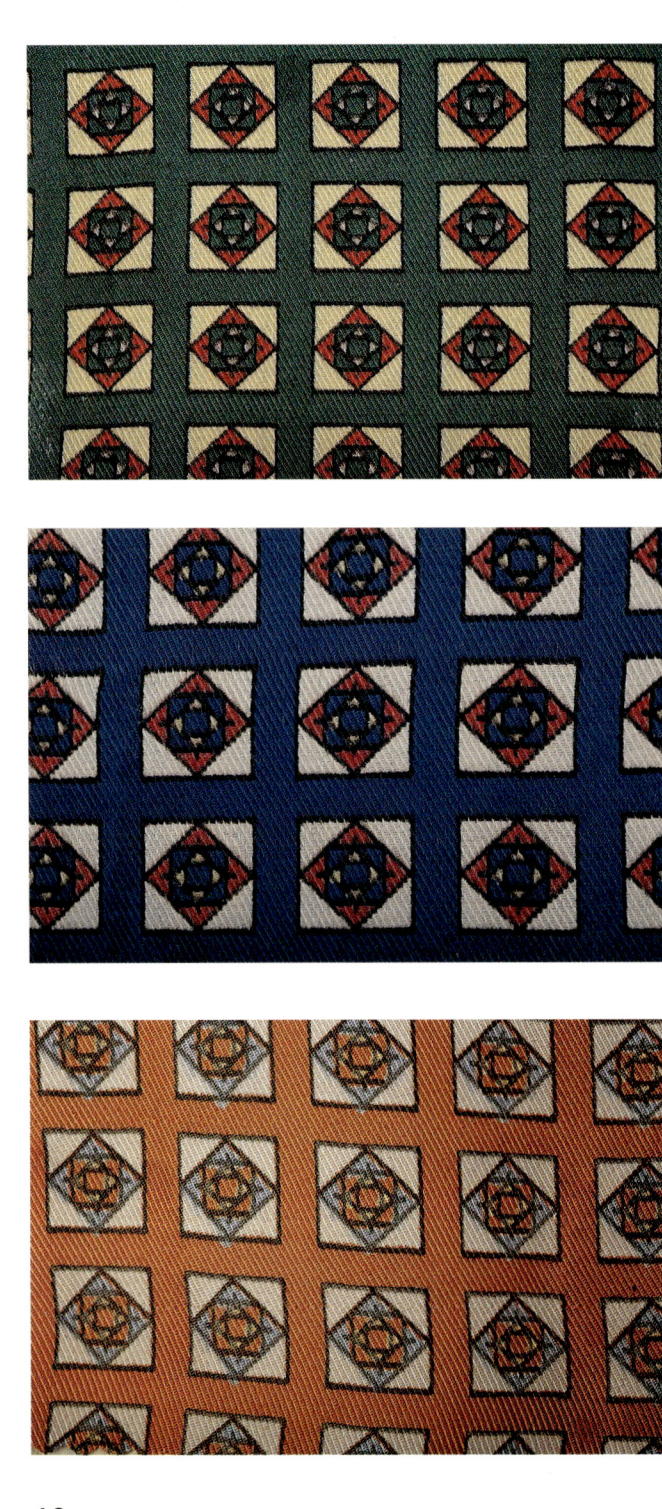

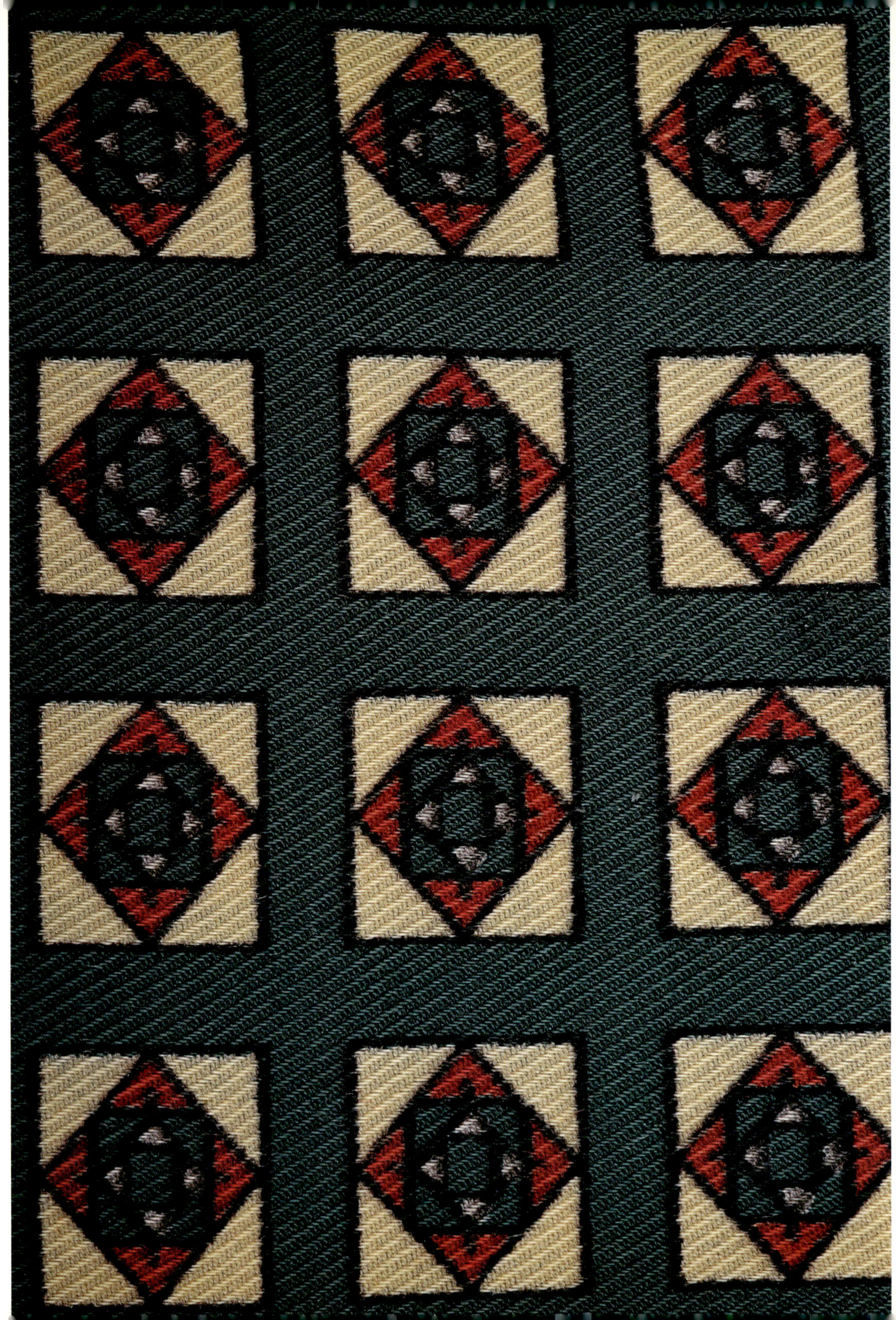

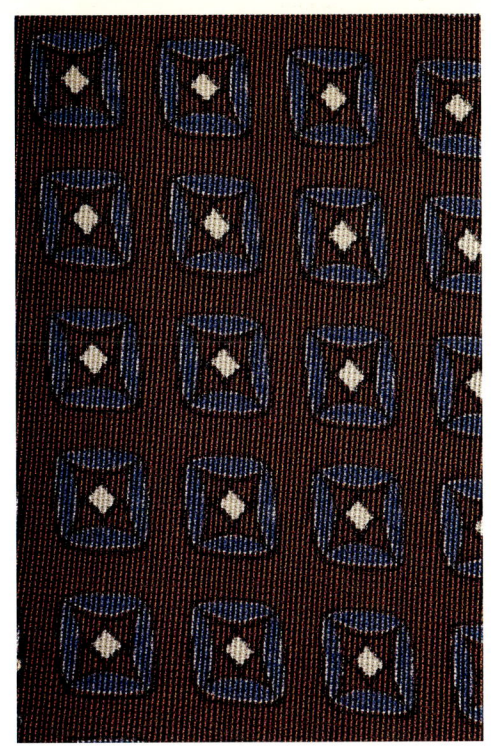

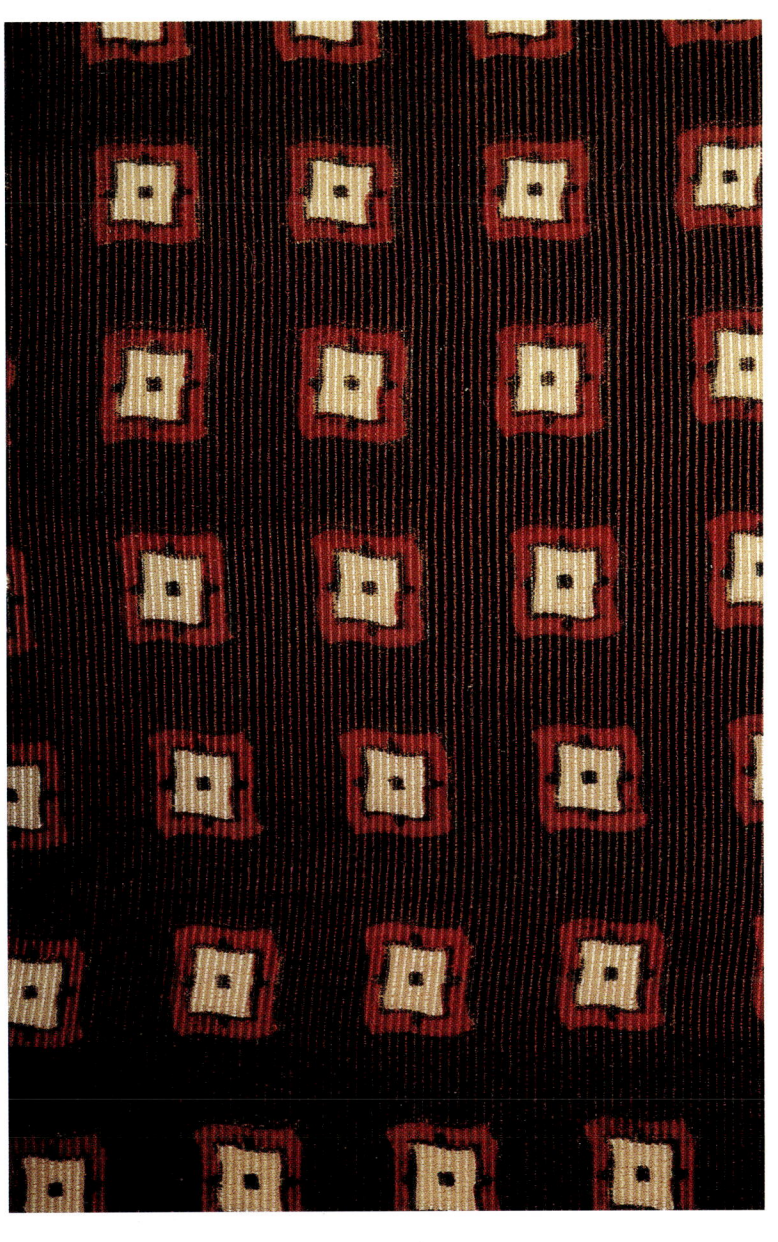

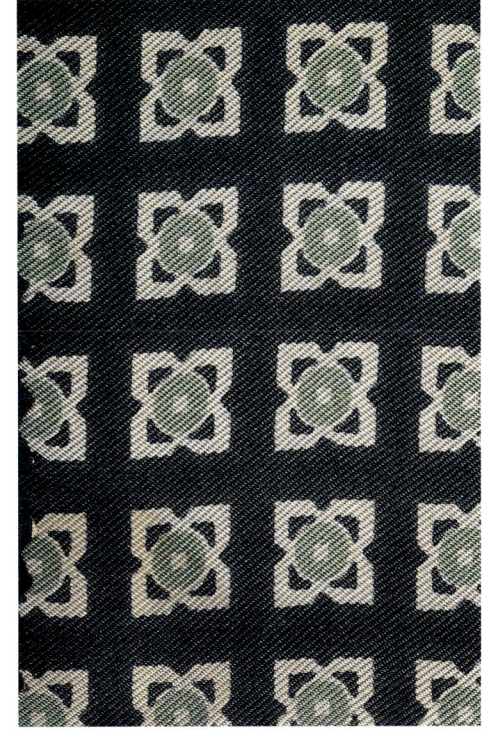

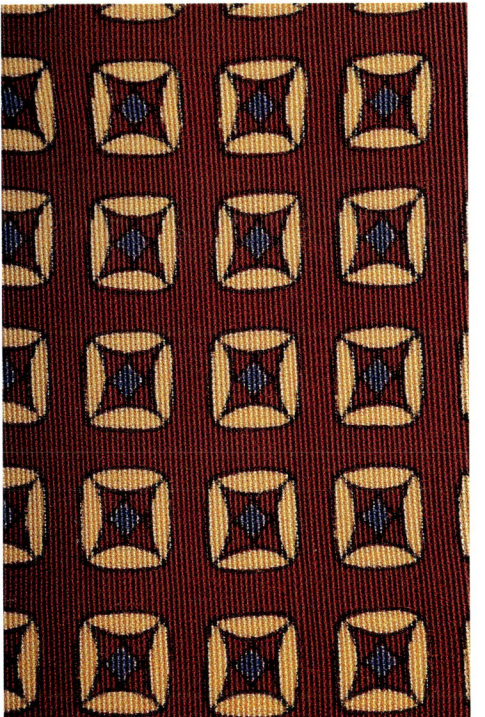

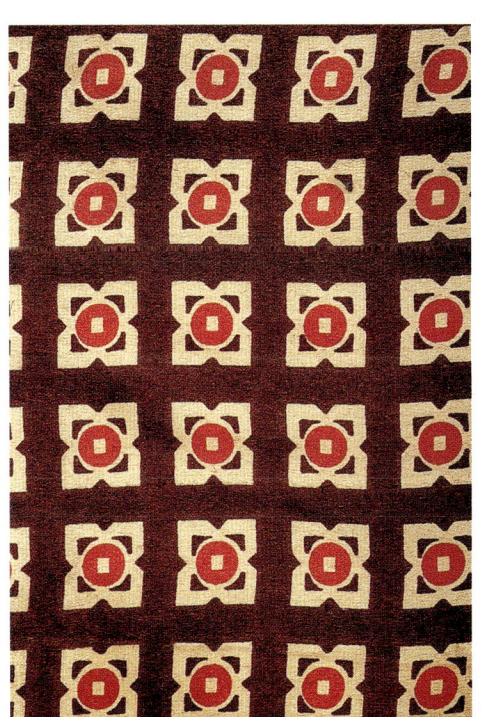

11

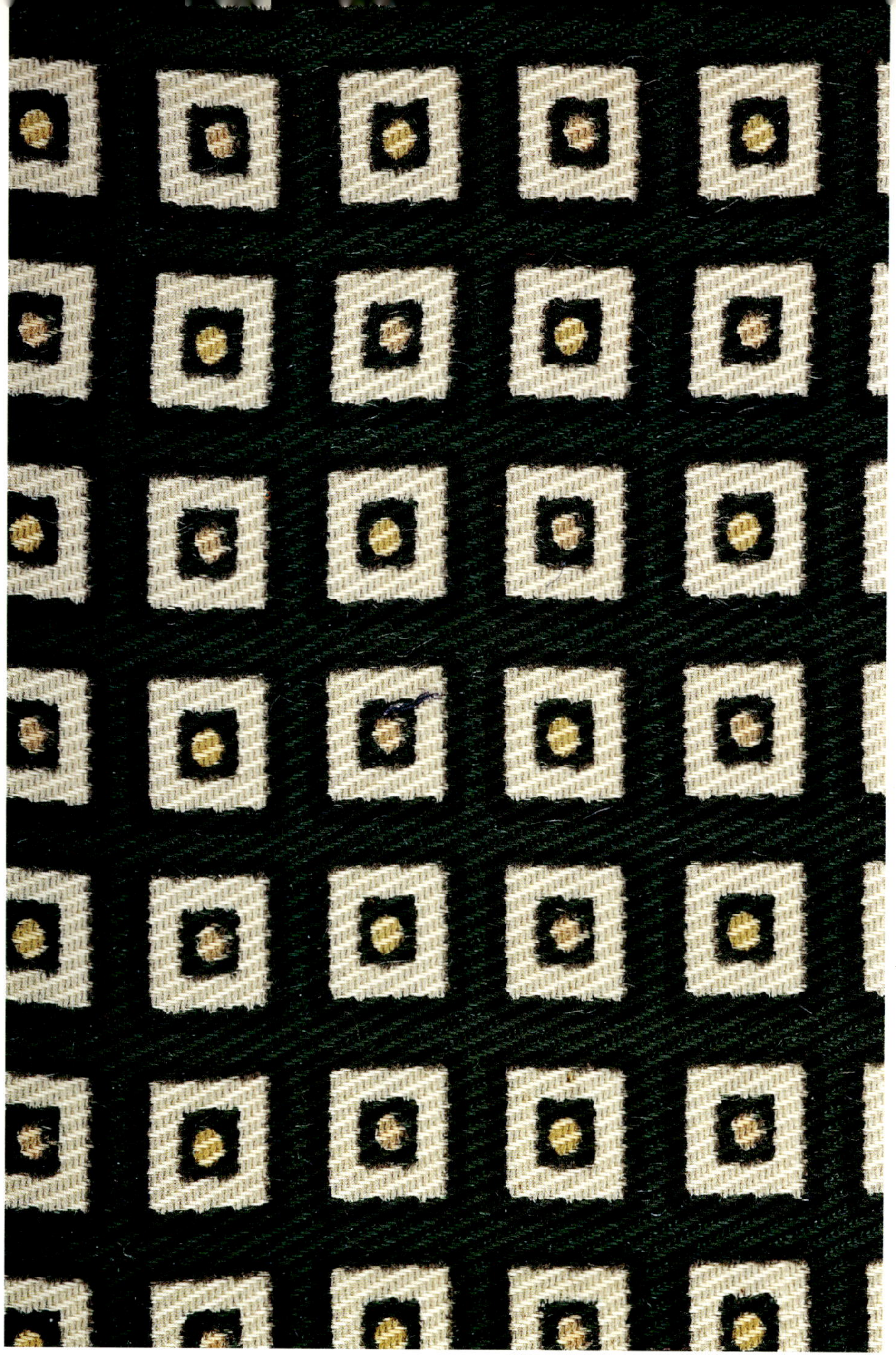

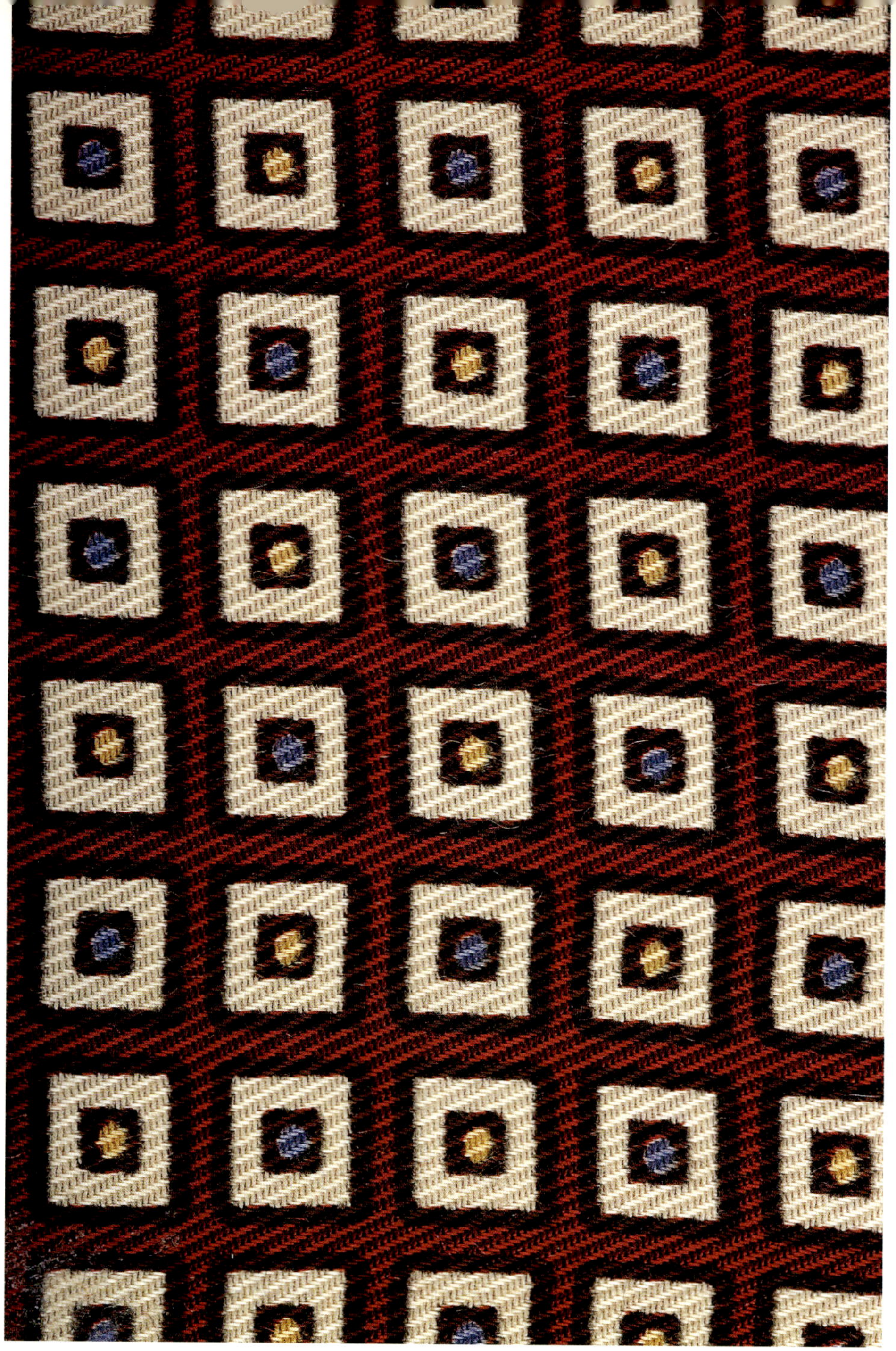

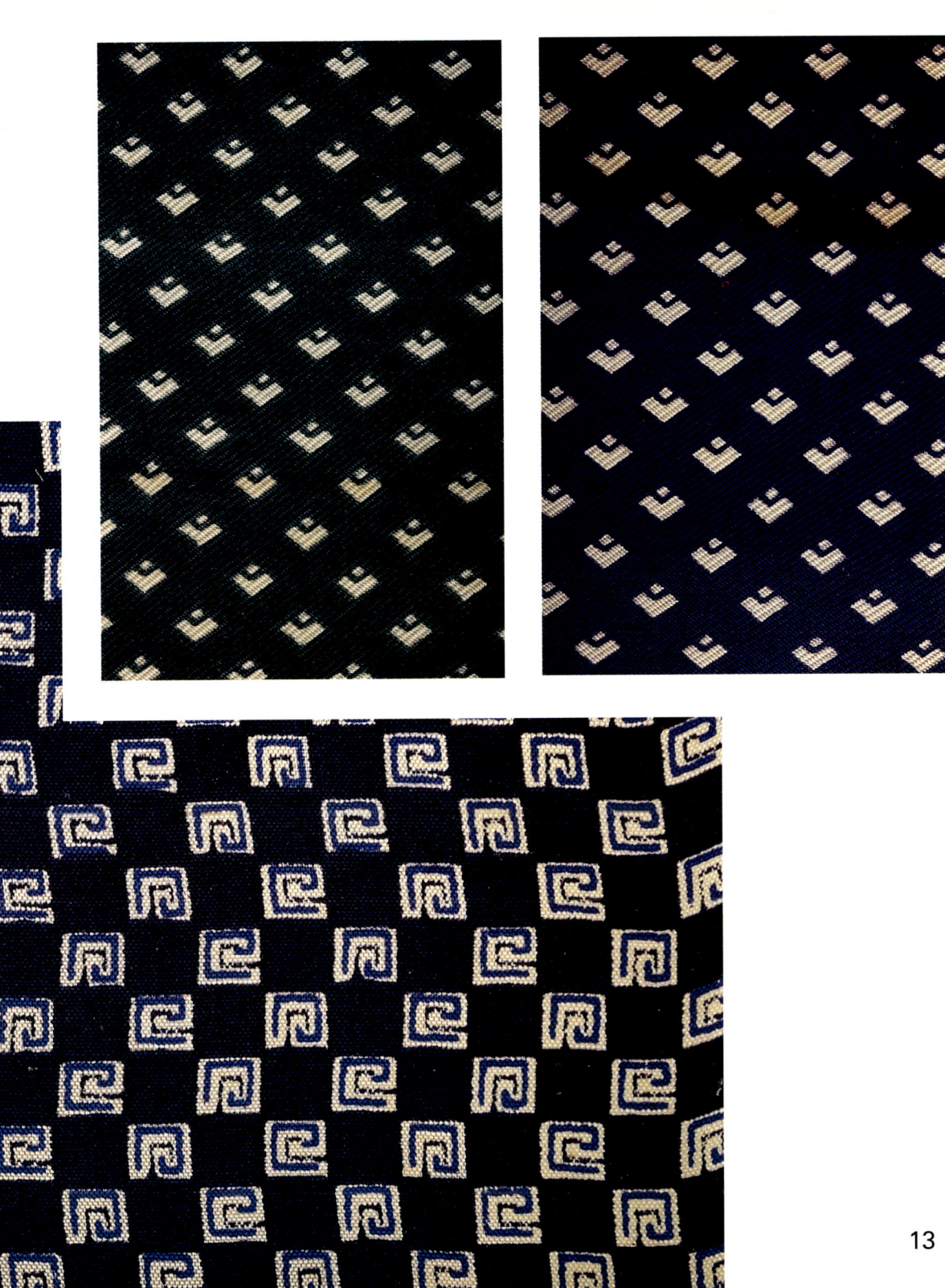

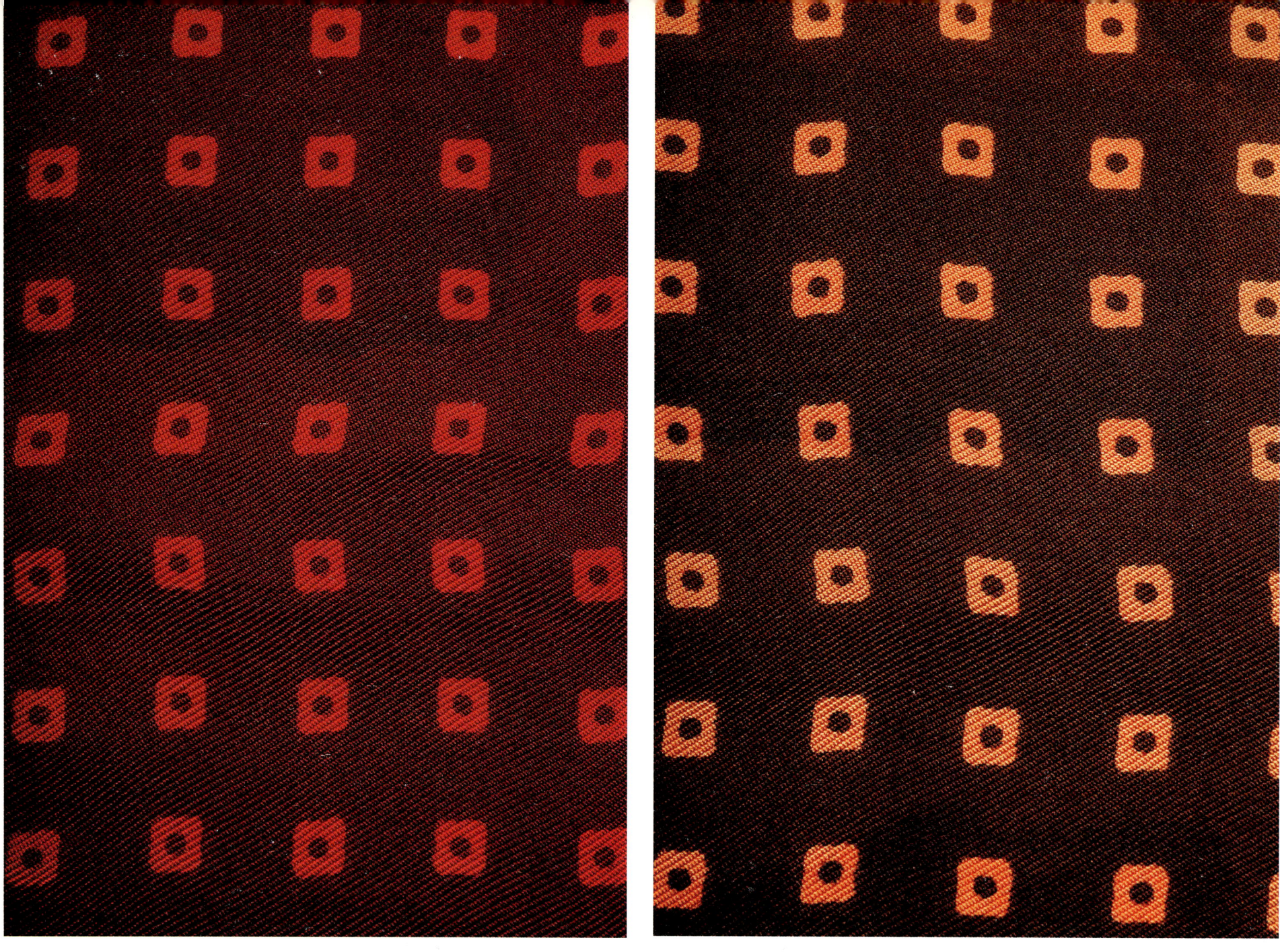

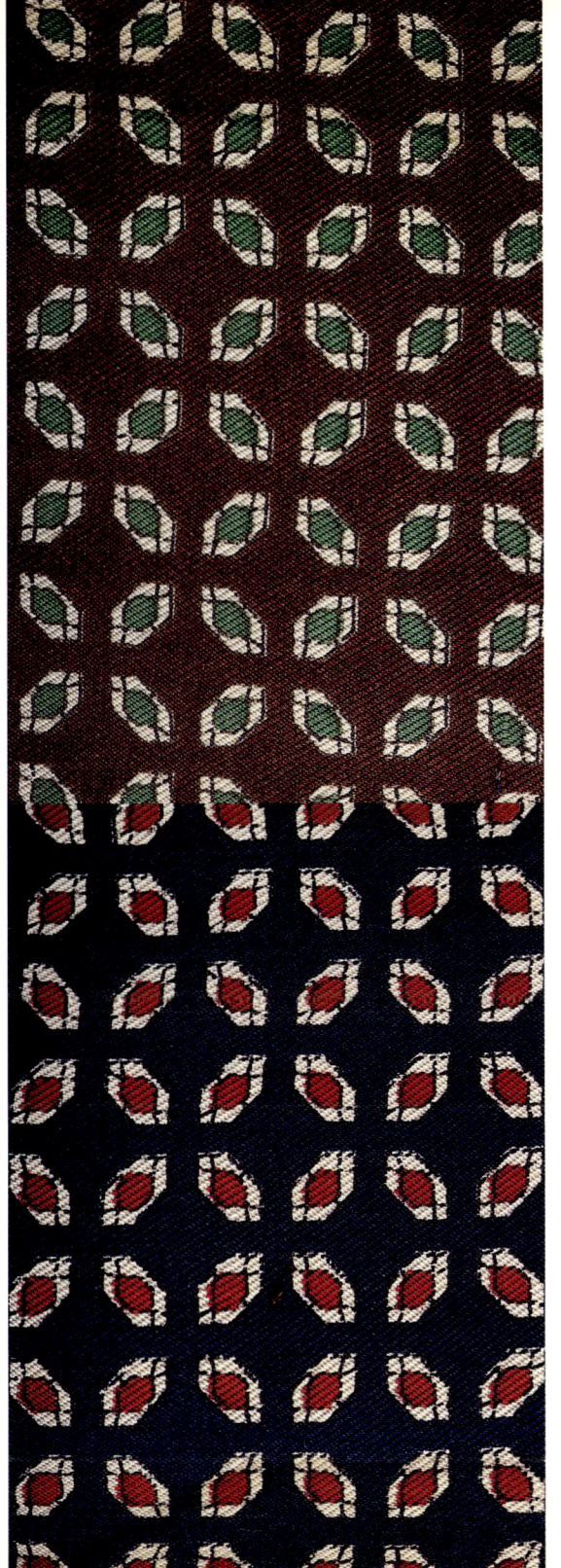

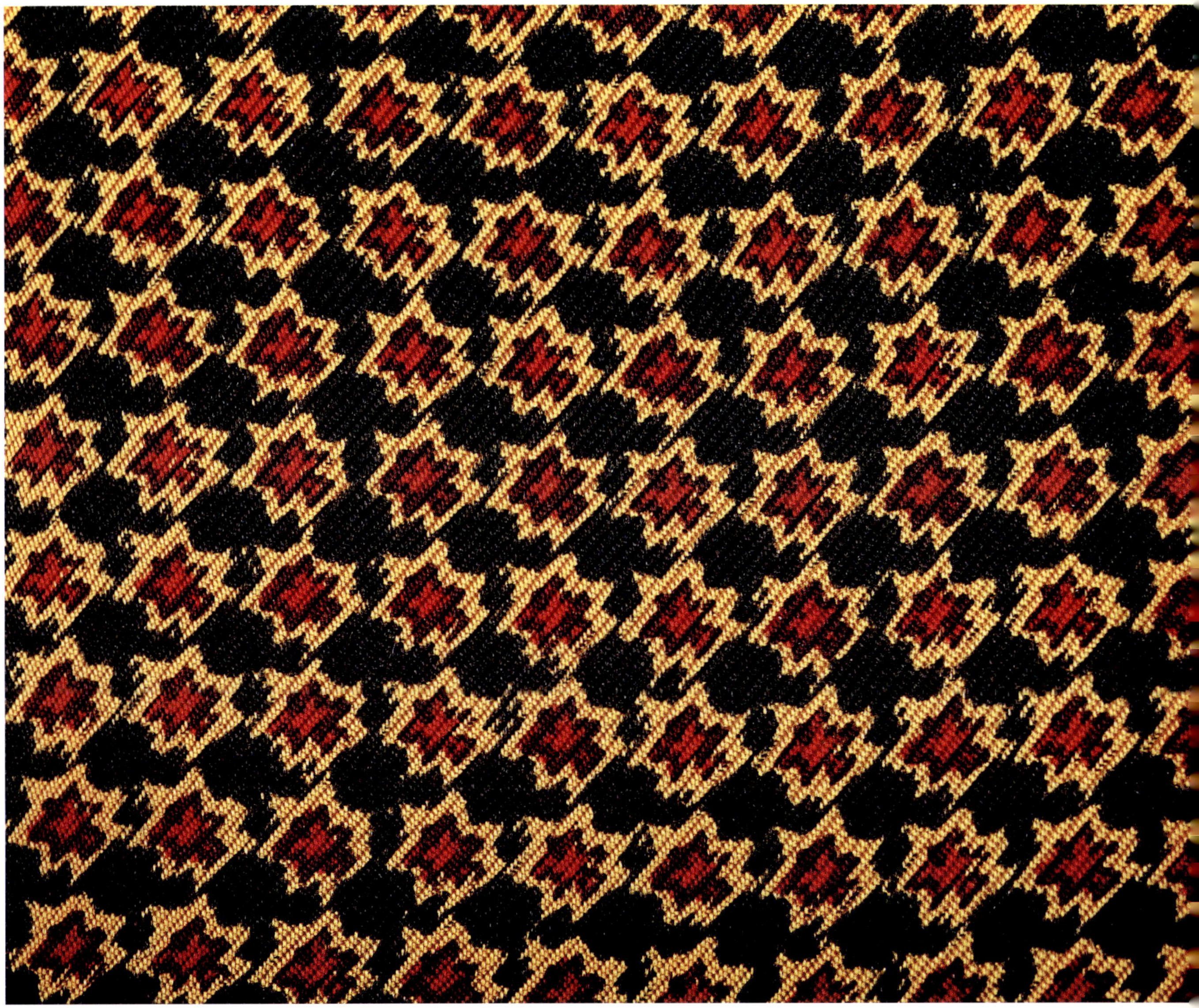

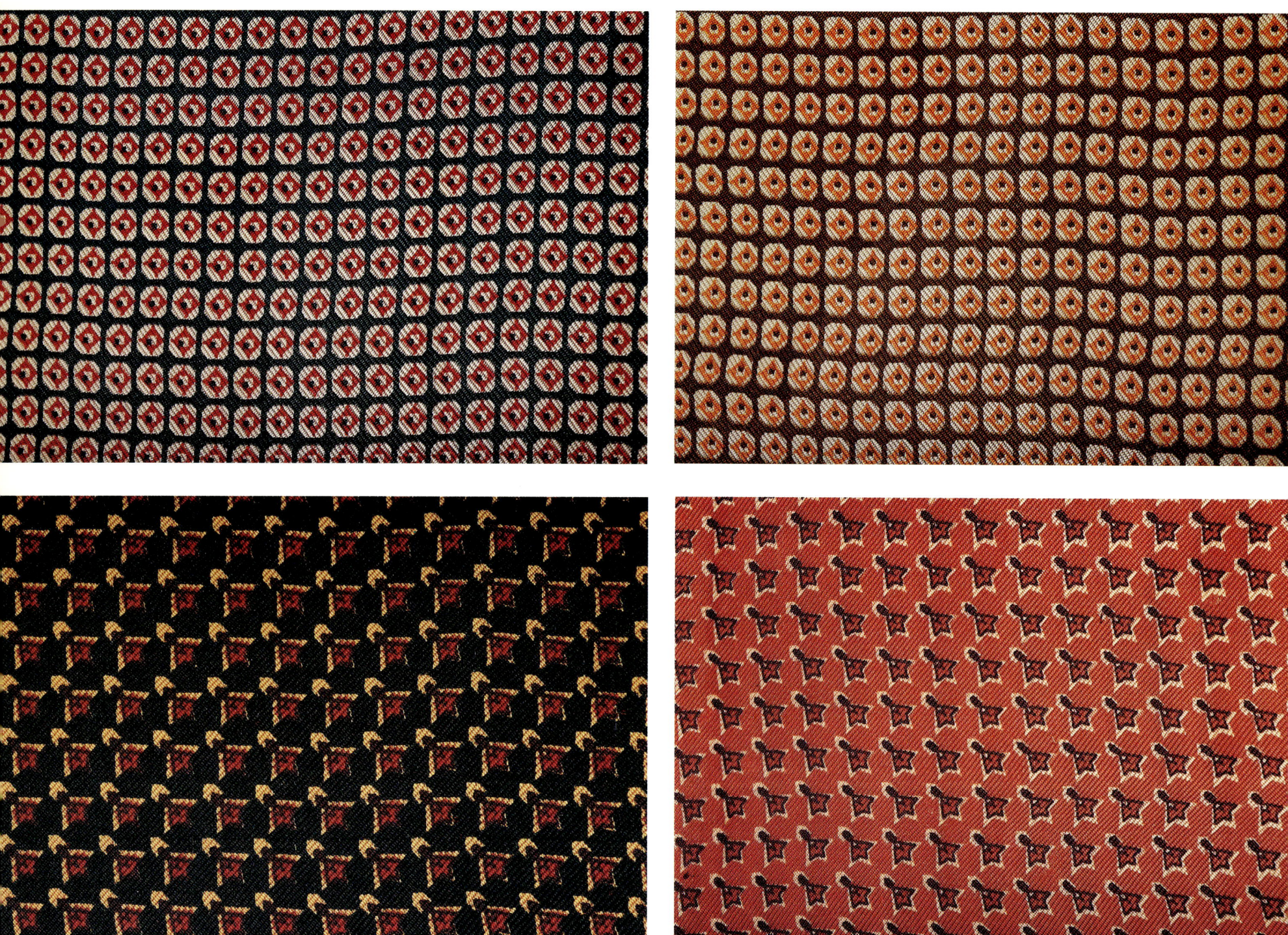

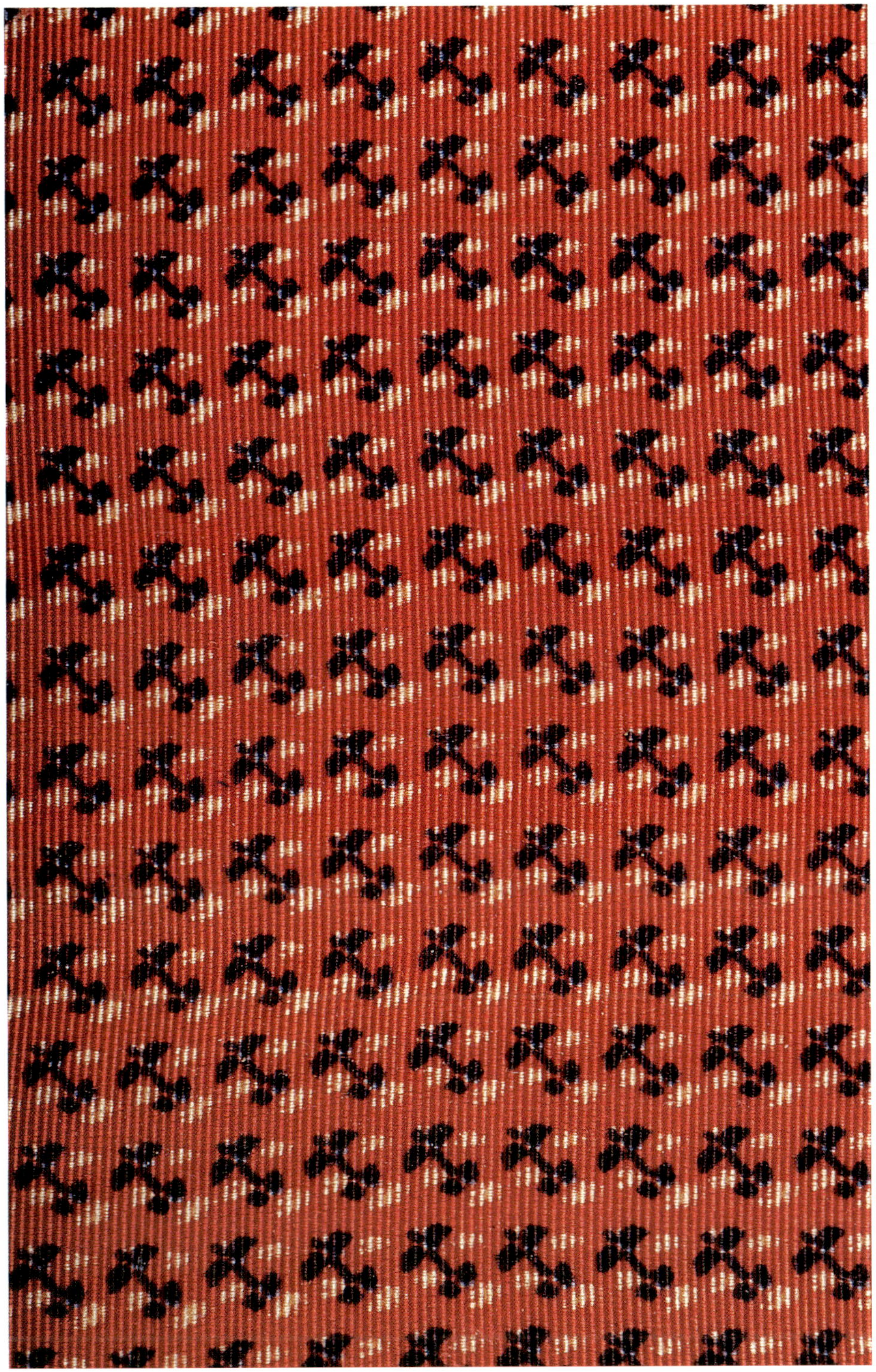

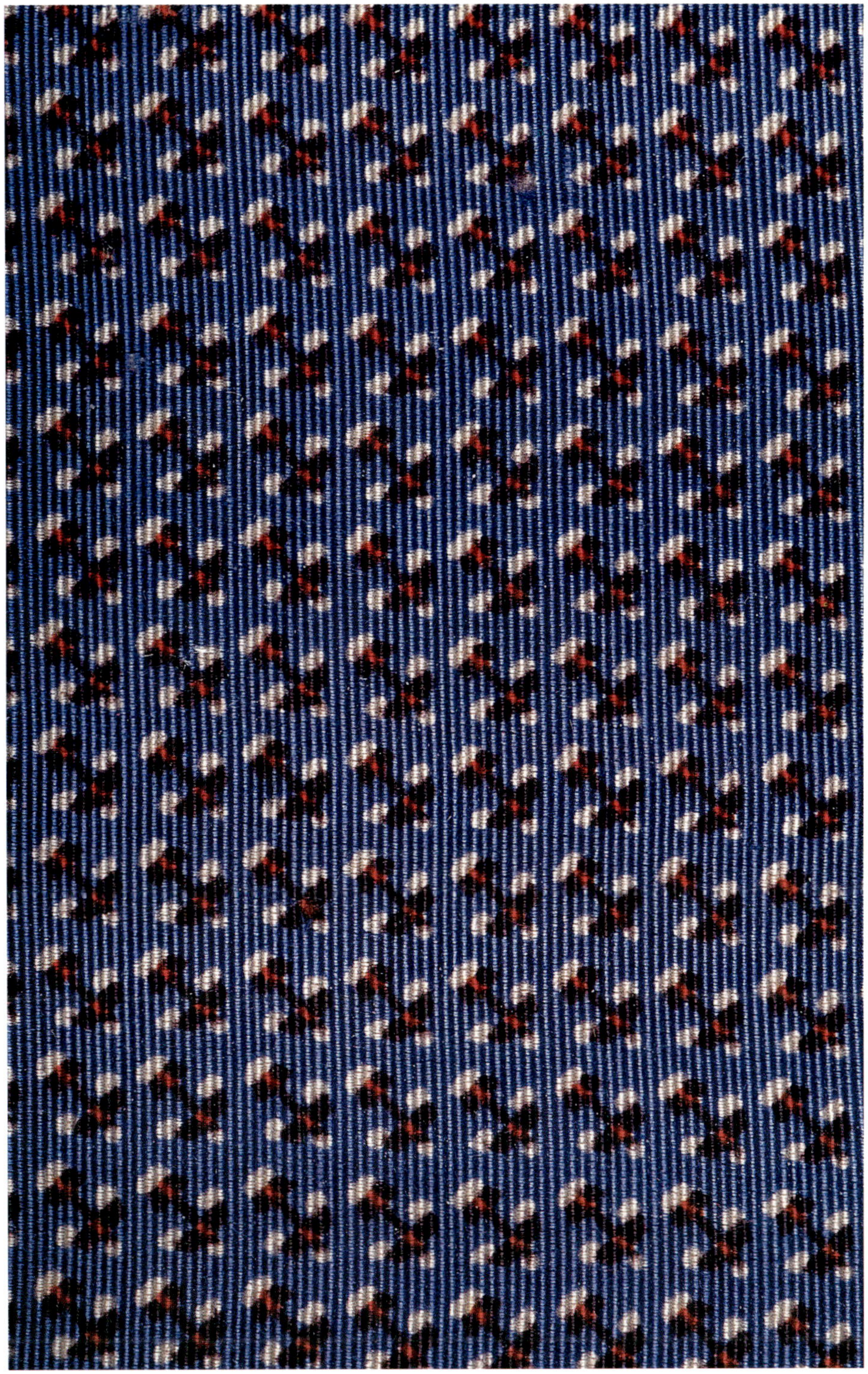

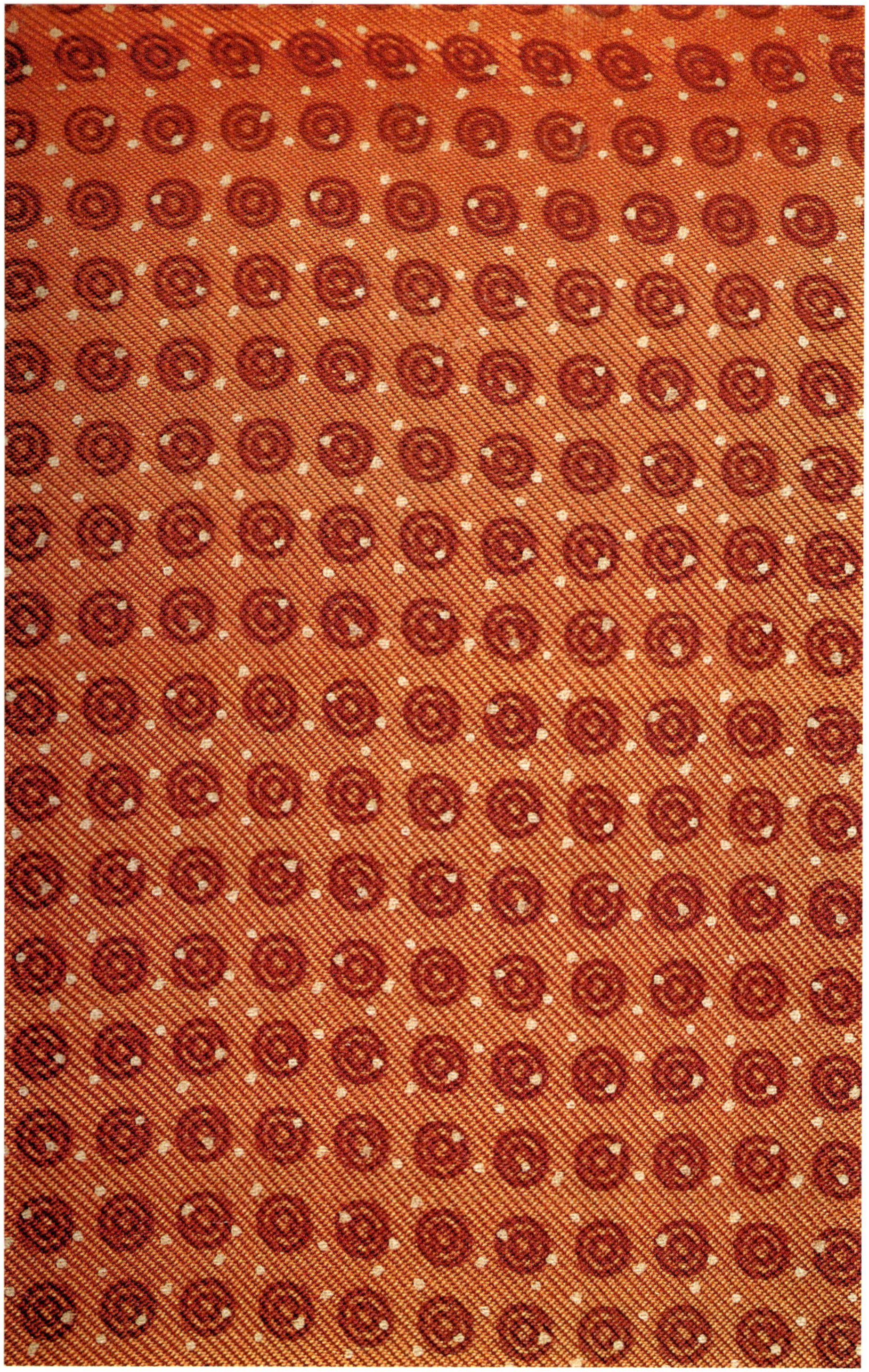

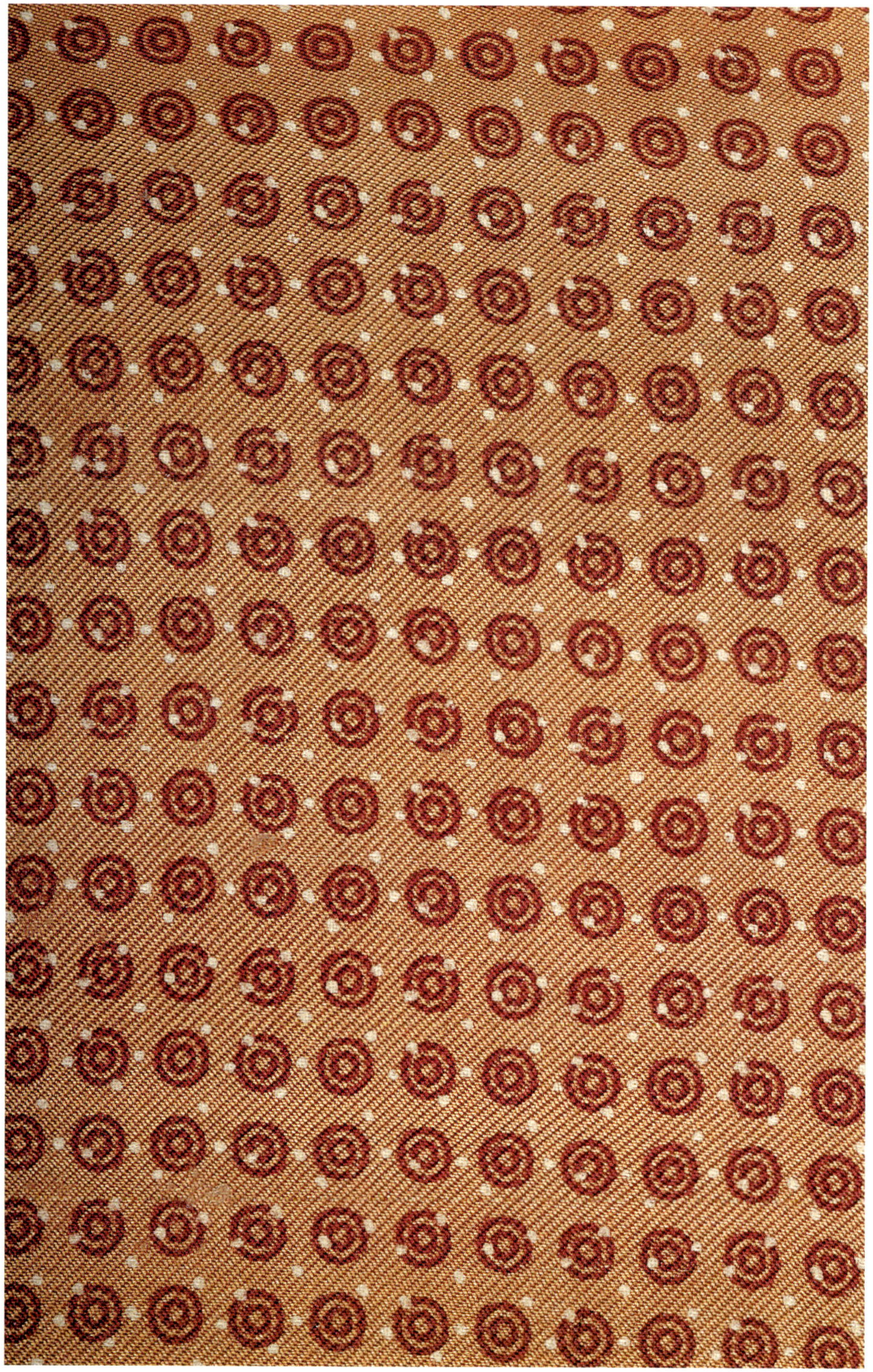

Scattered

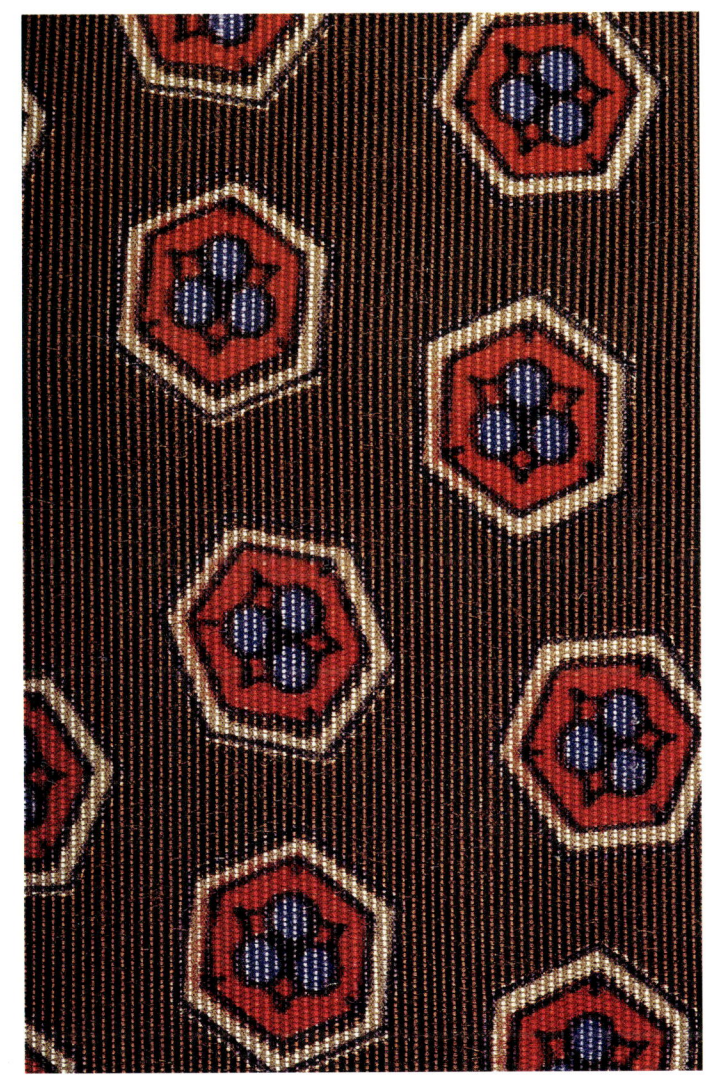

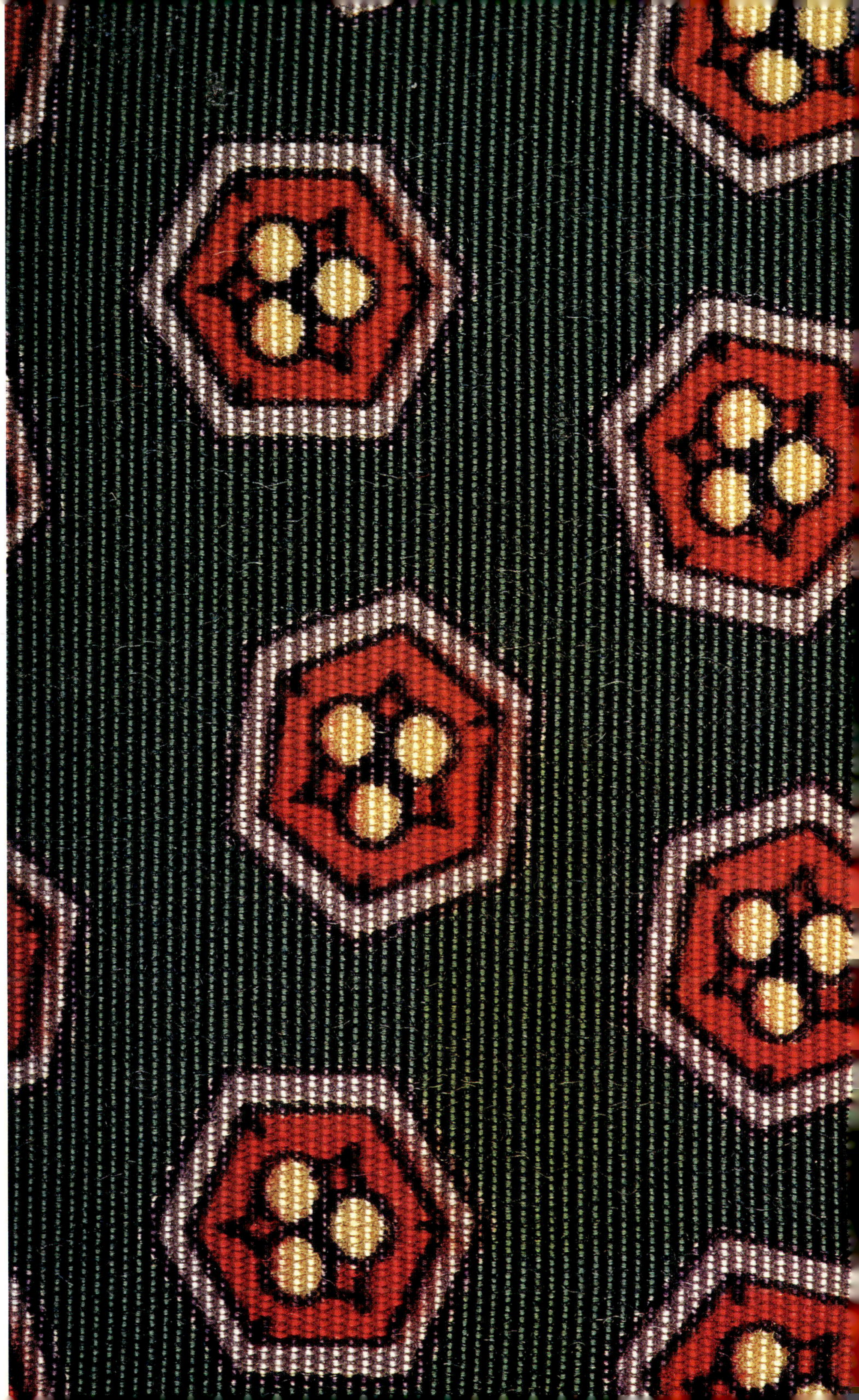

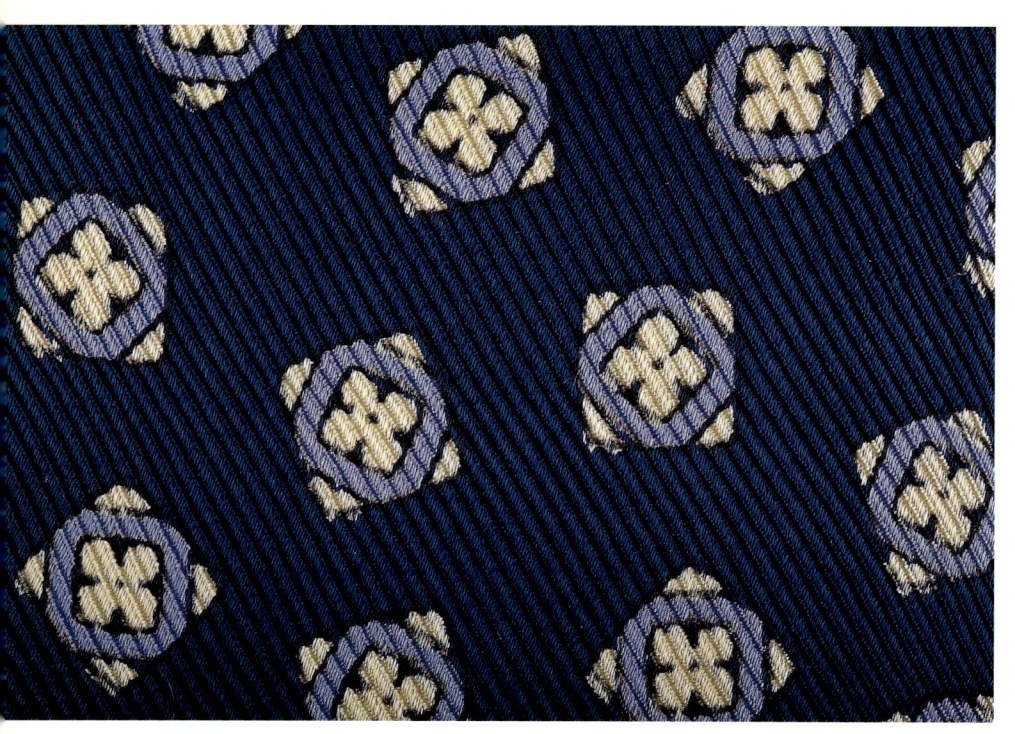
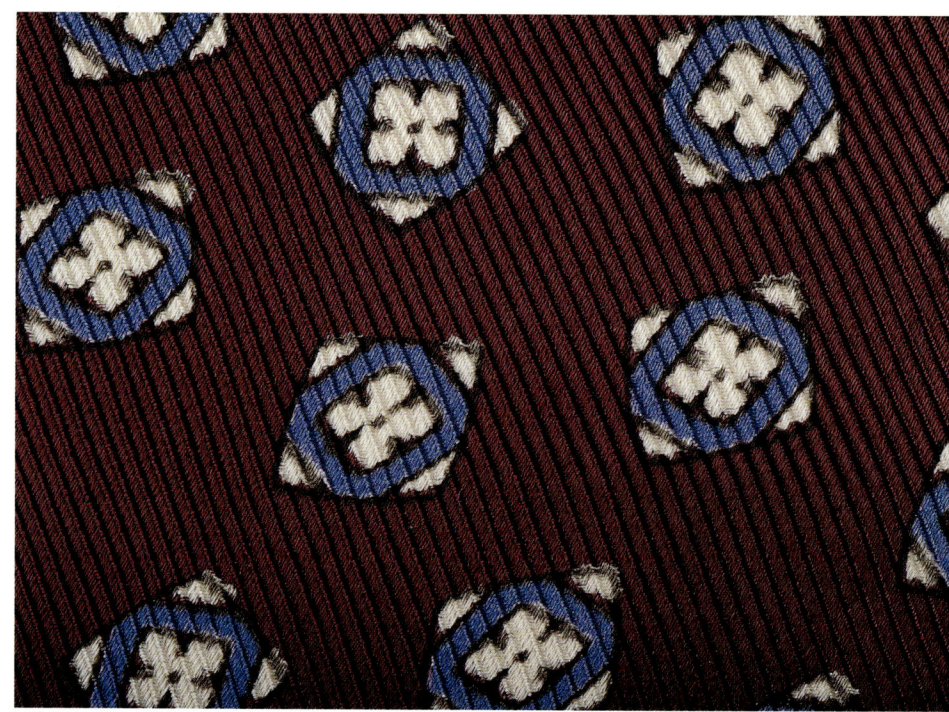
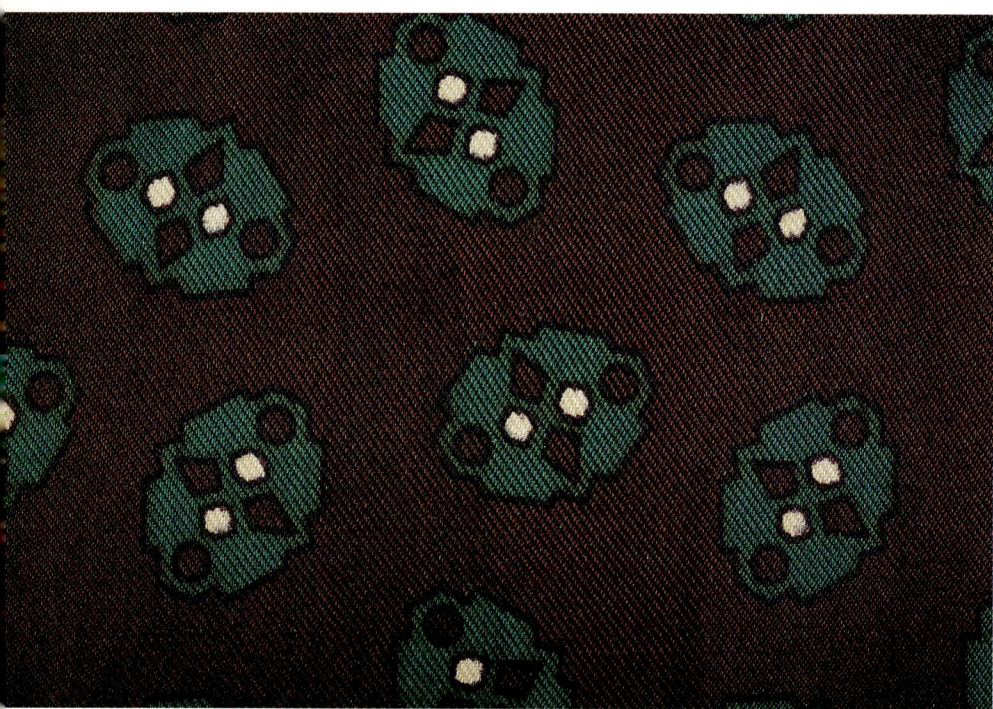
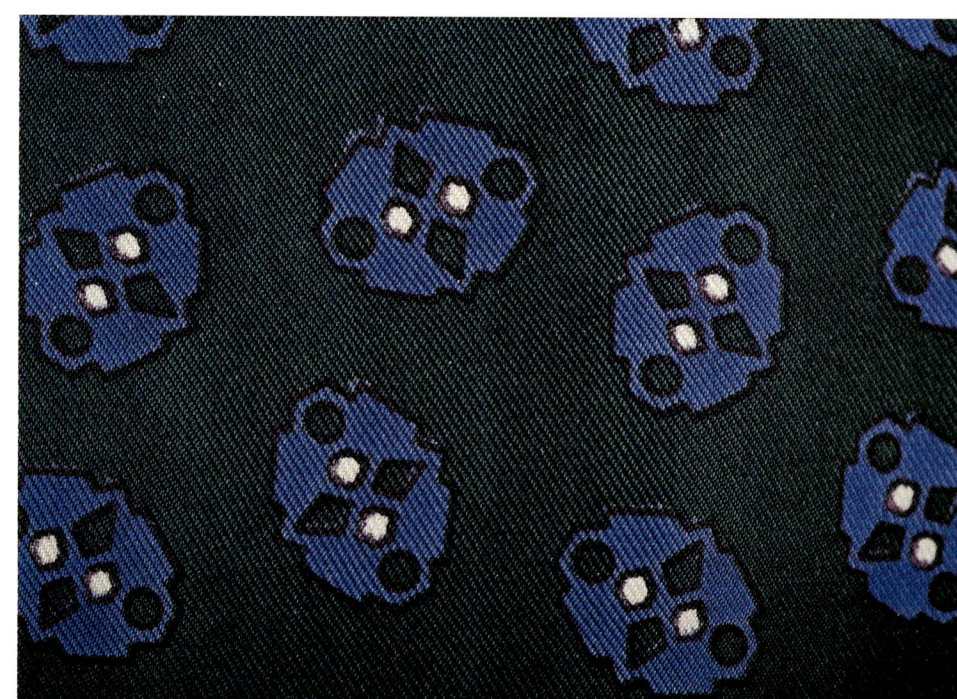

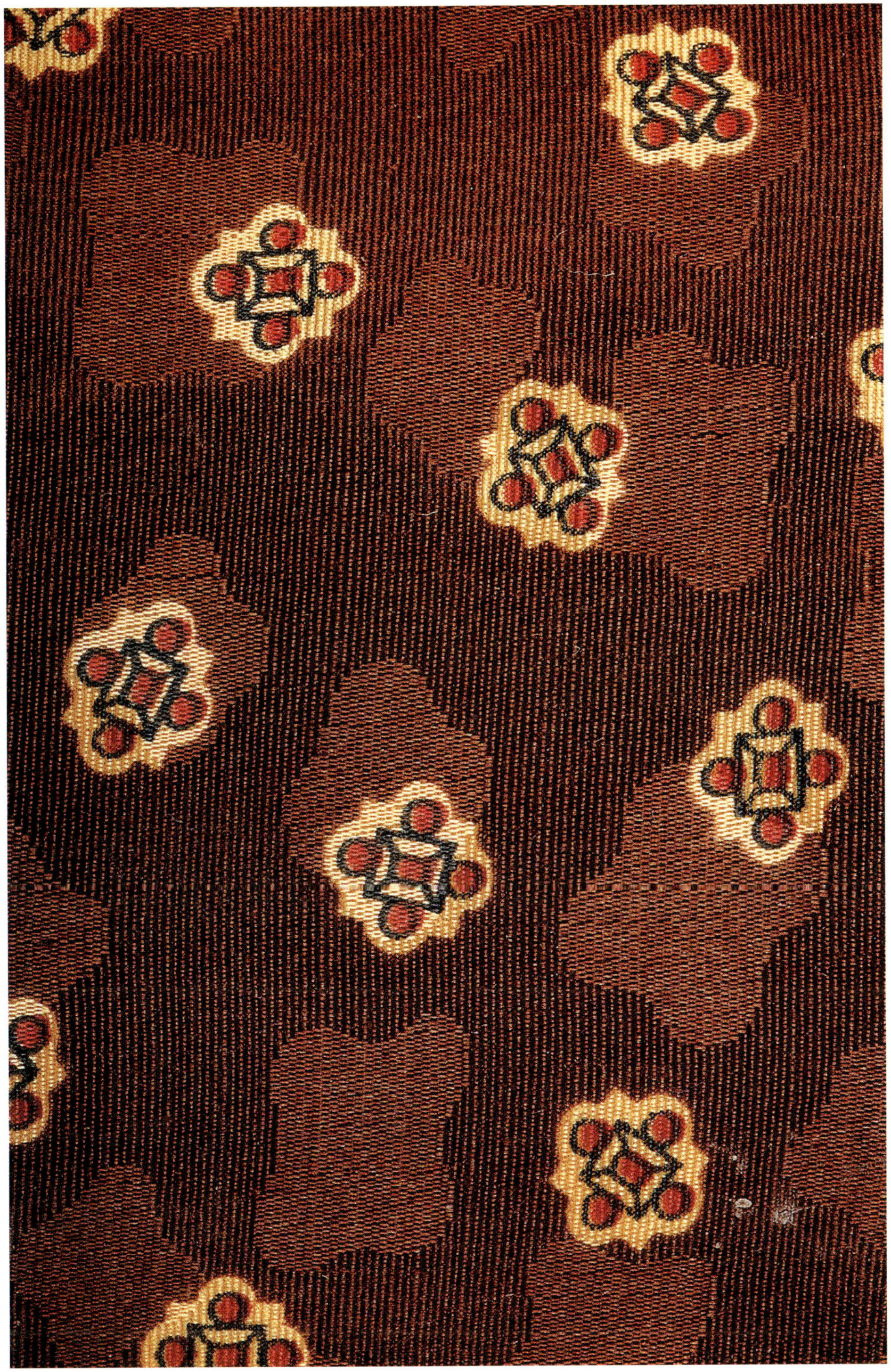
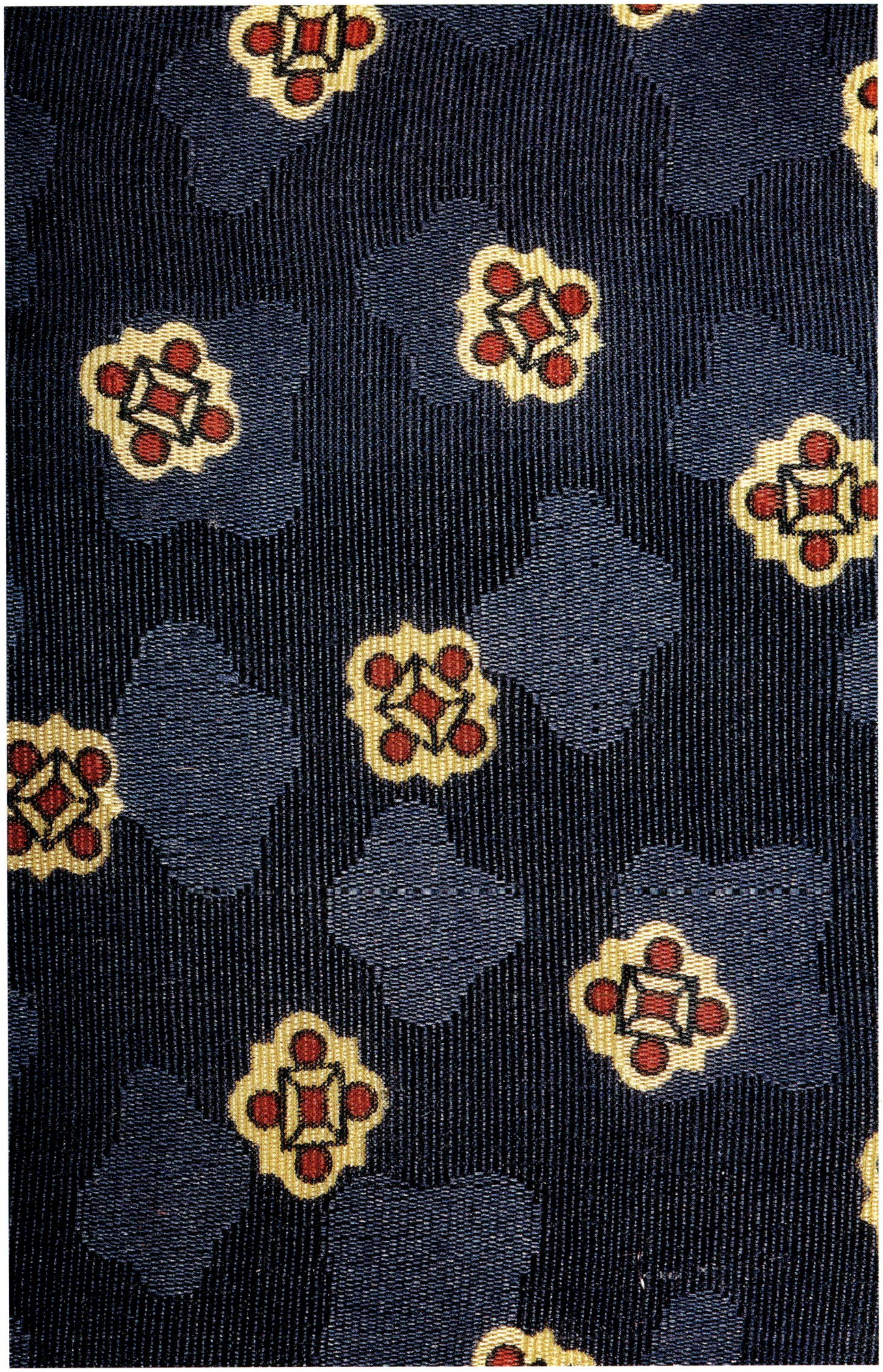

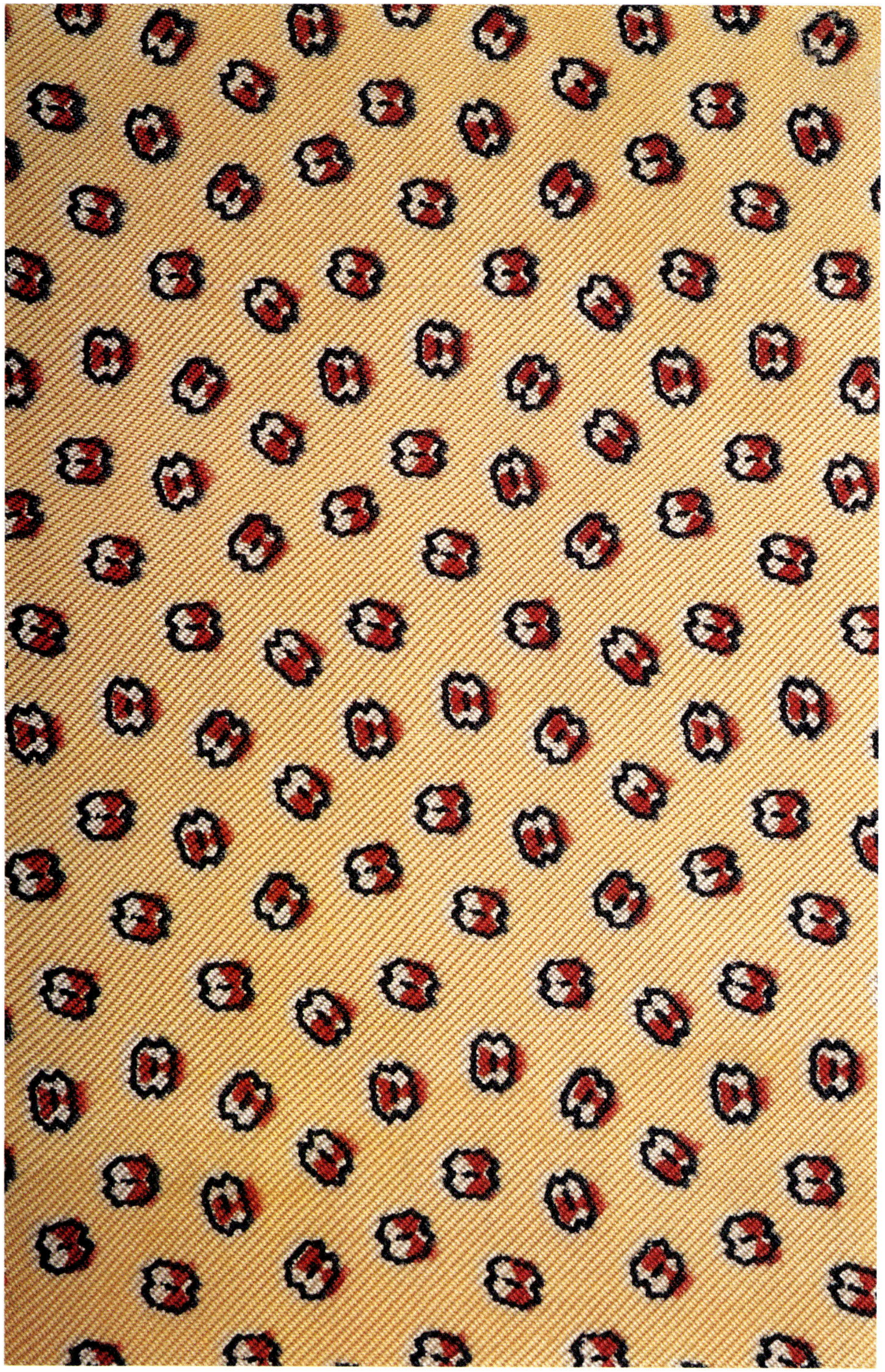

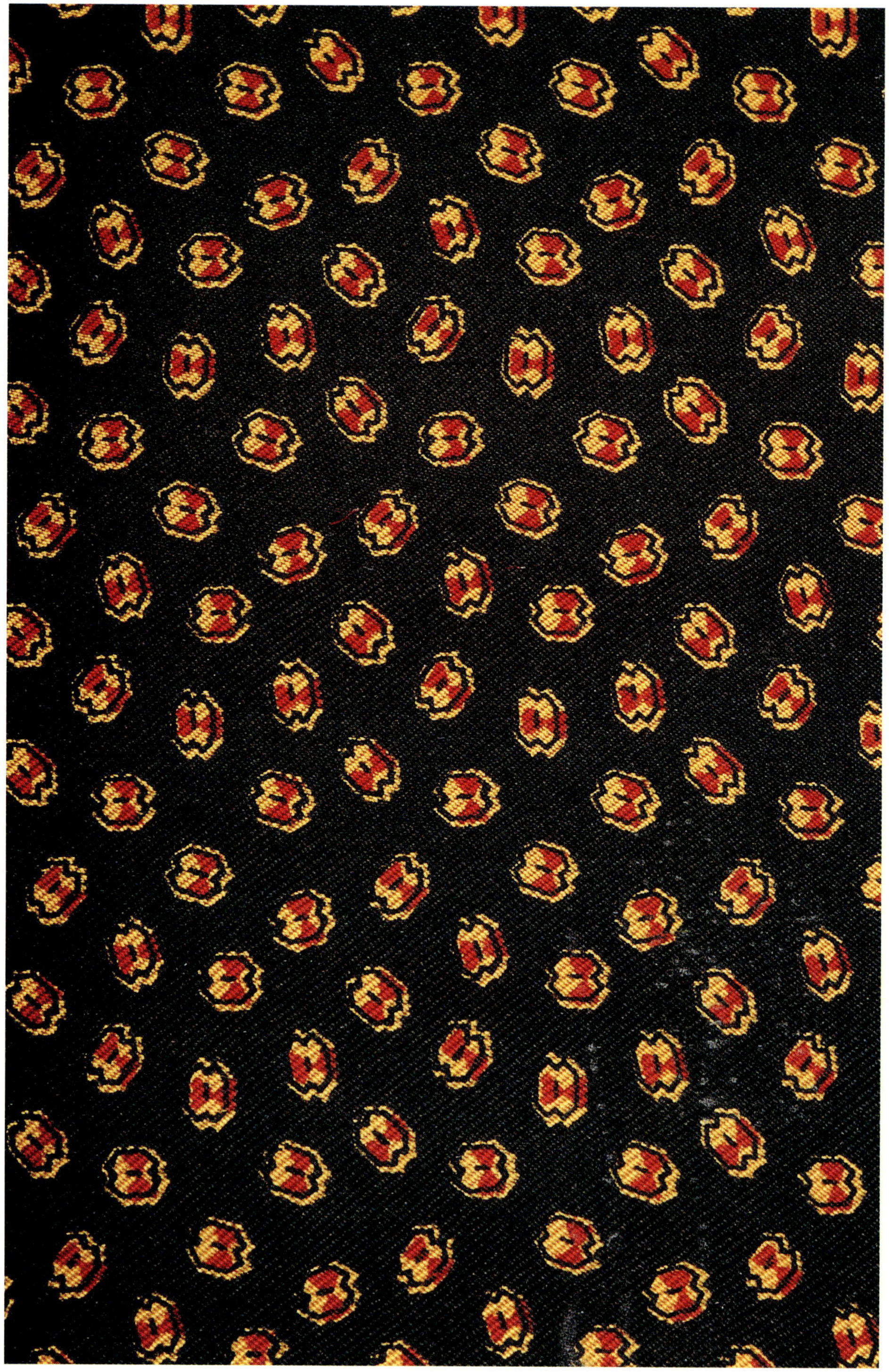

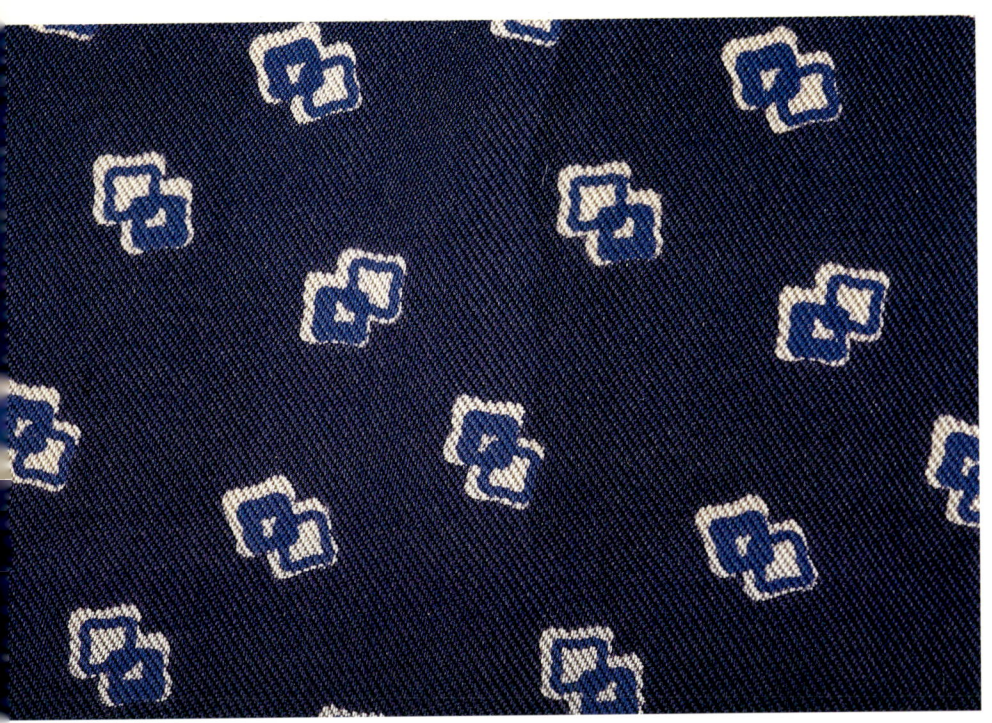
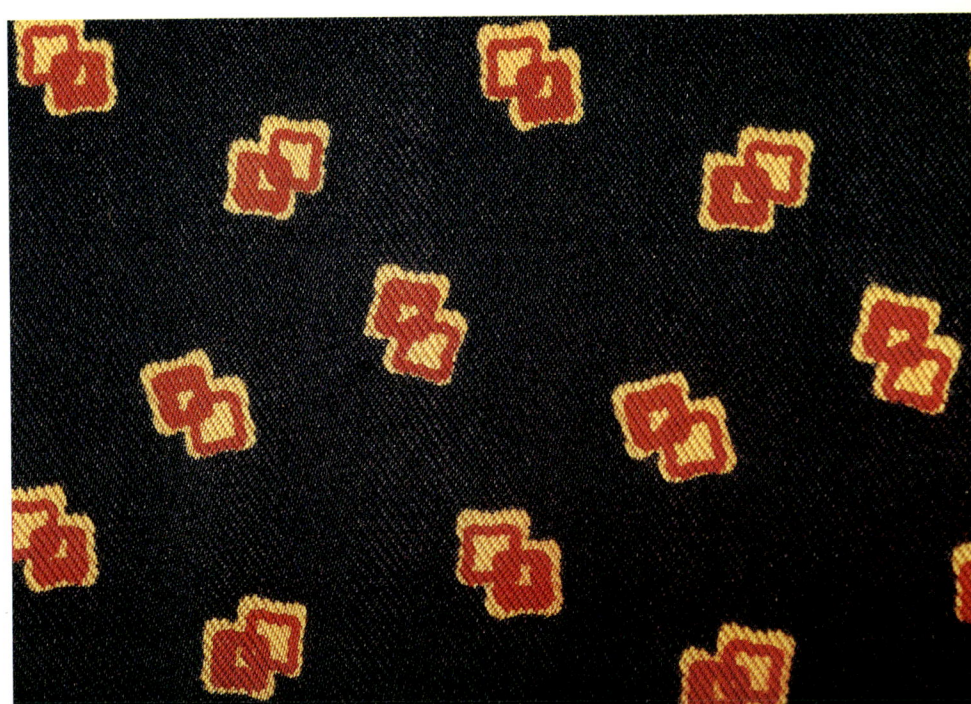
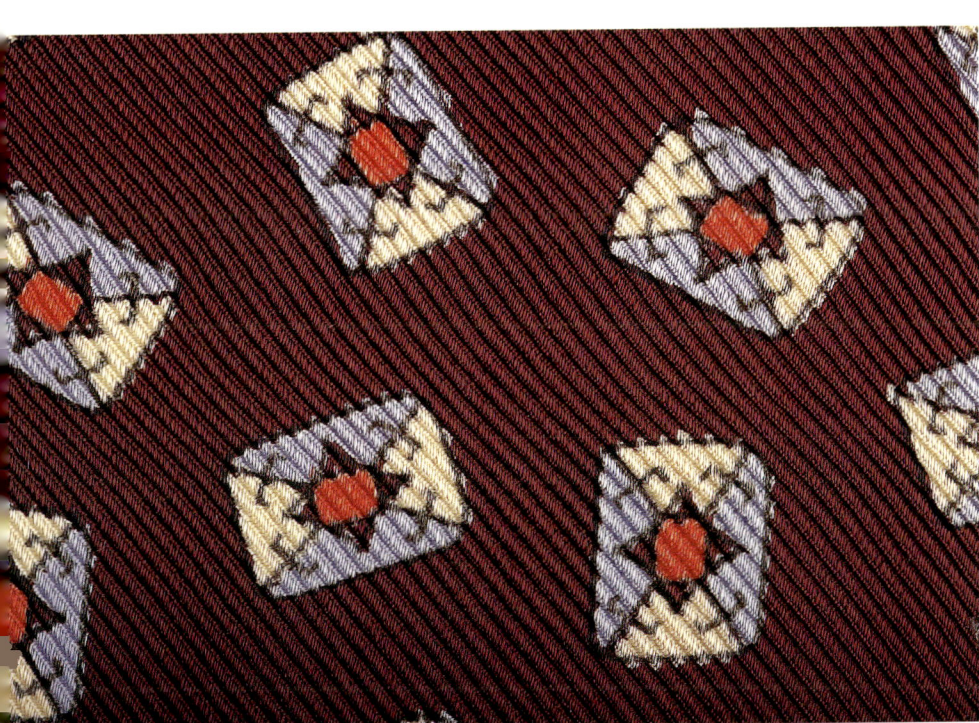
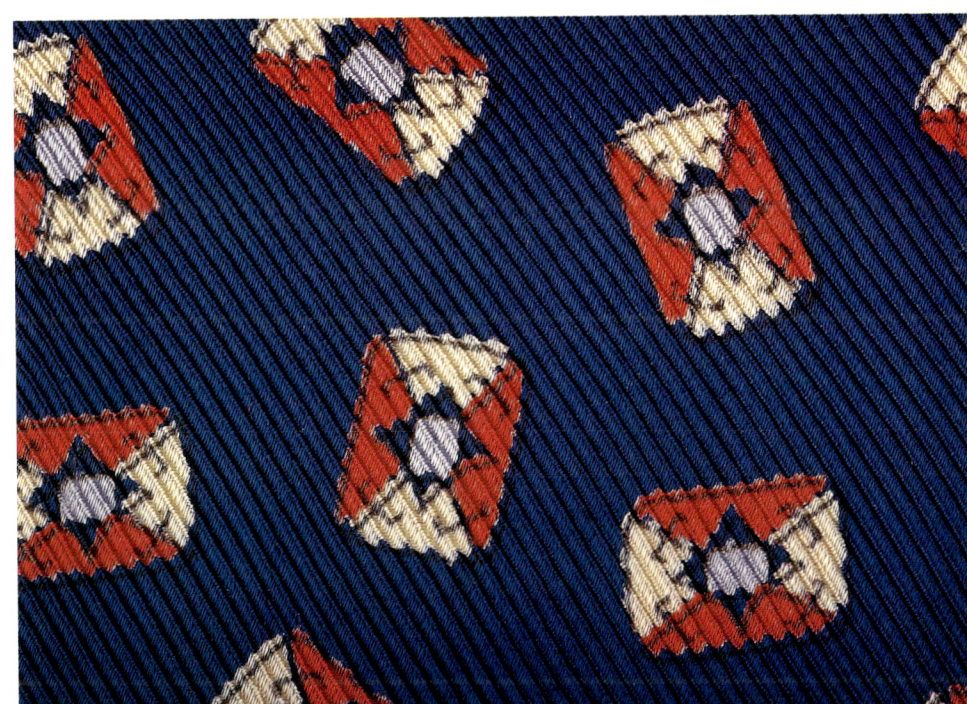

23

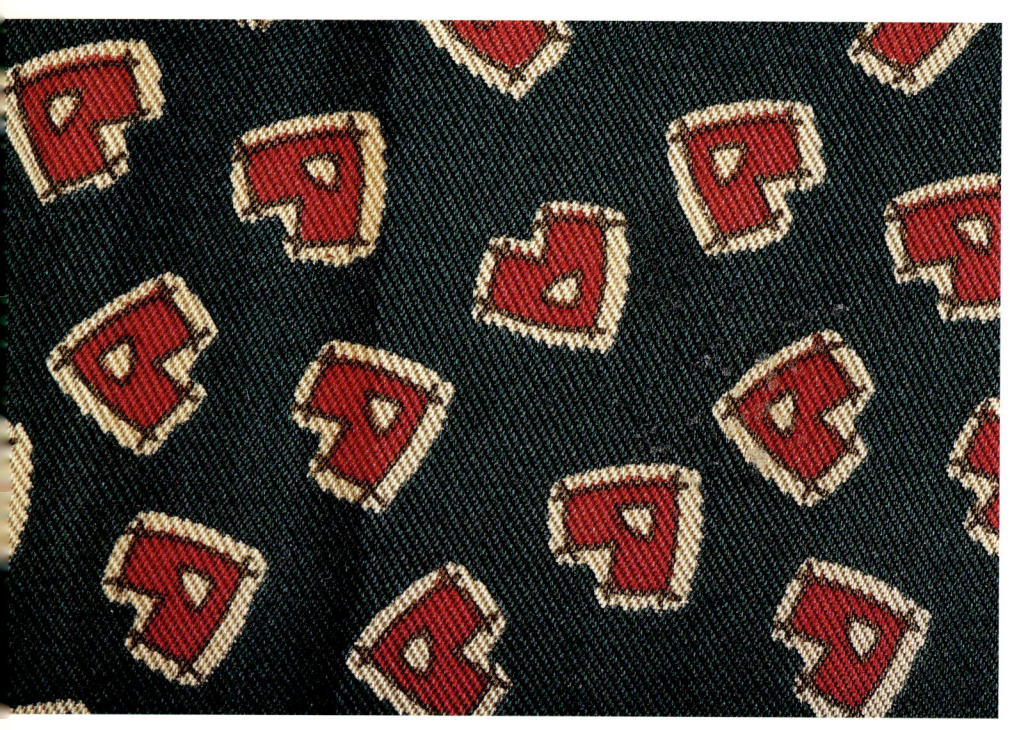
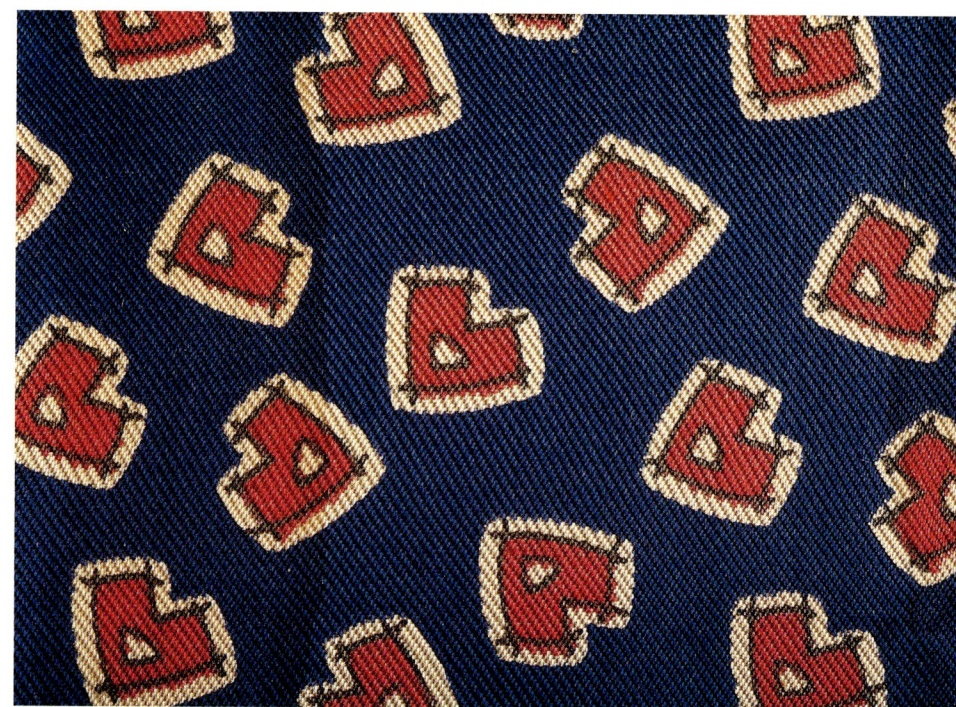
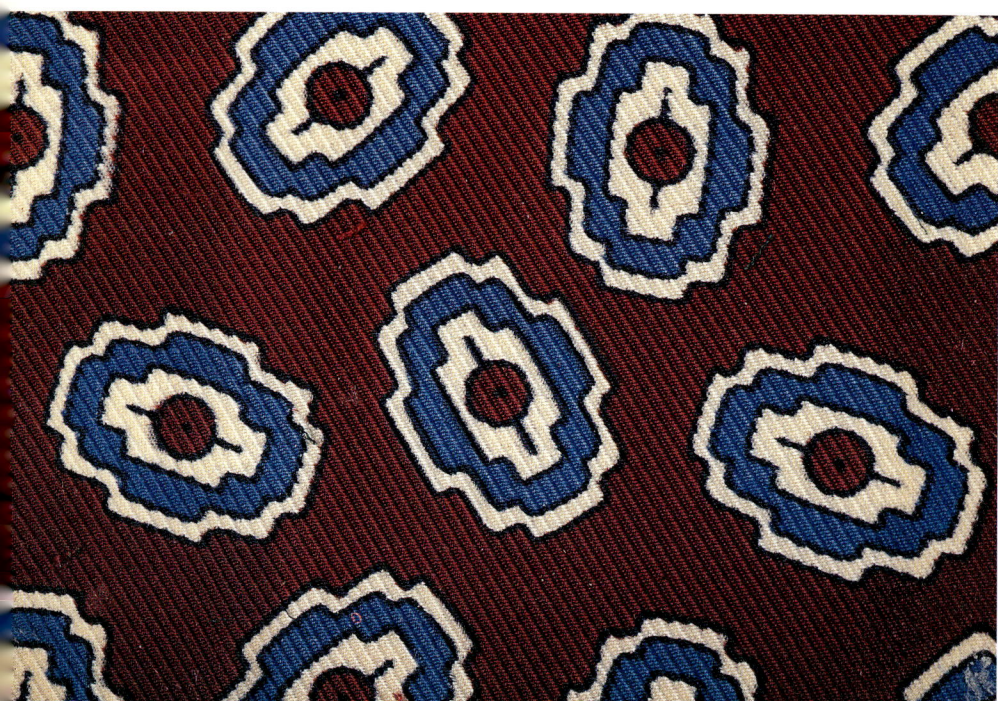
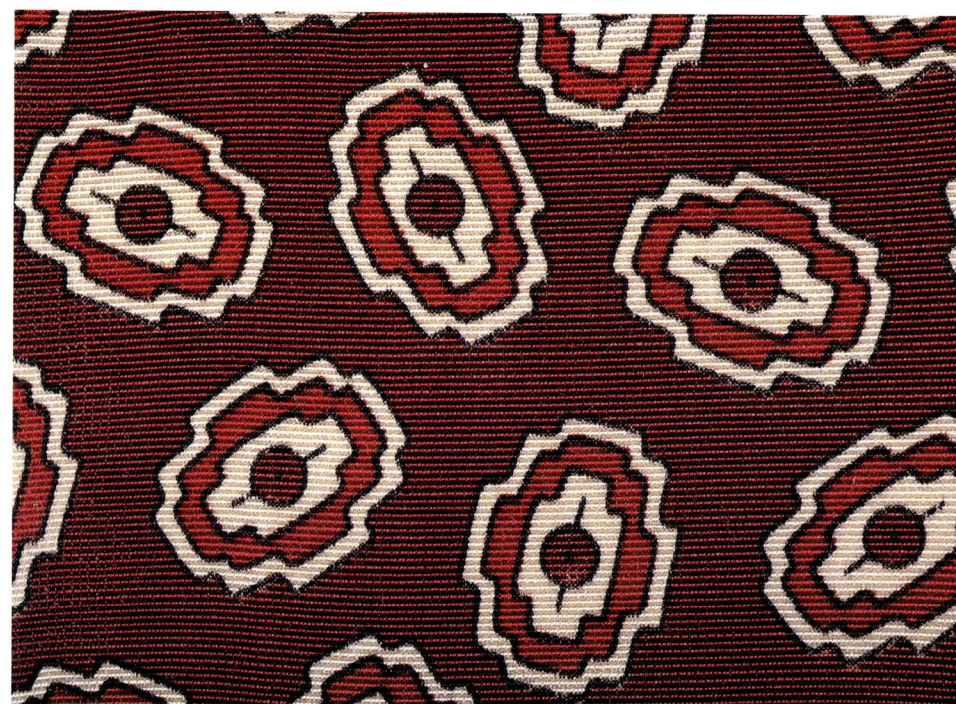

24

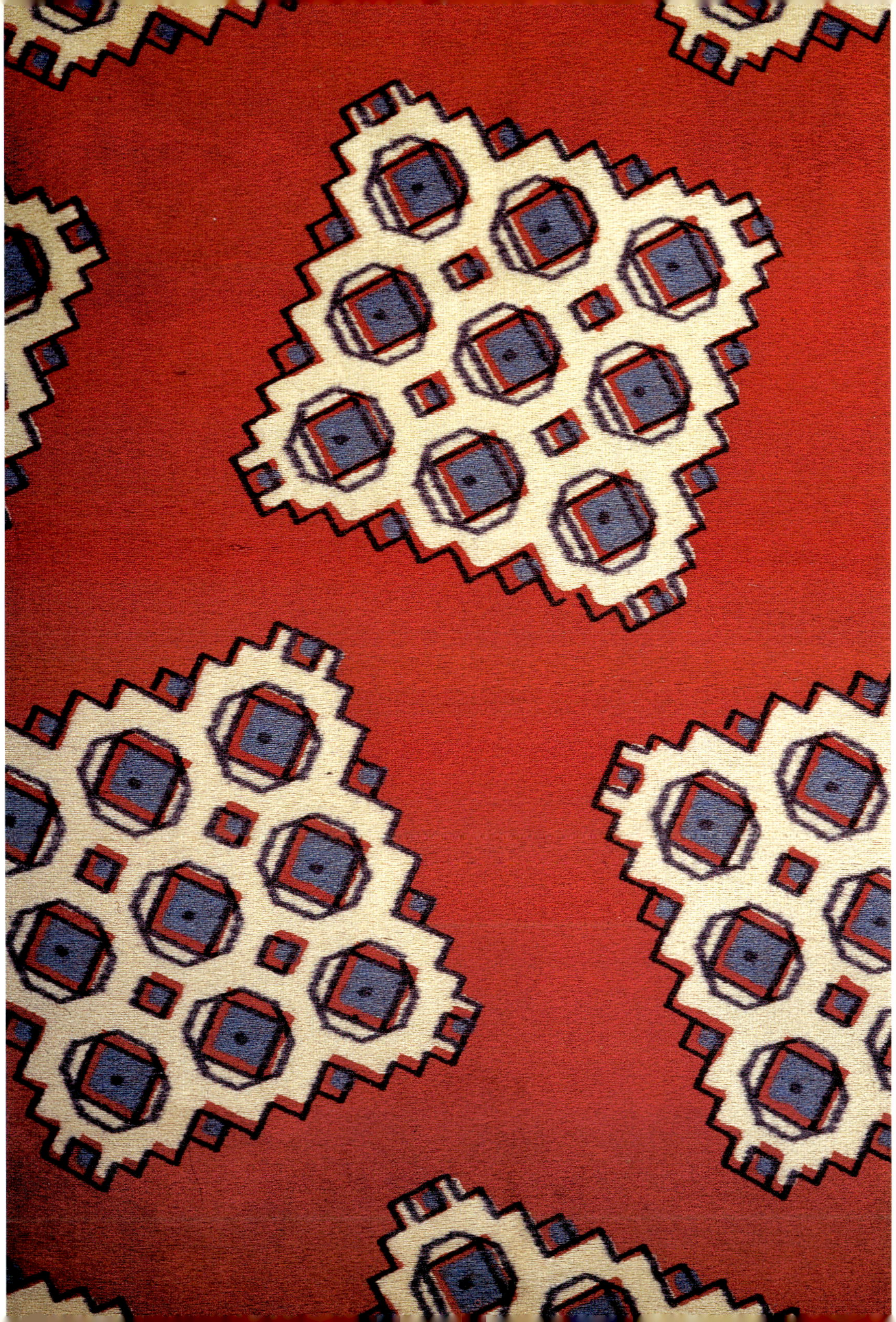

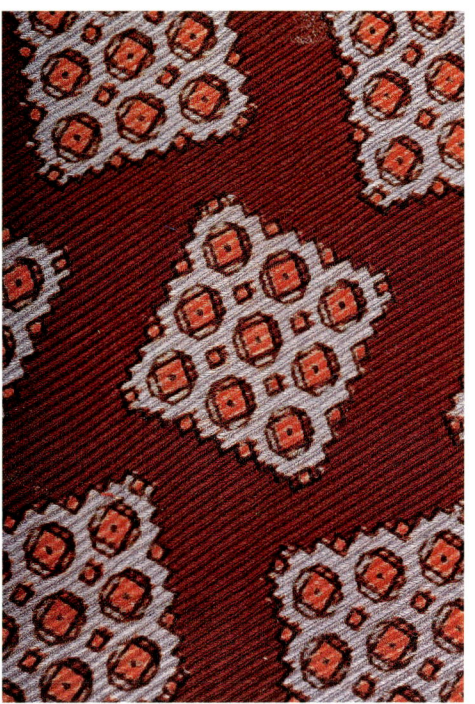

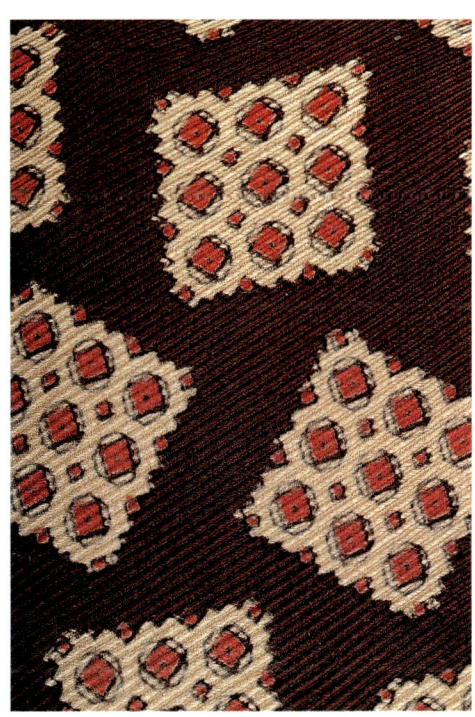

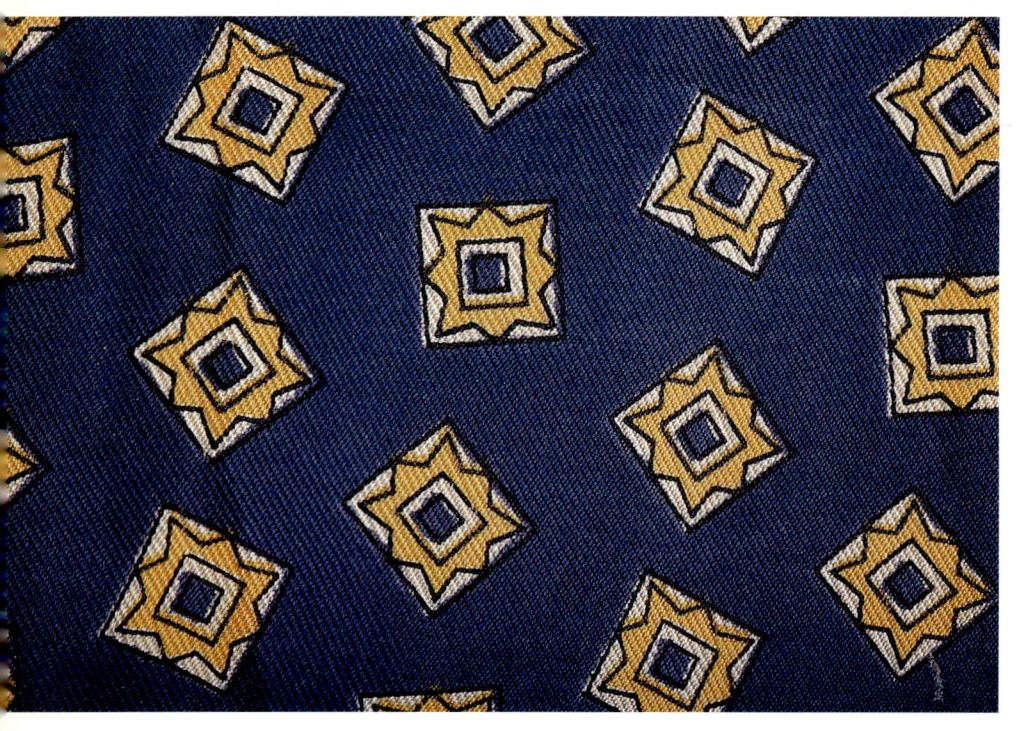
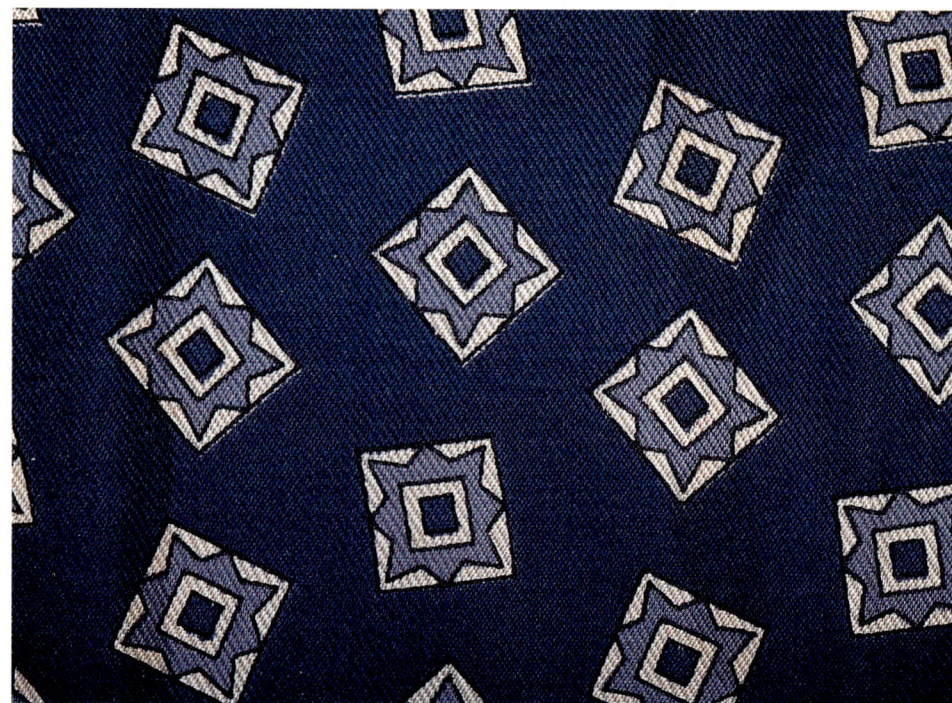
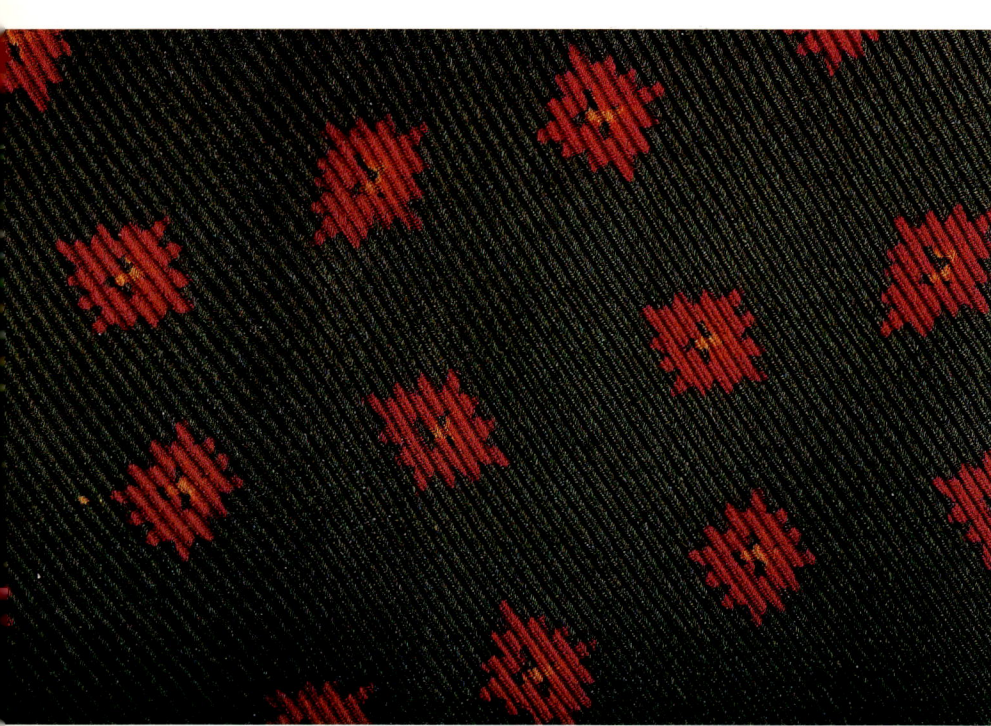
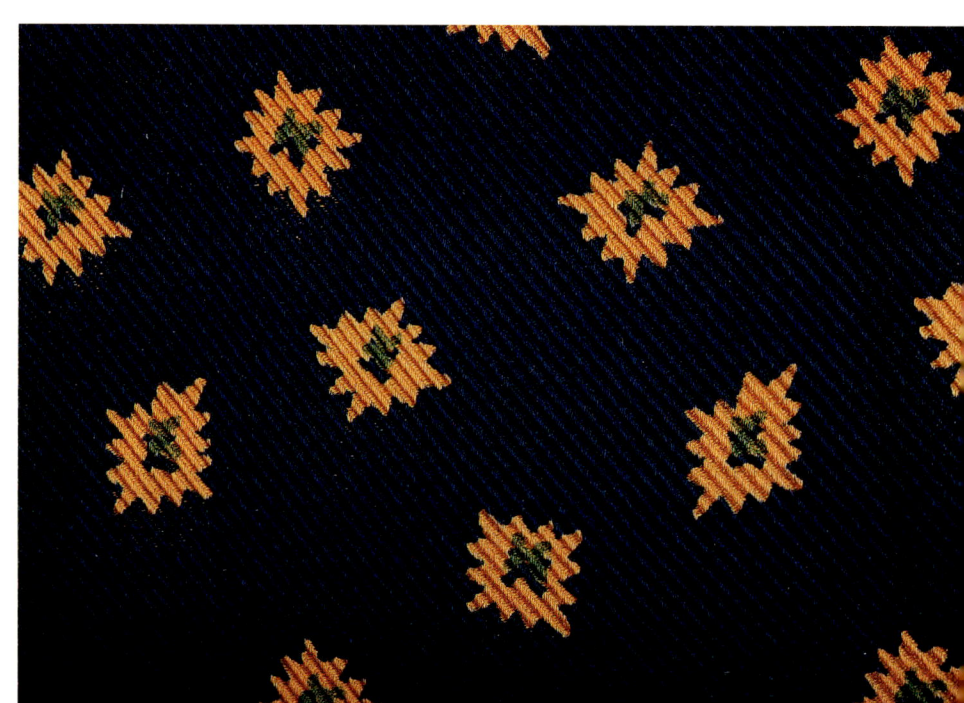

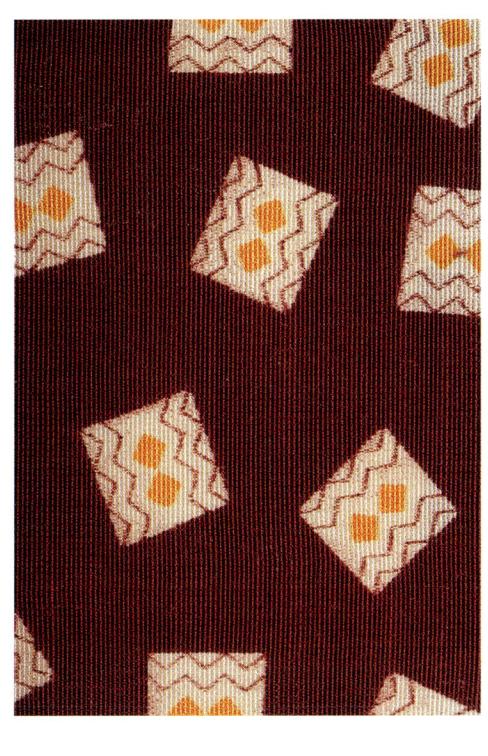
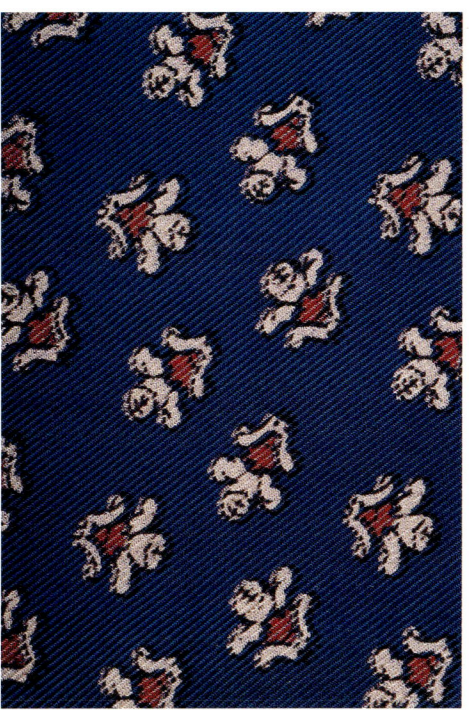
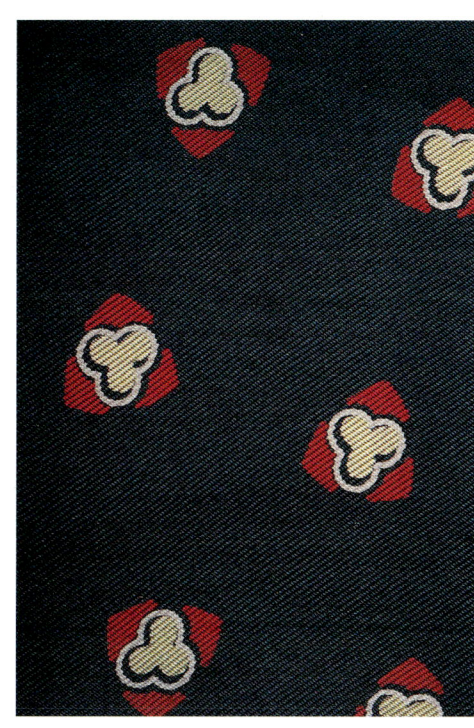
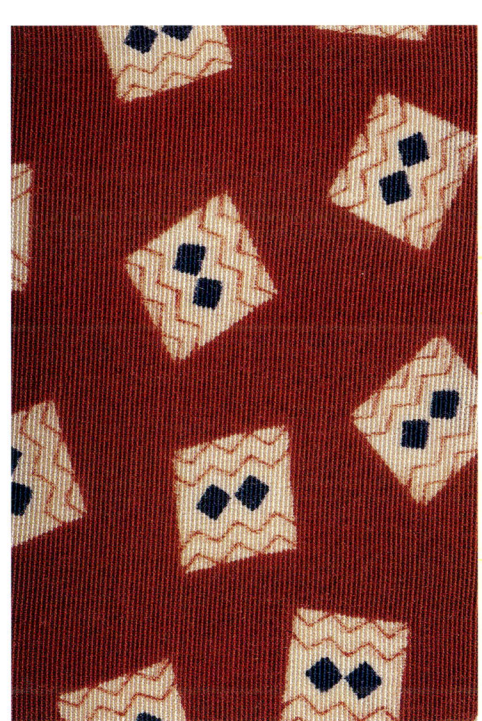
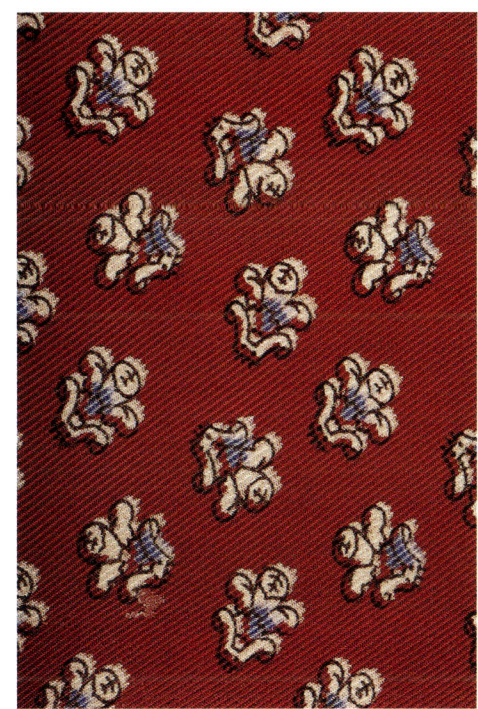
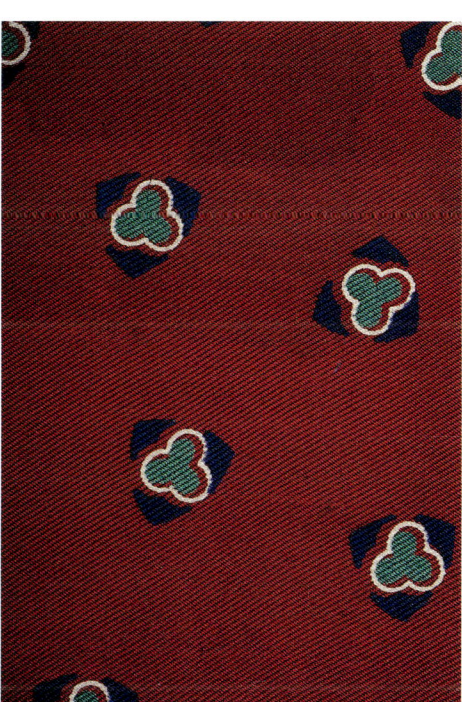

27

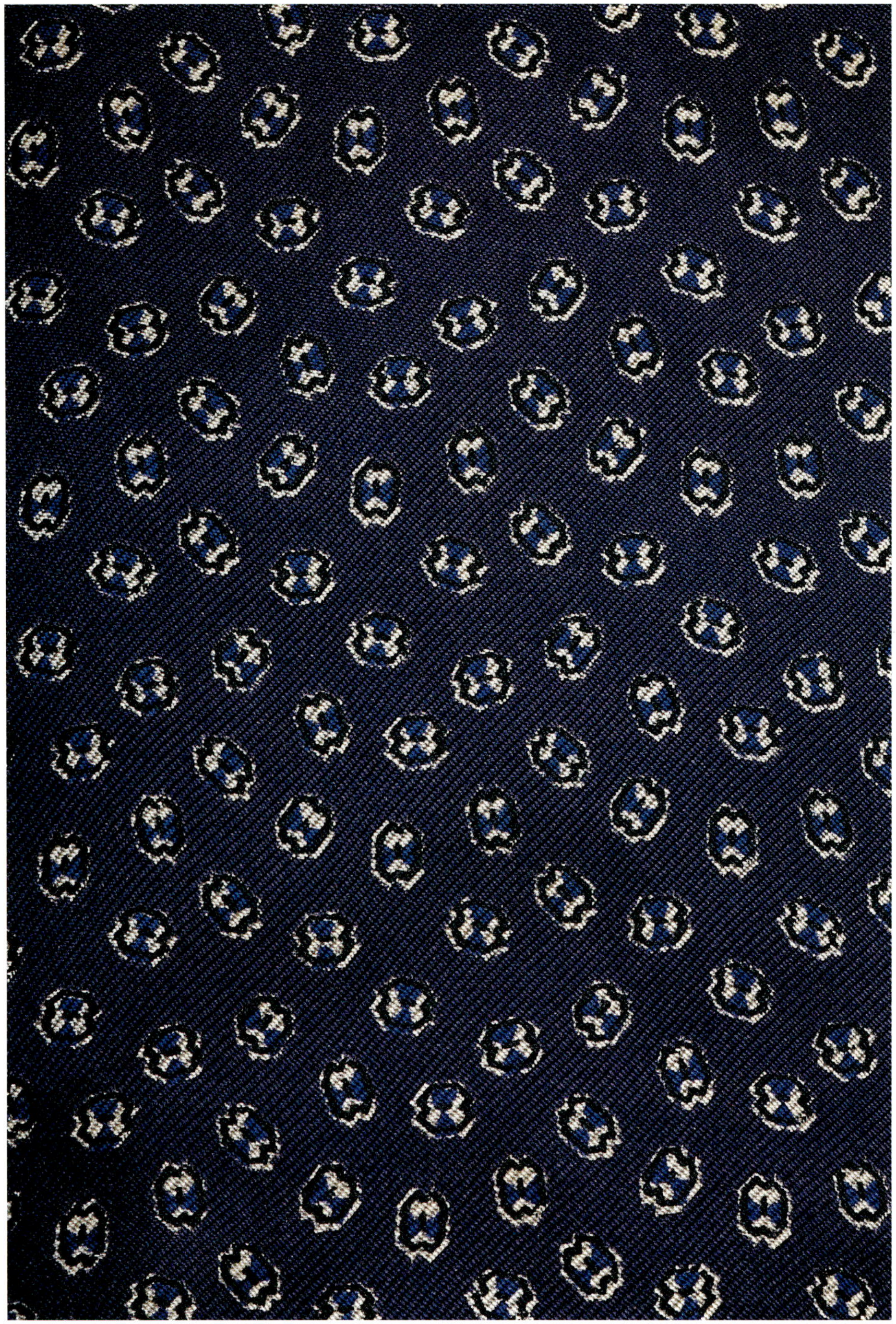

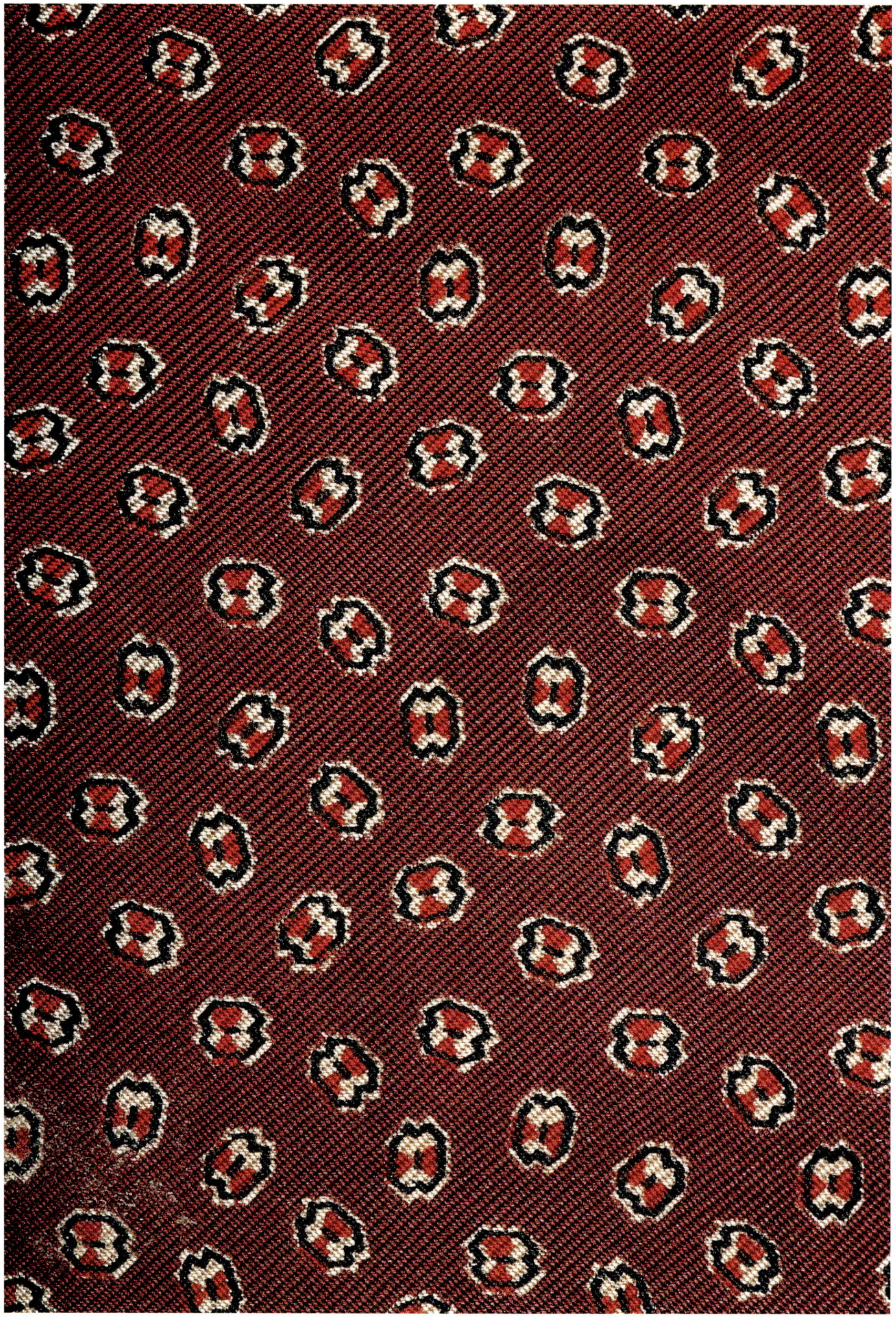

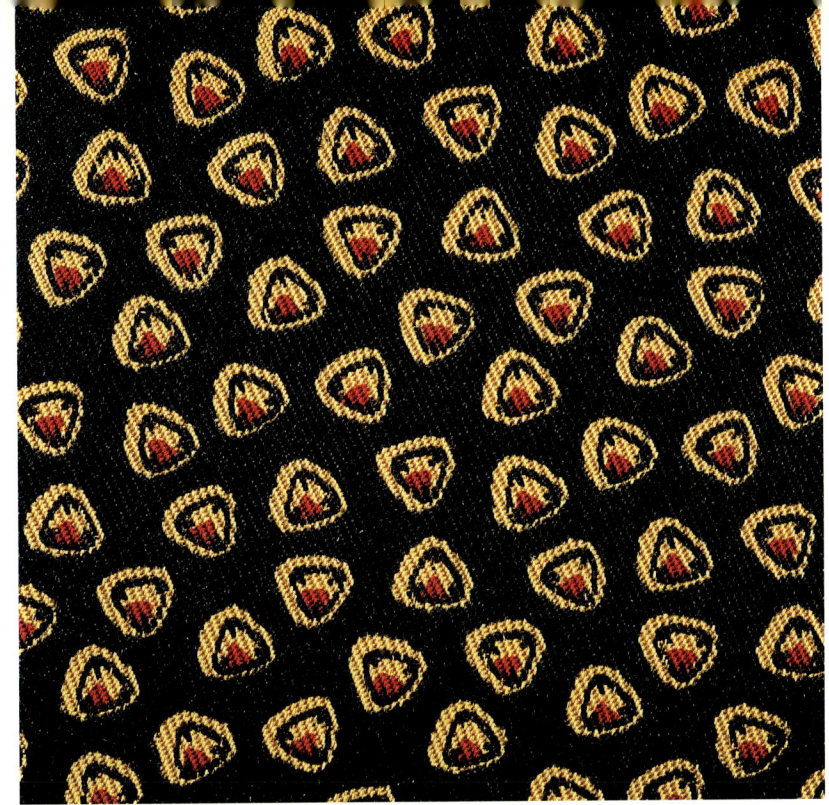
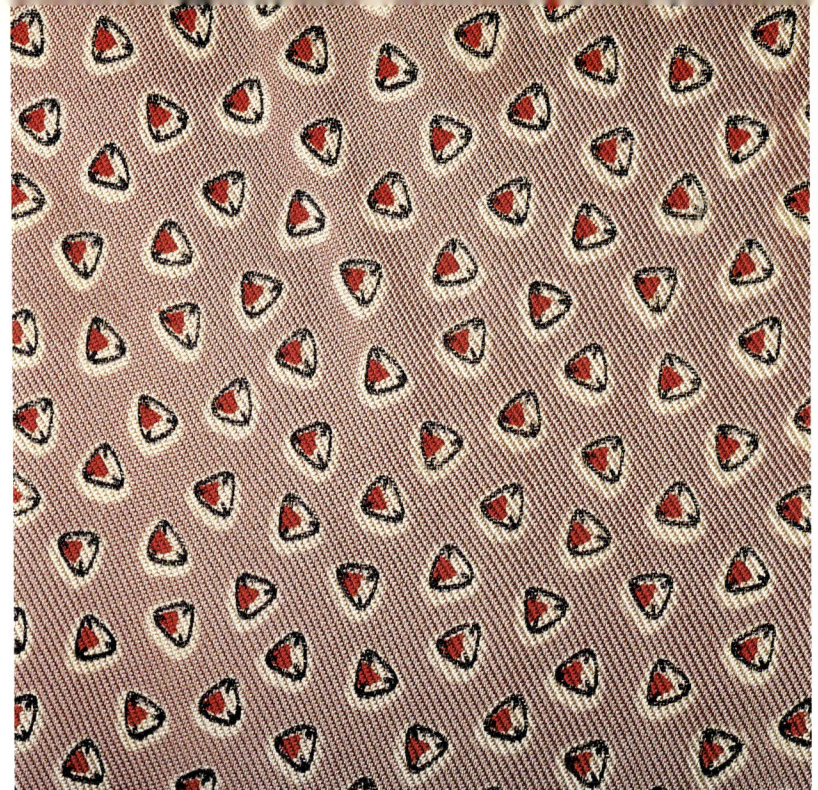
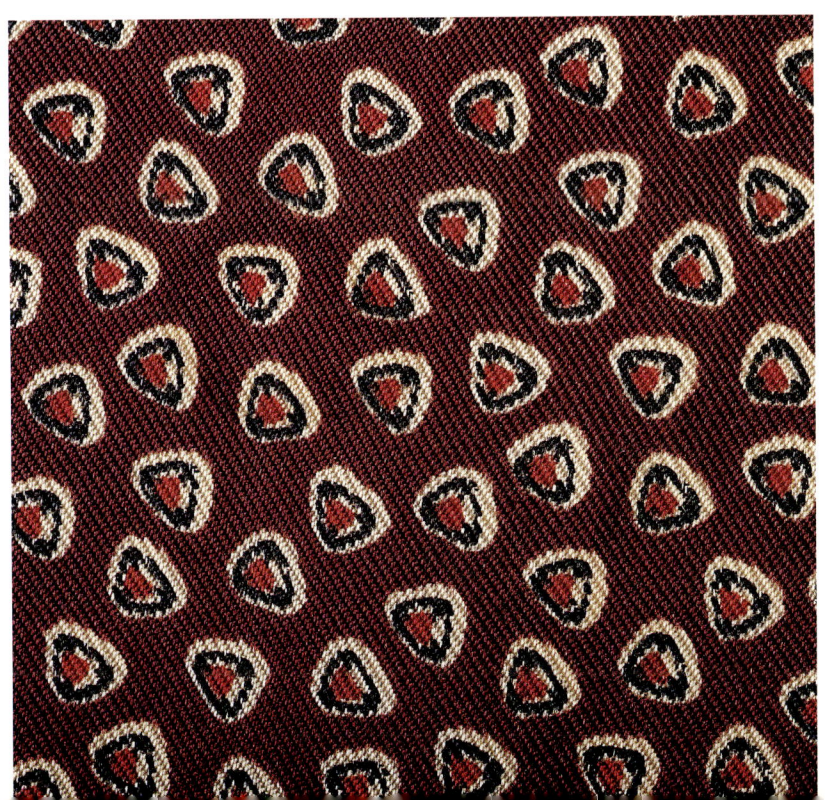

29

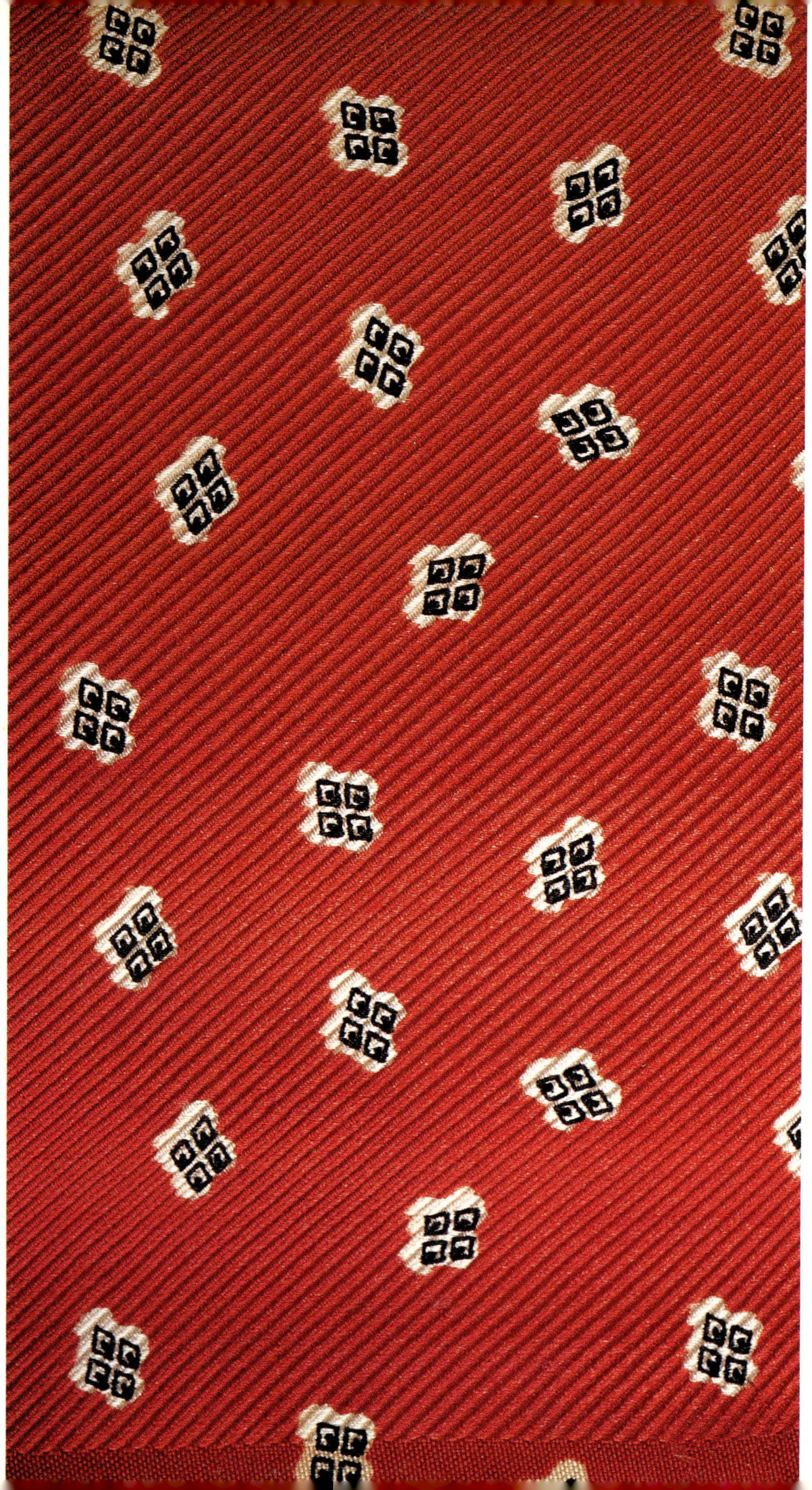

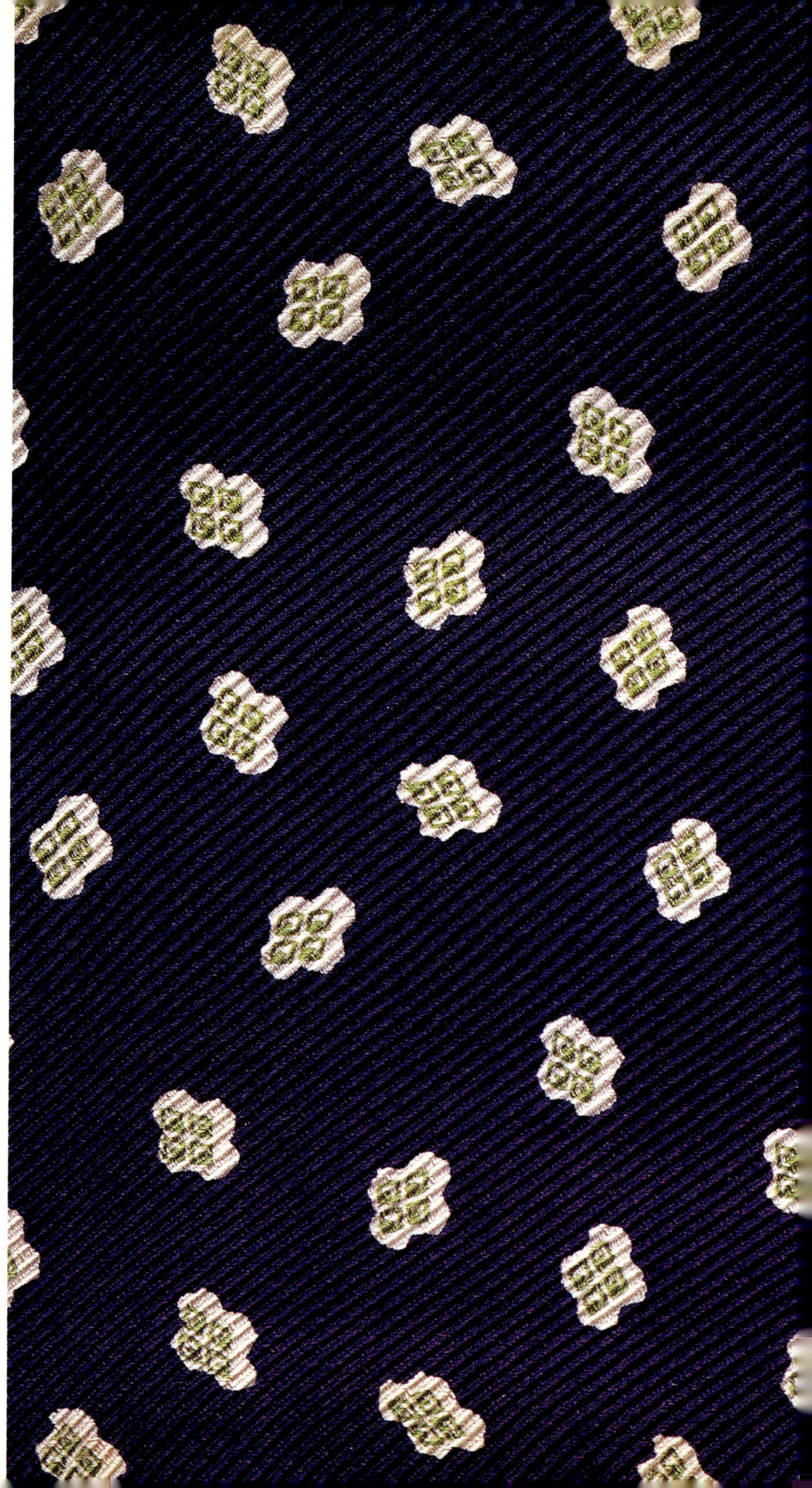

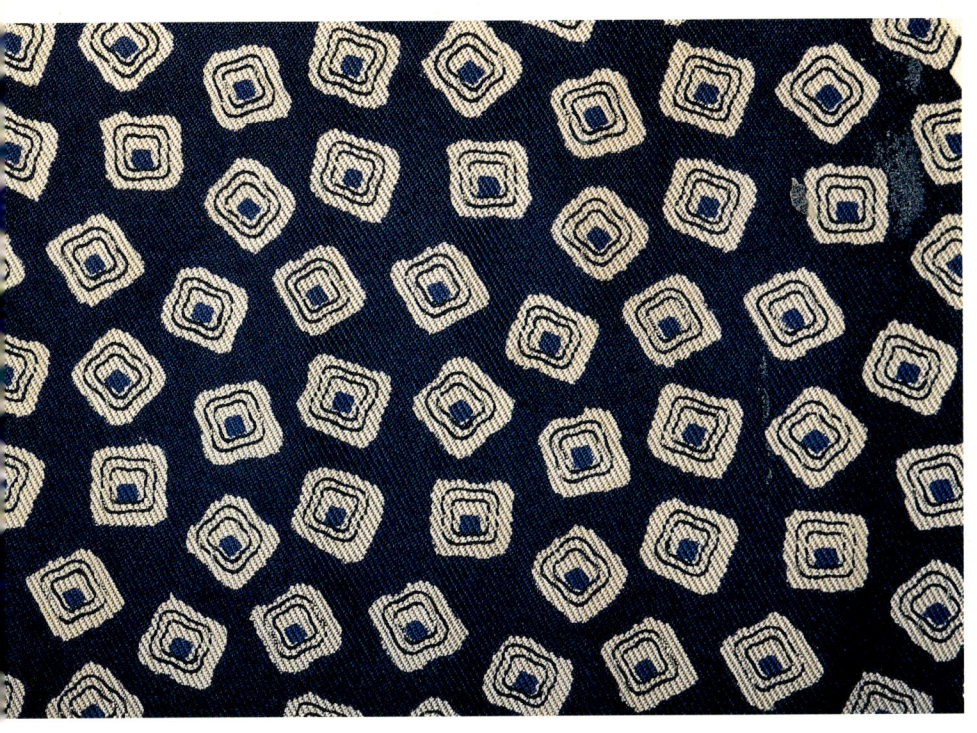
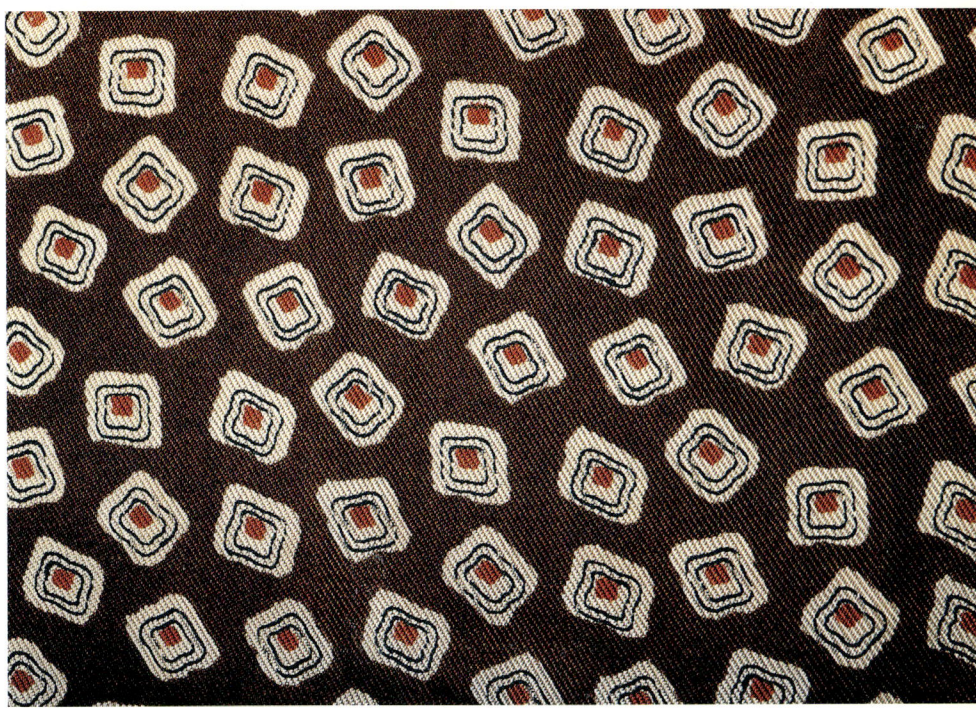
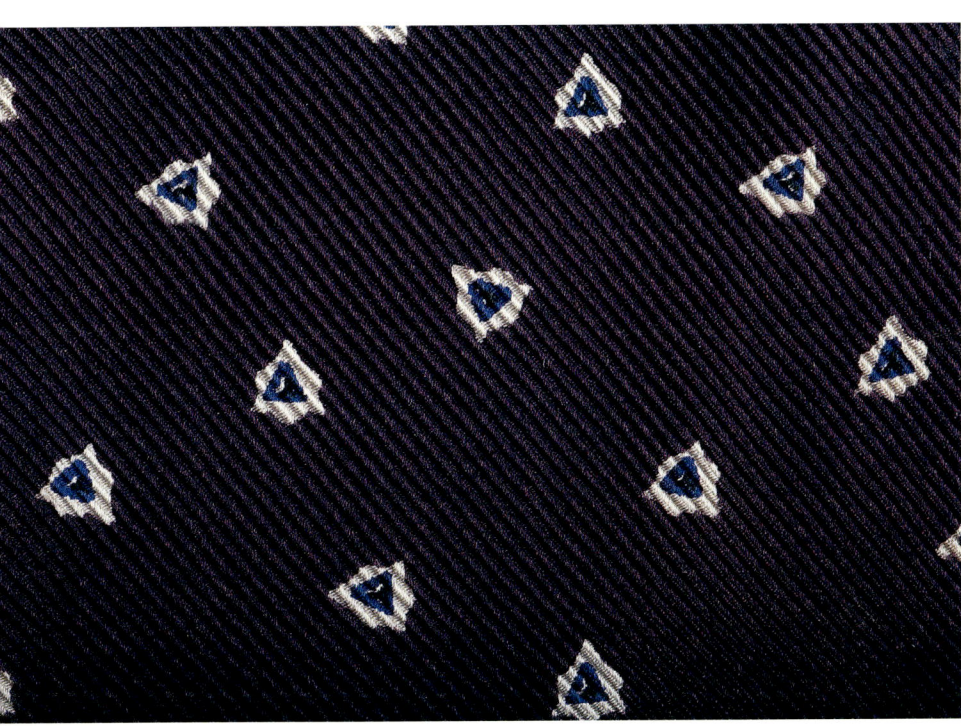
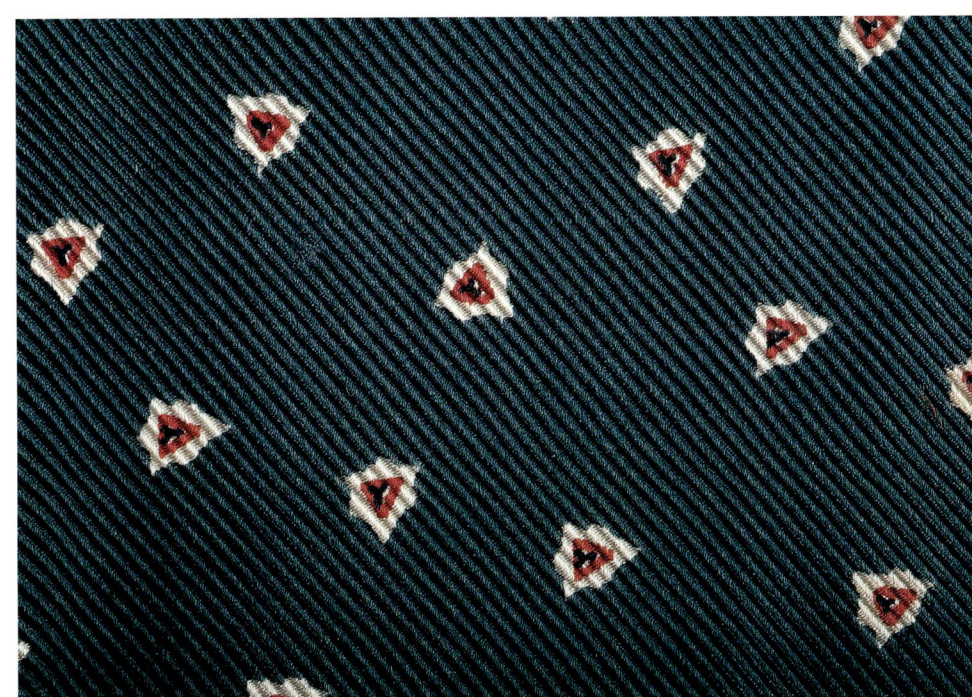

31

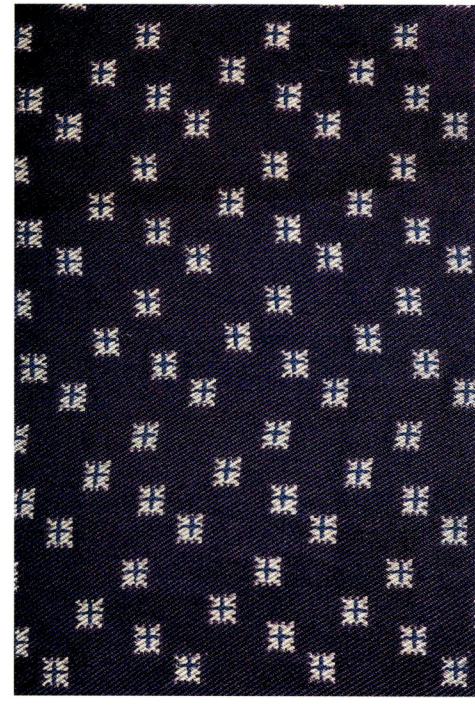

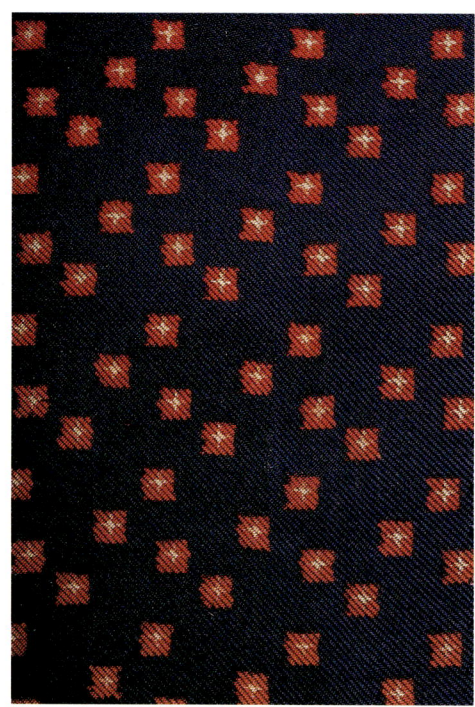

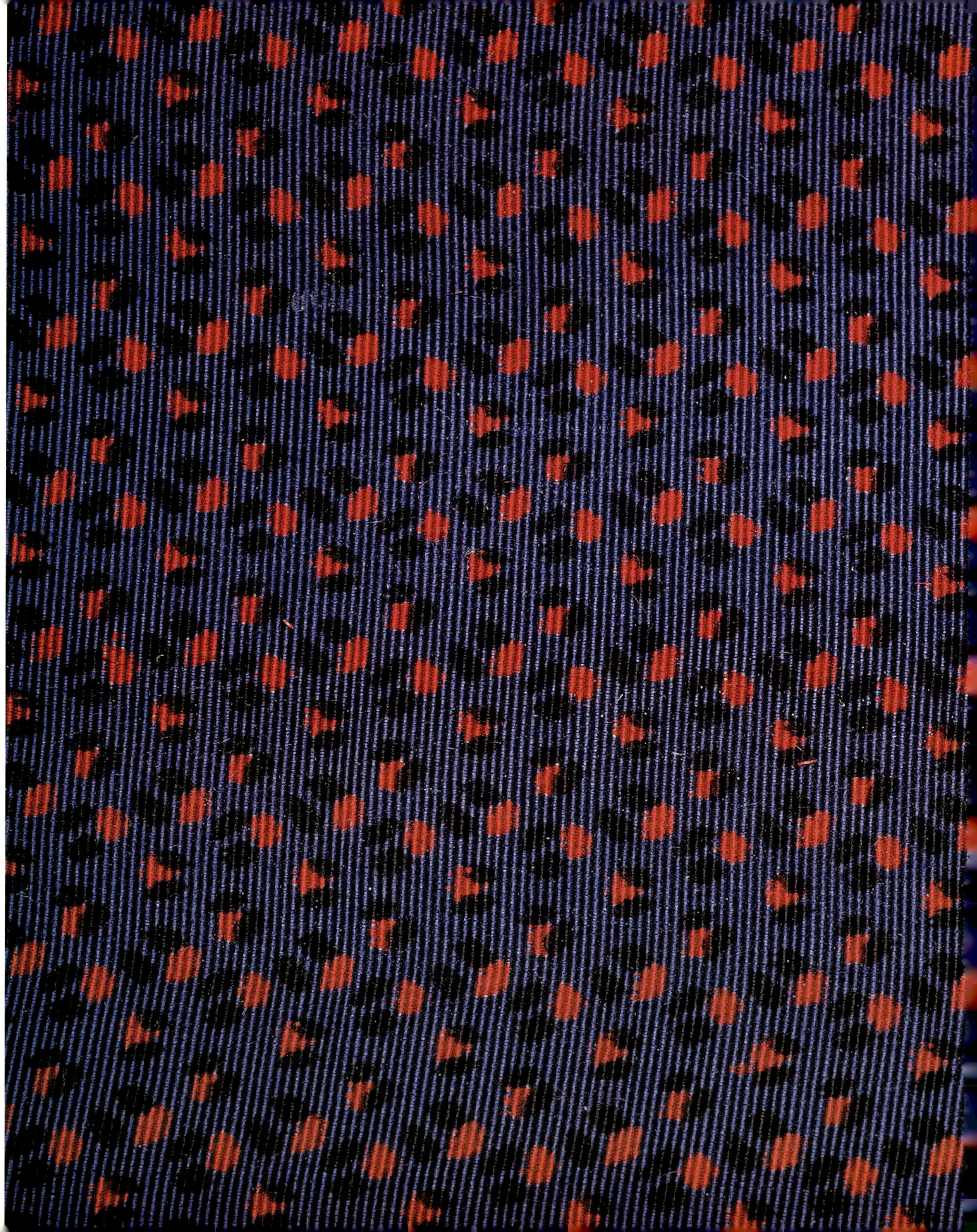

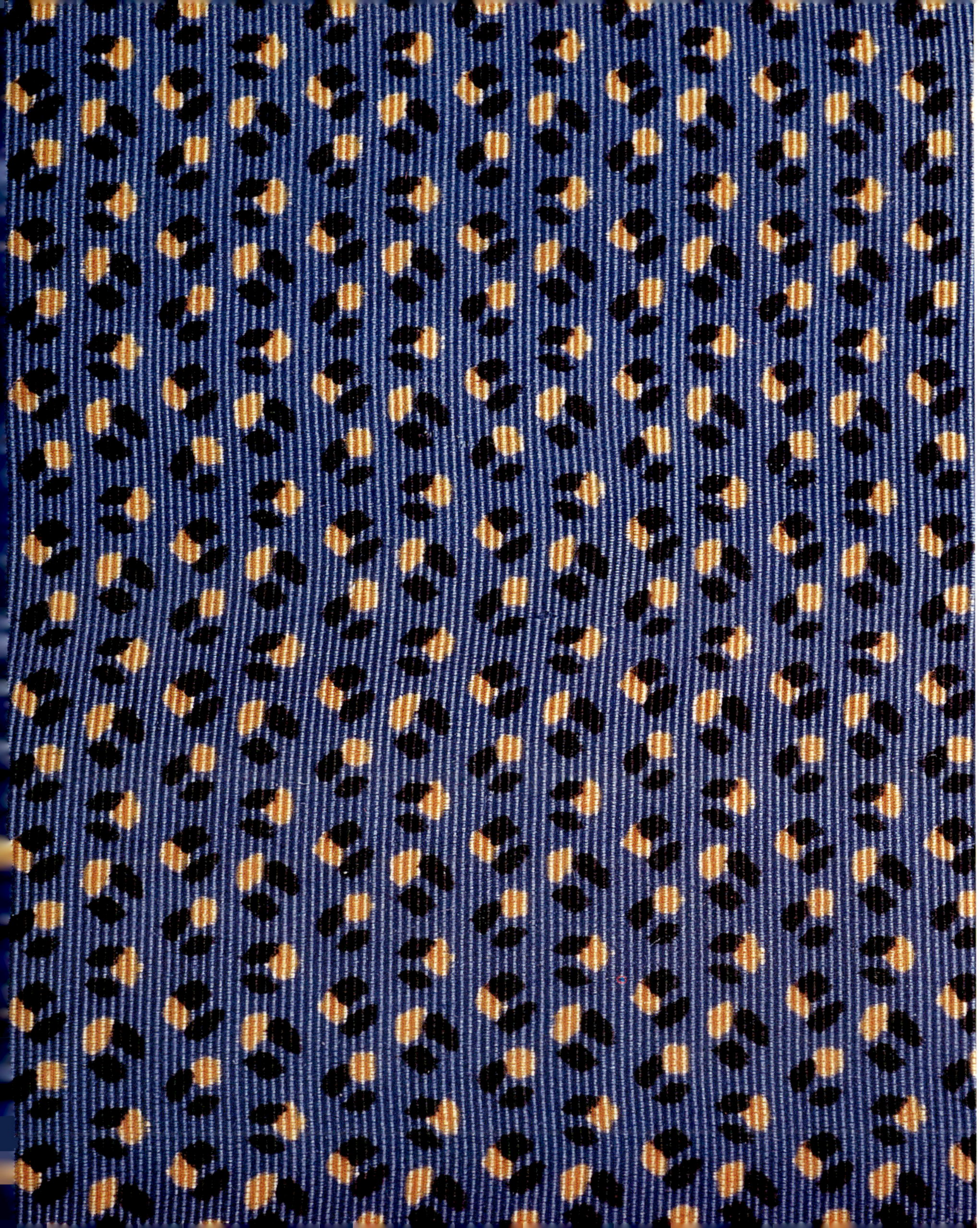

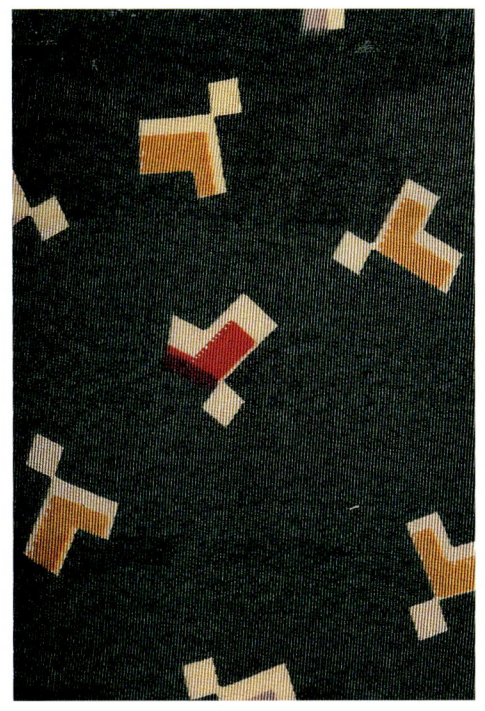

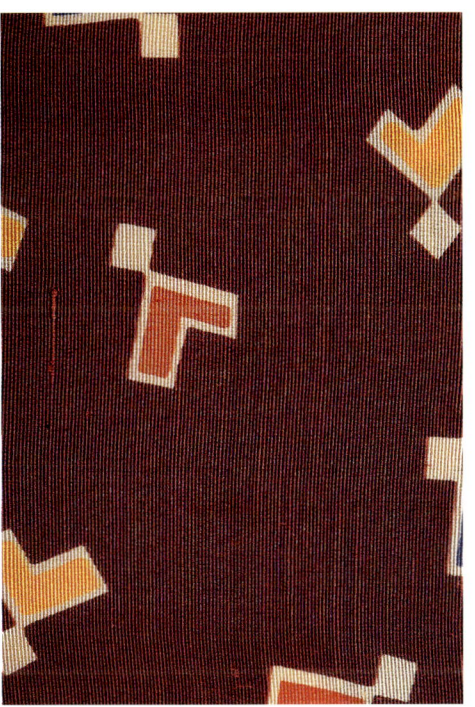

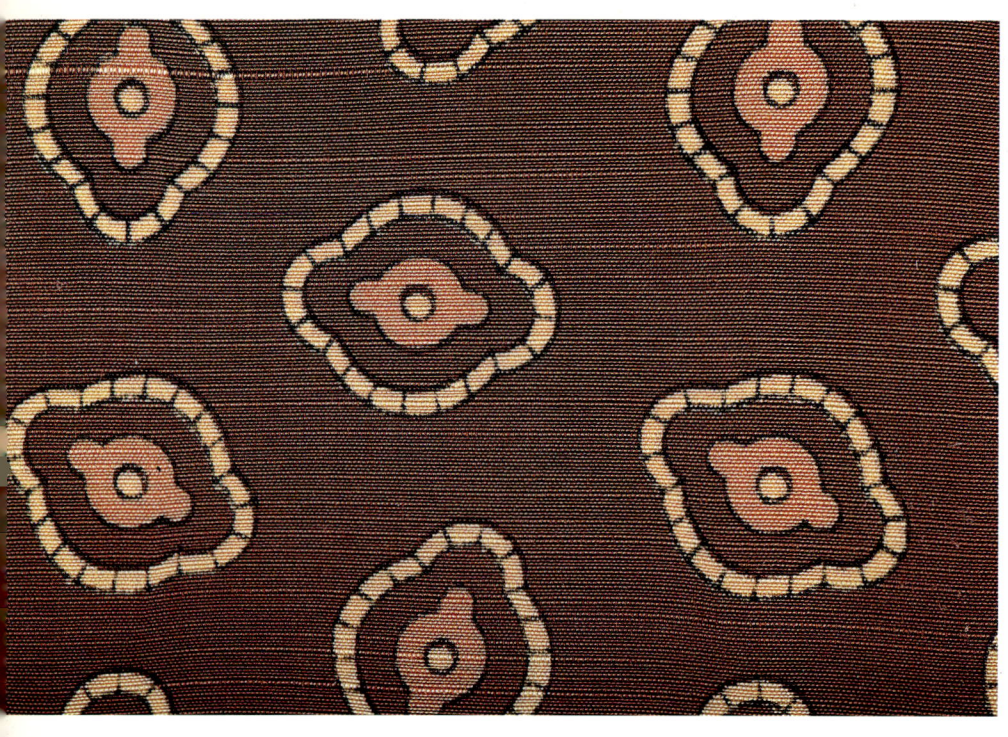
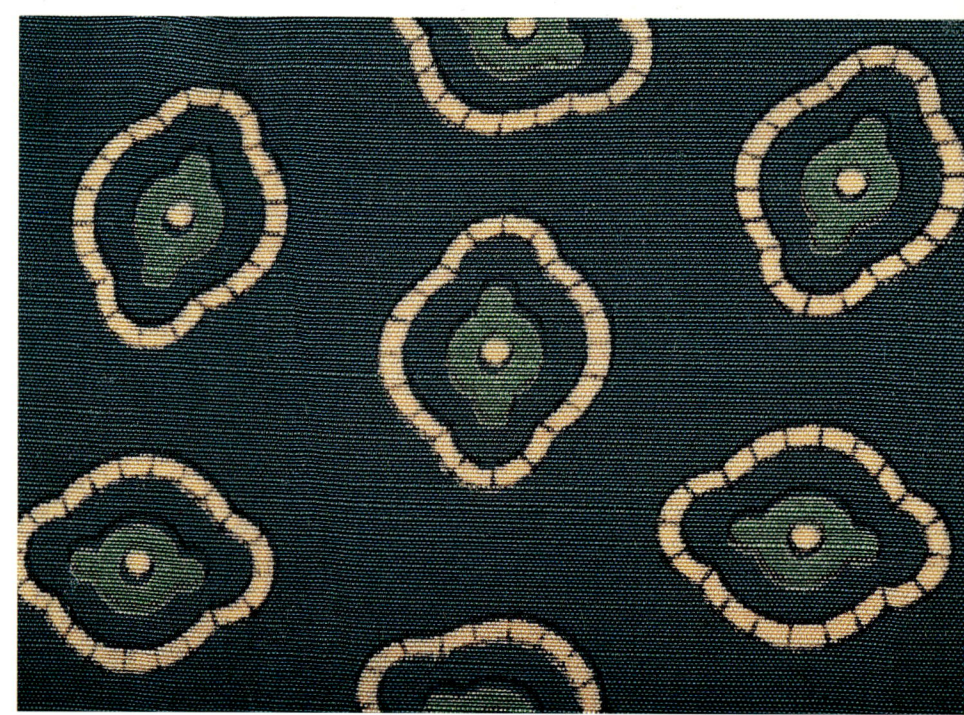
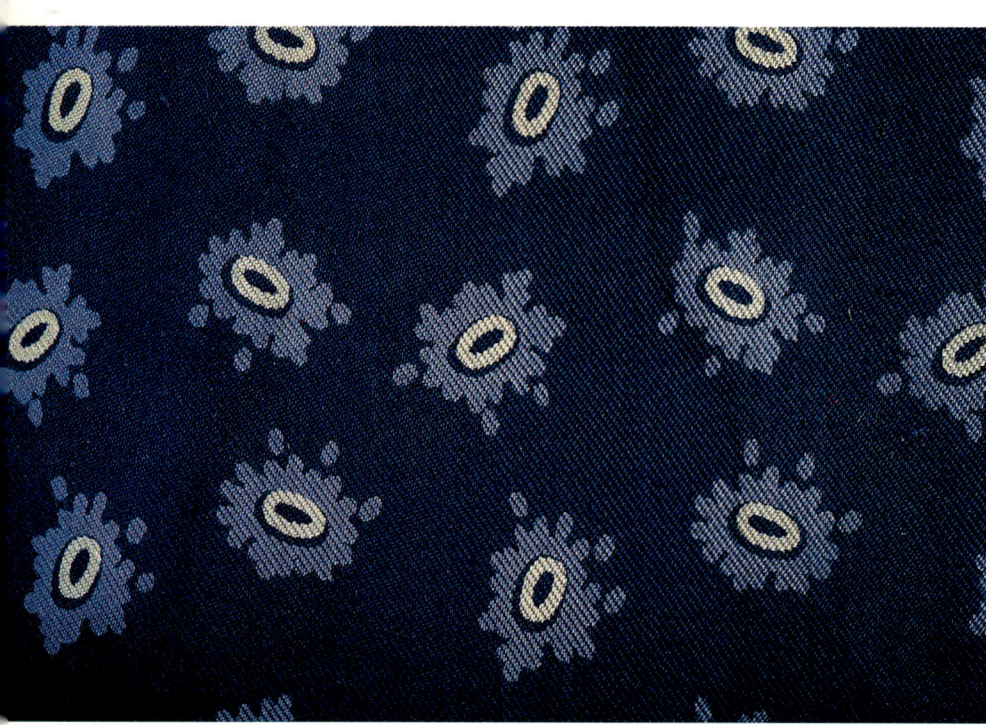
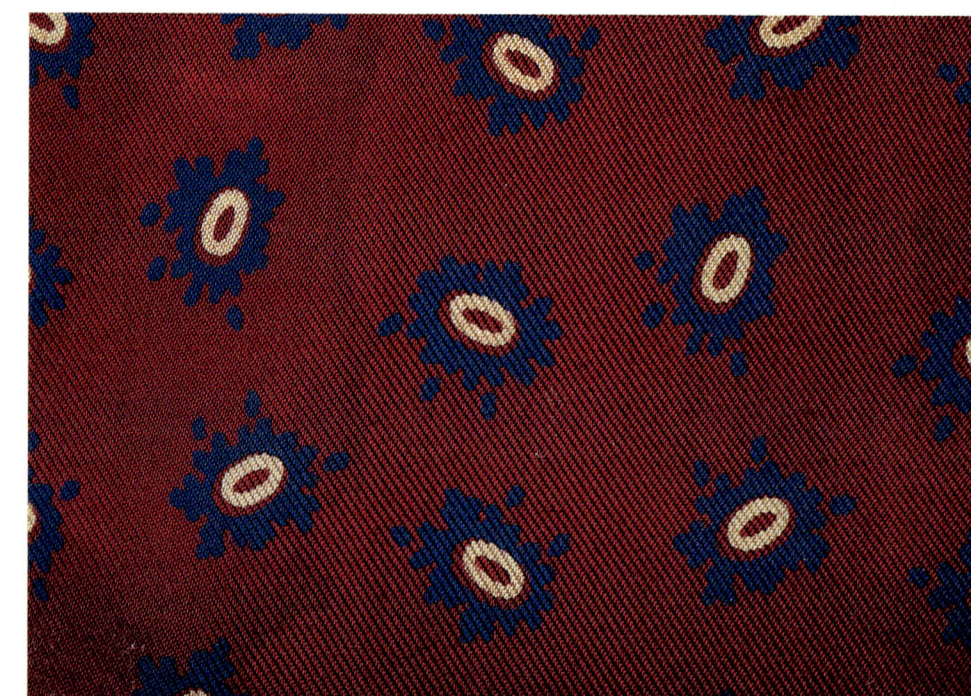

34

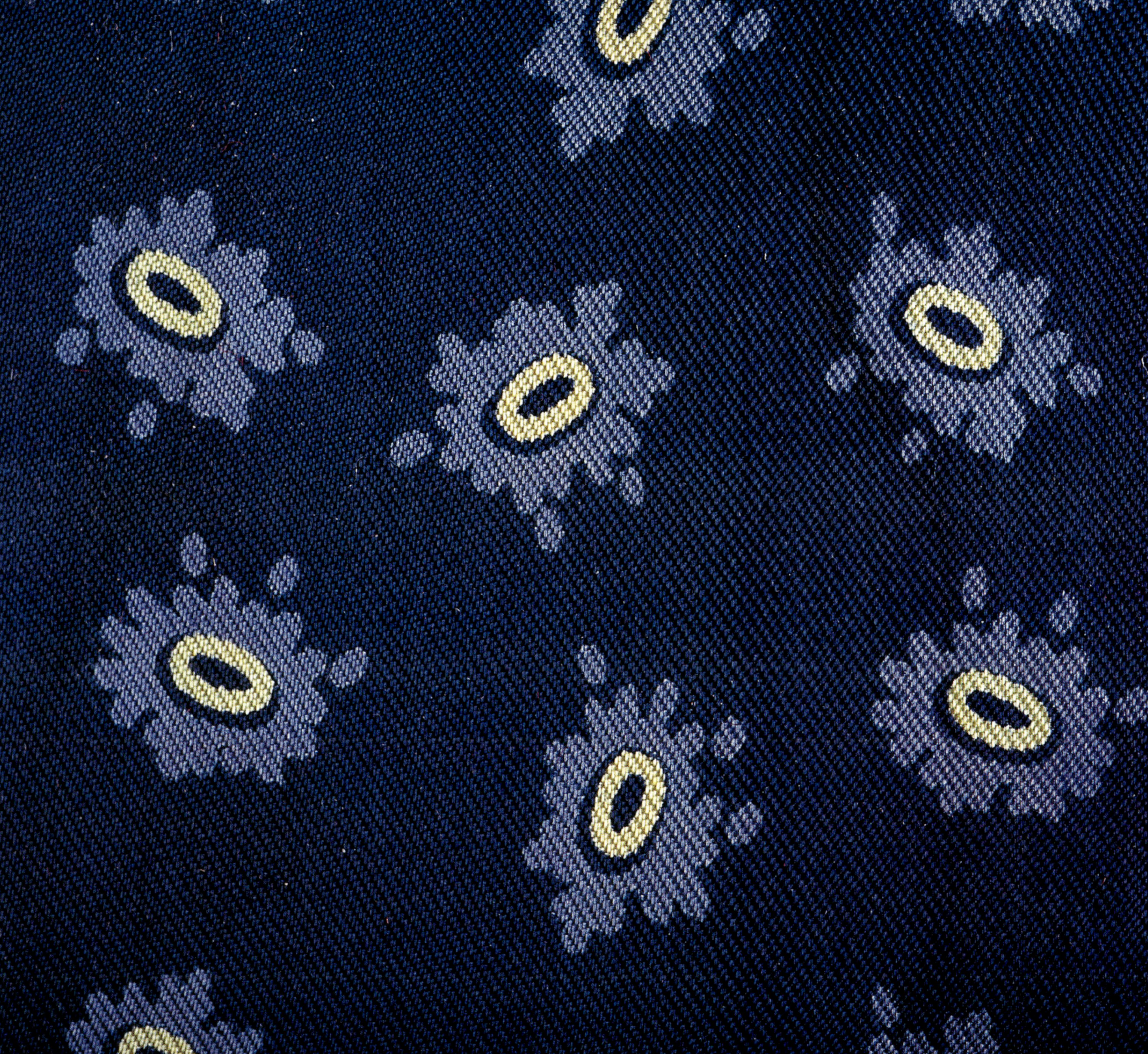

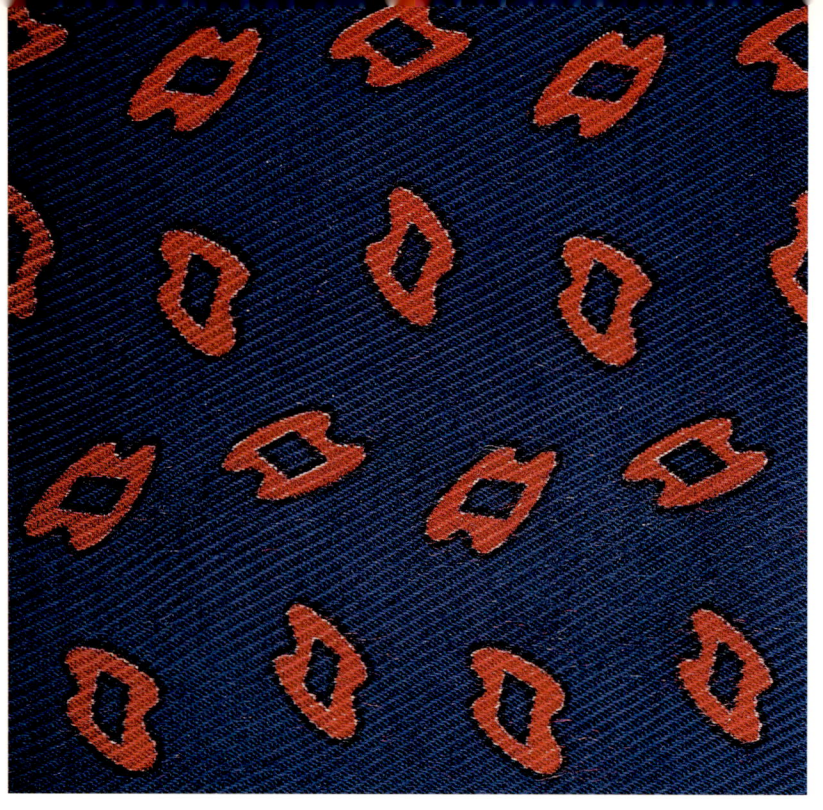
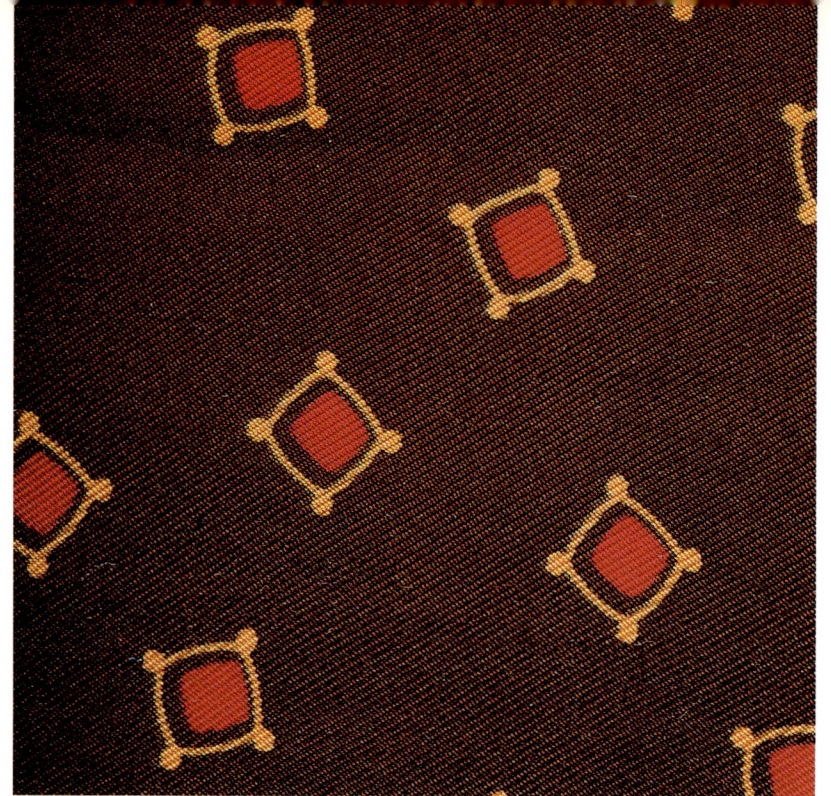
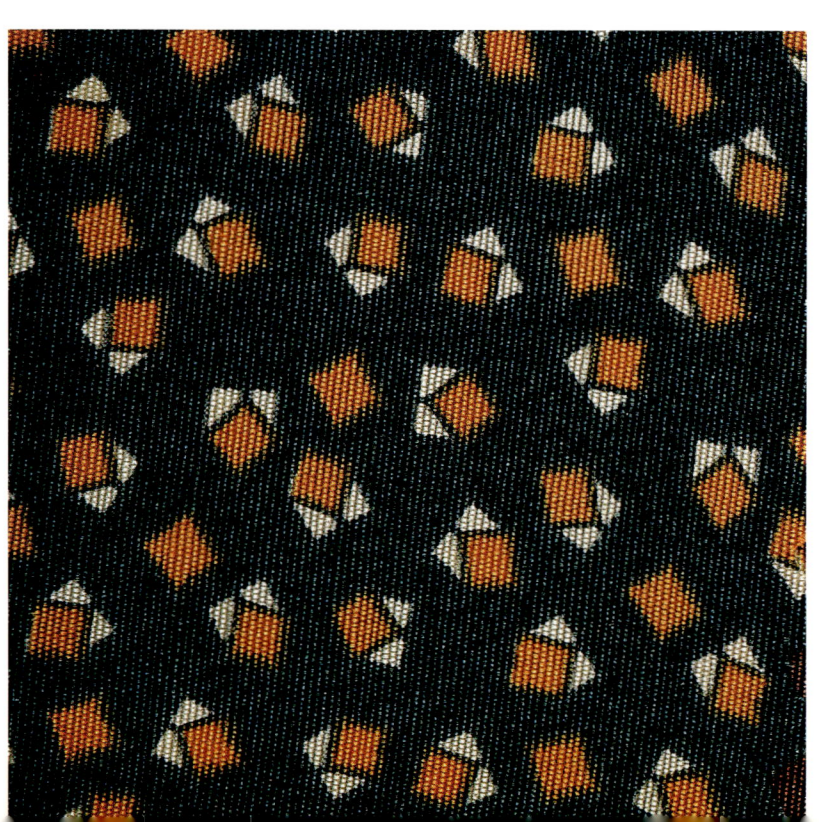
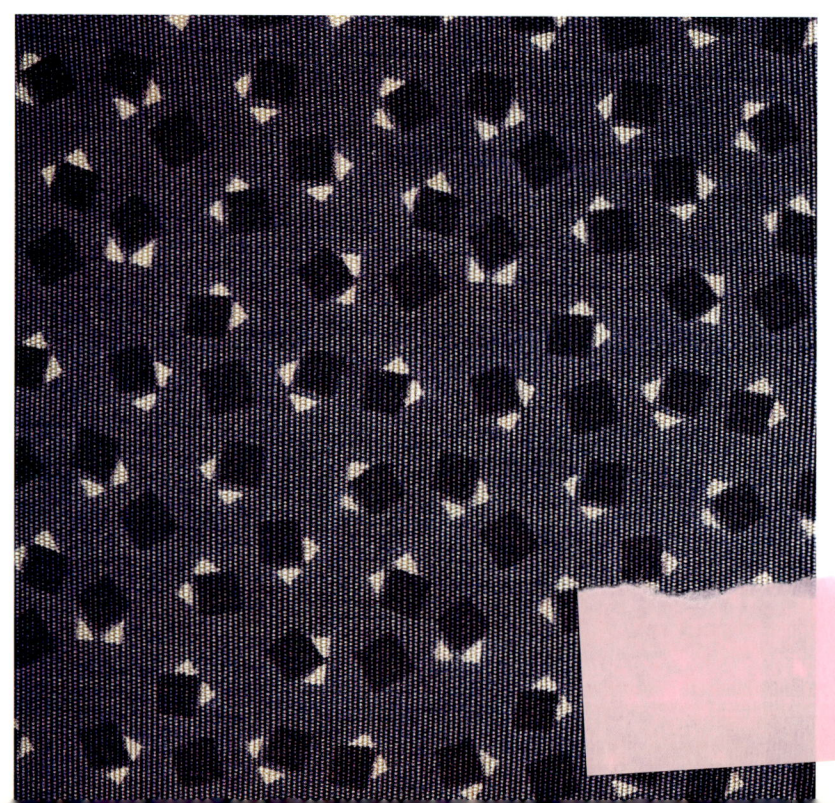

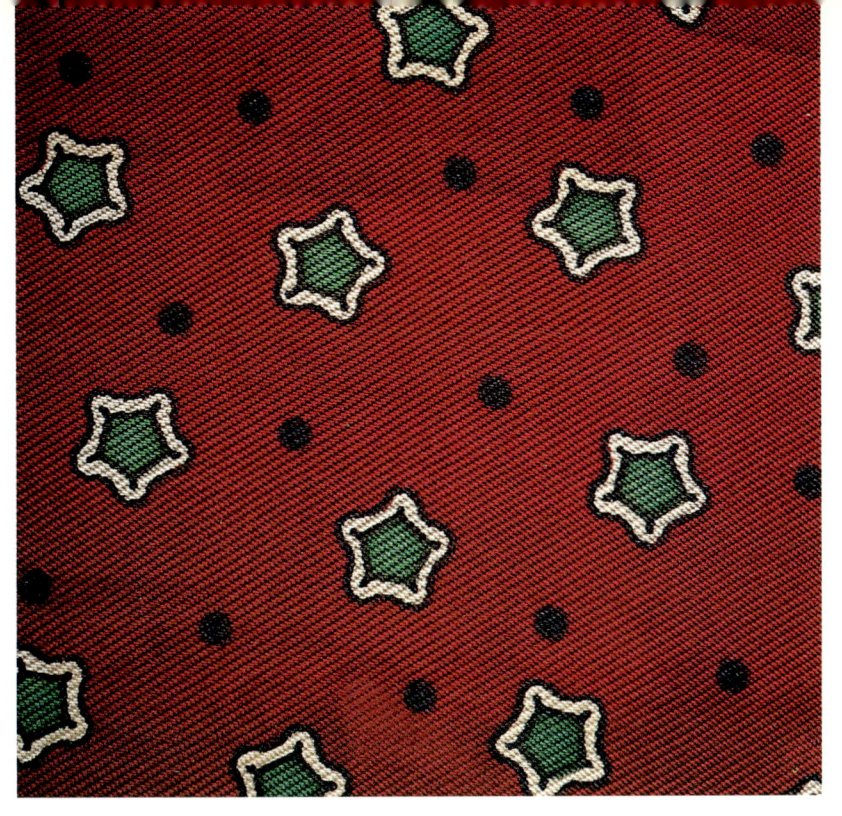
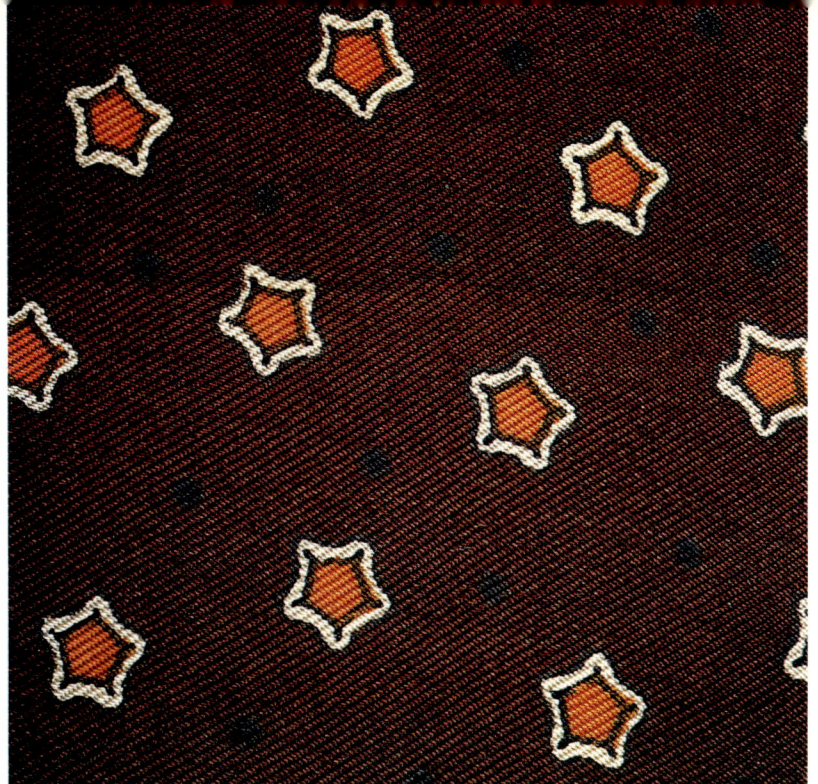
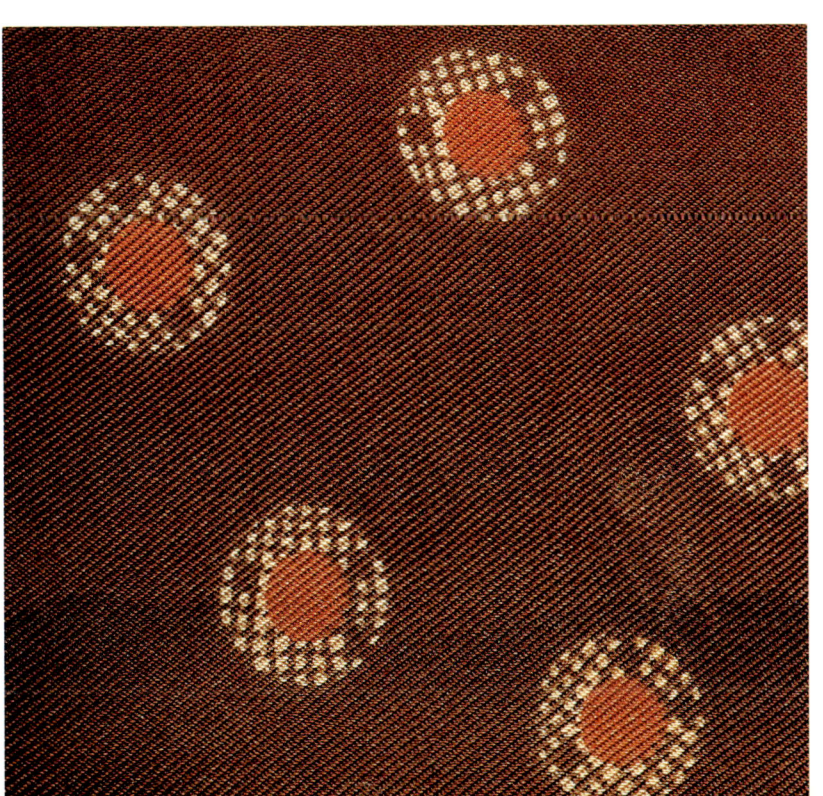
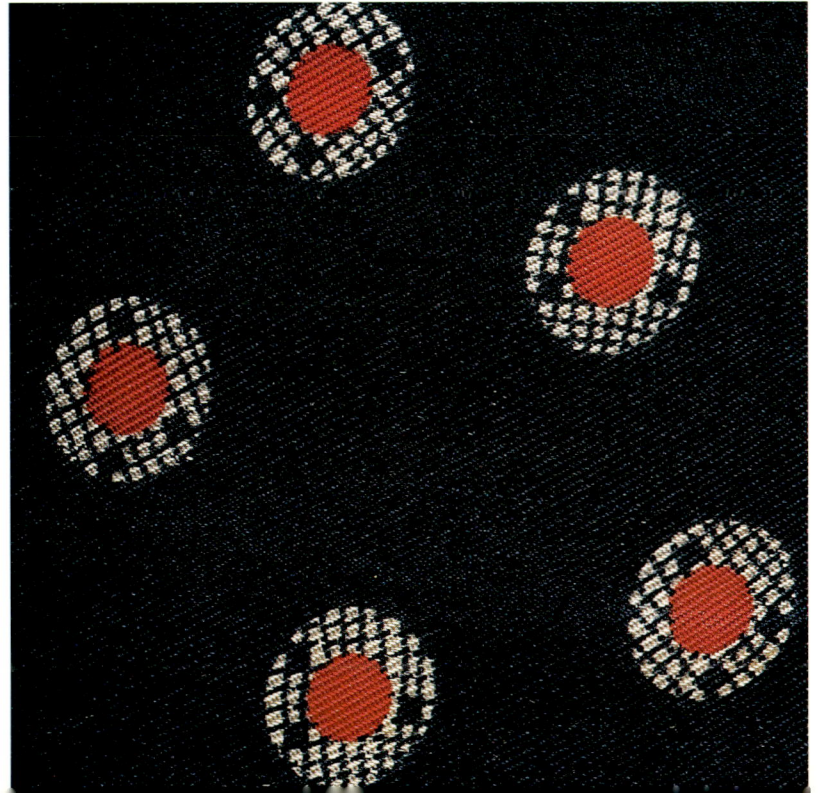

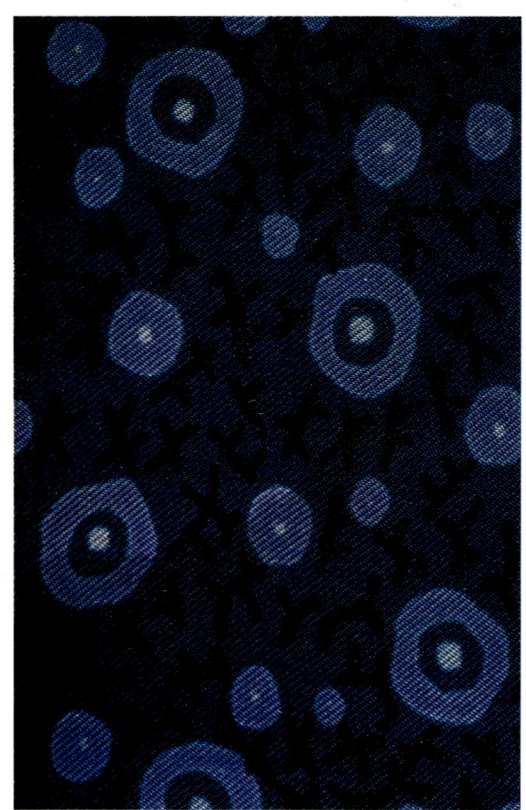
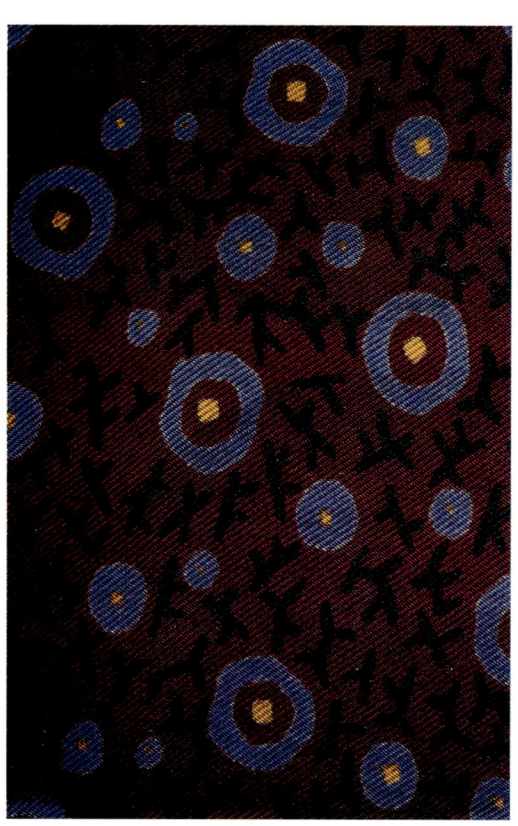

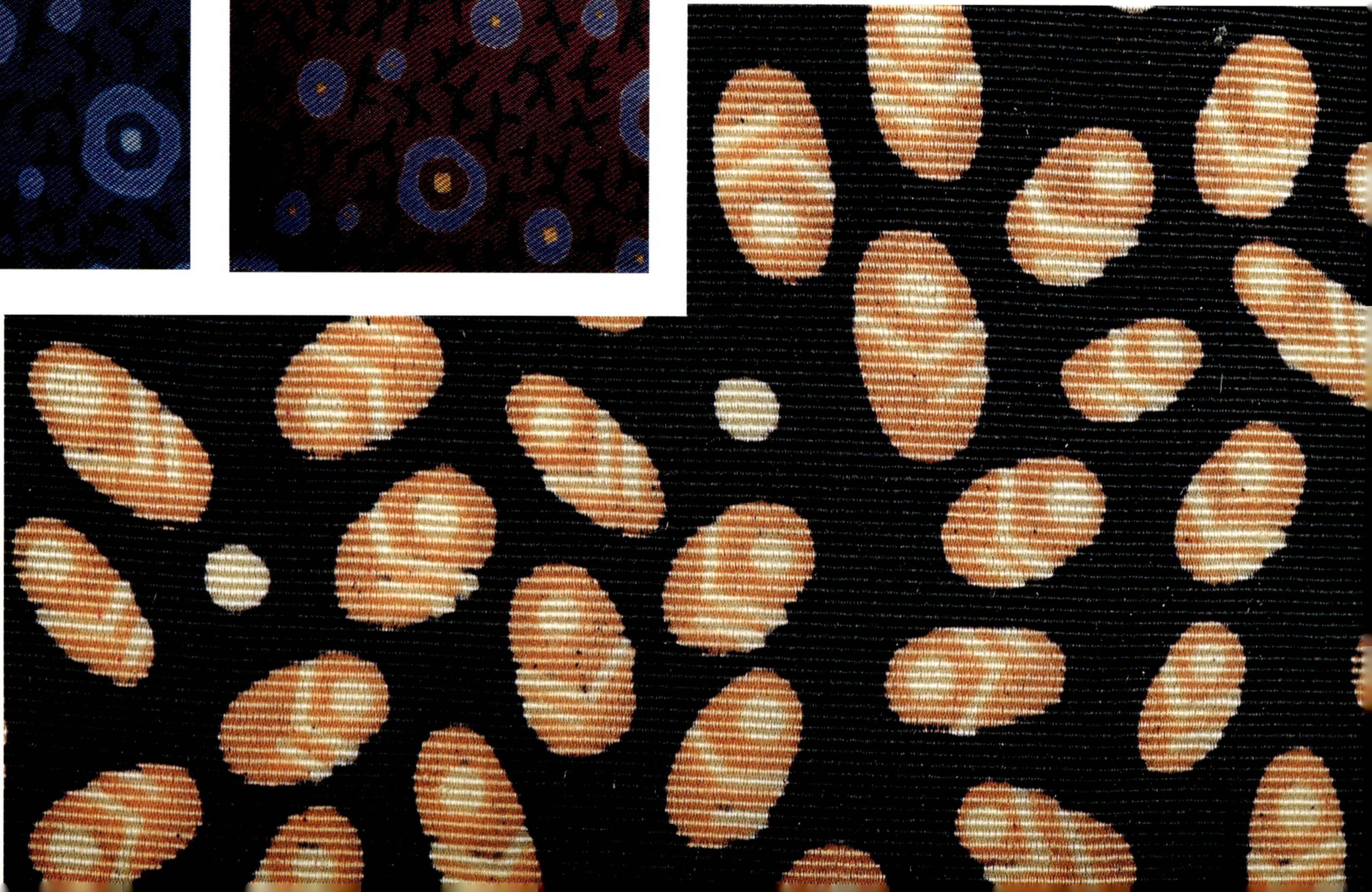

Alternating

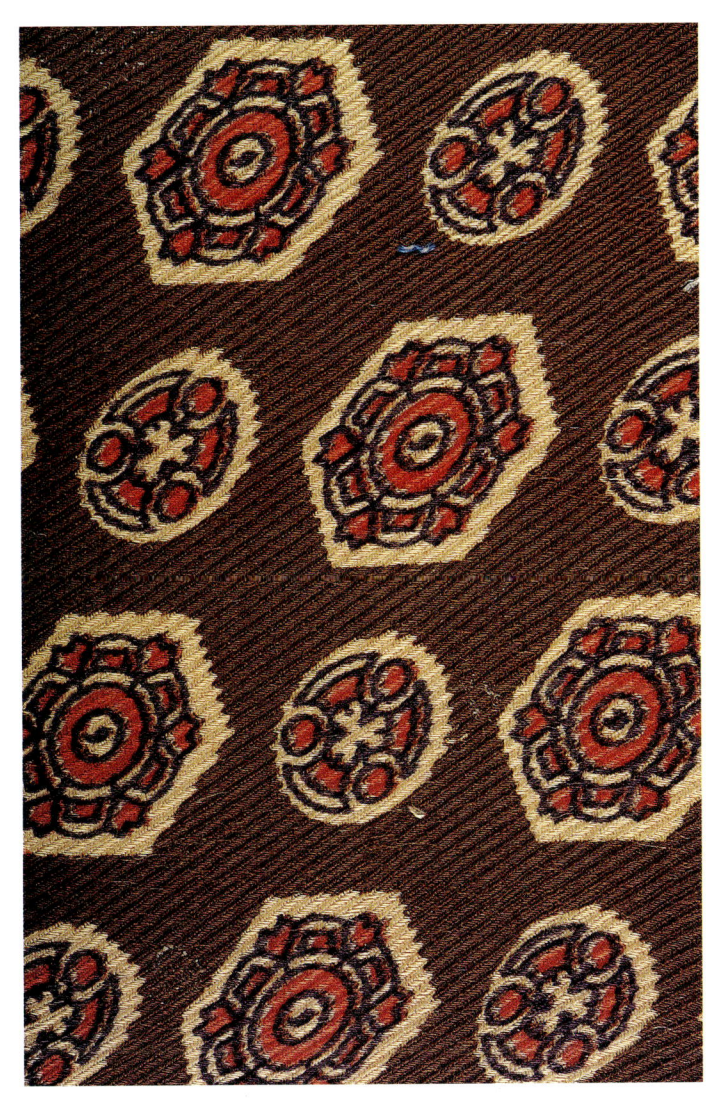

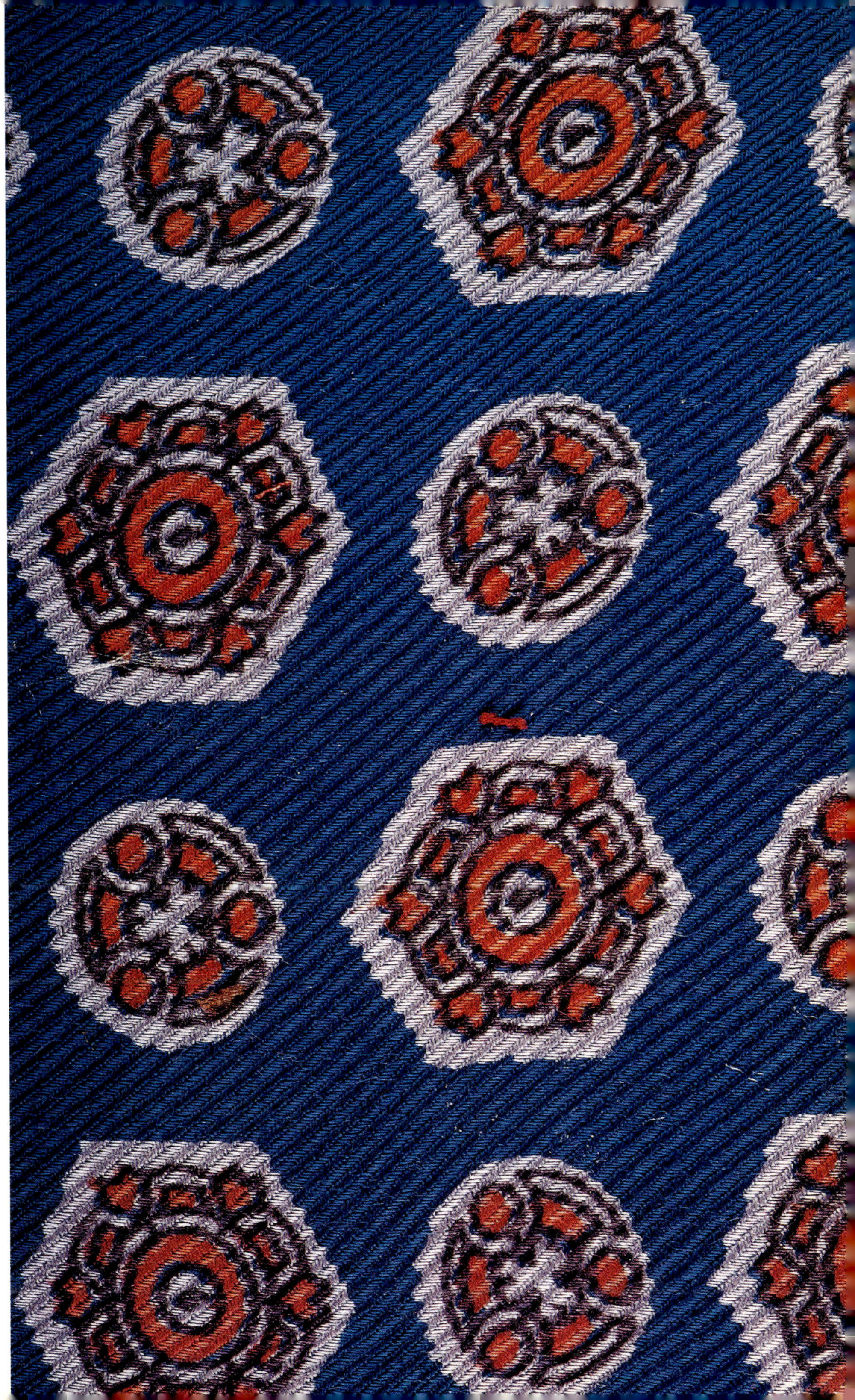

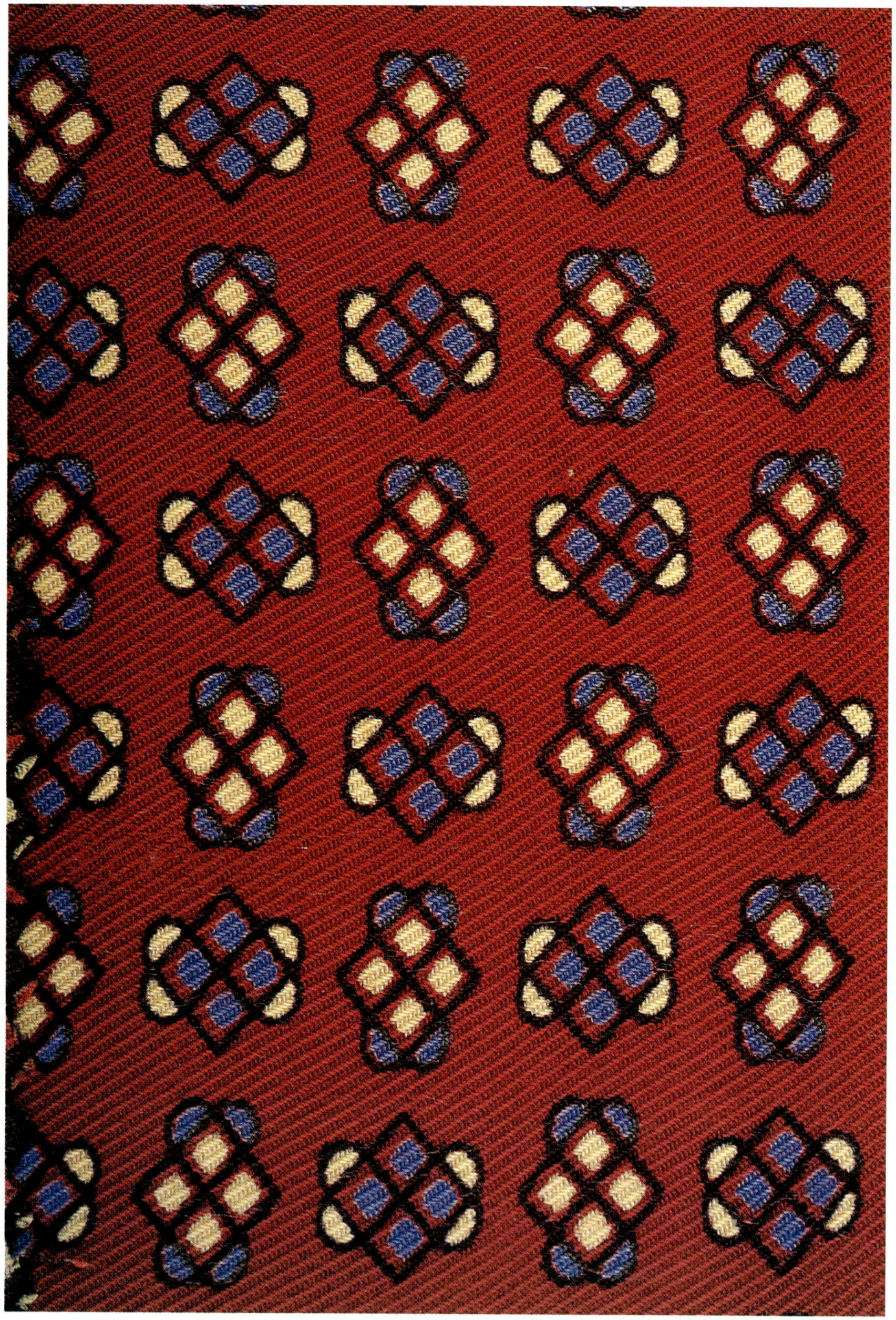

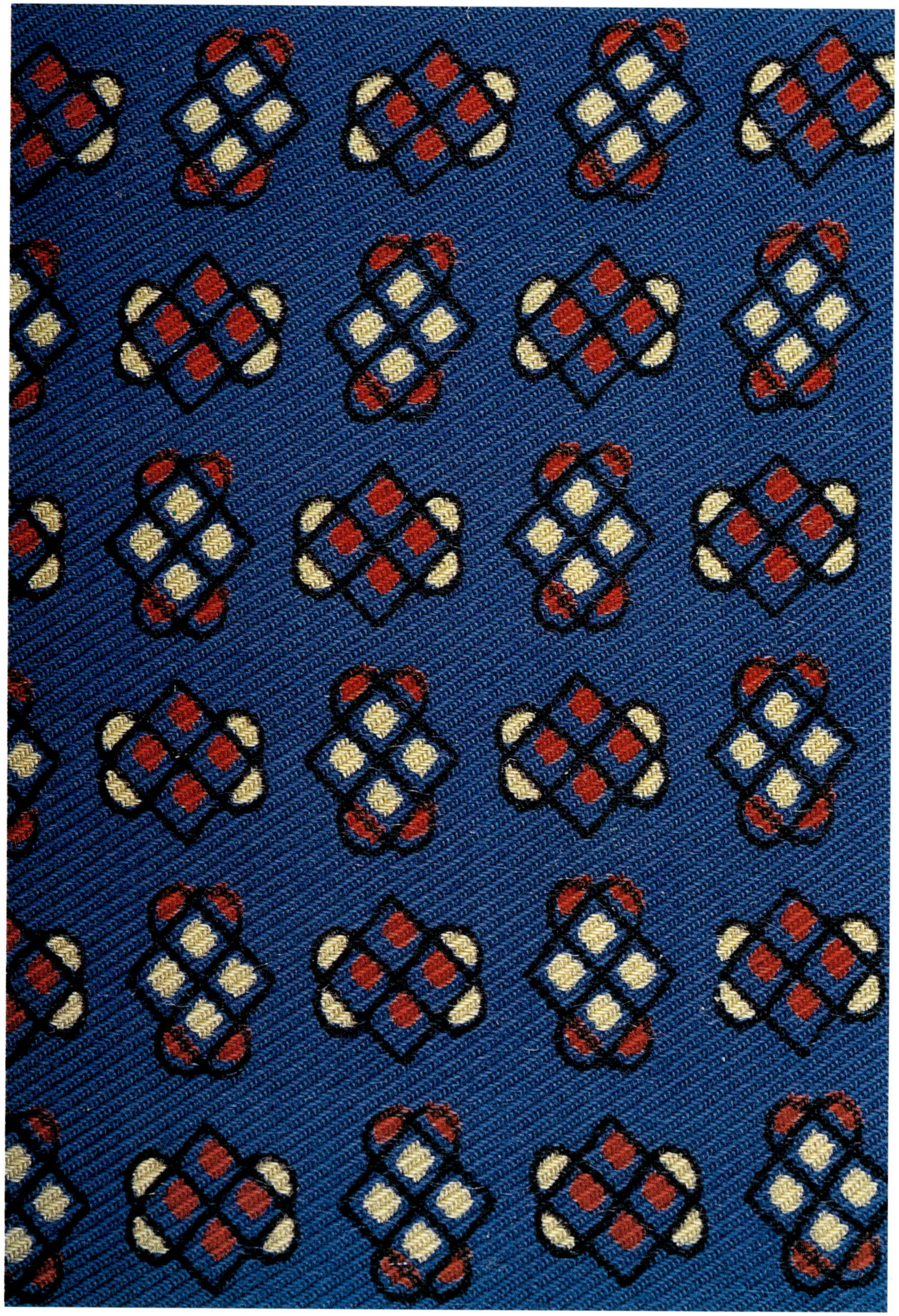

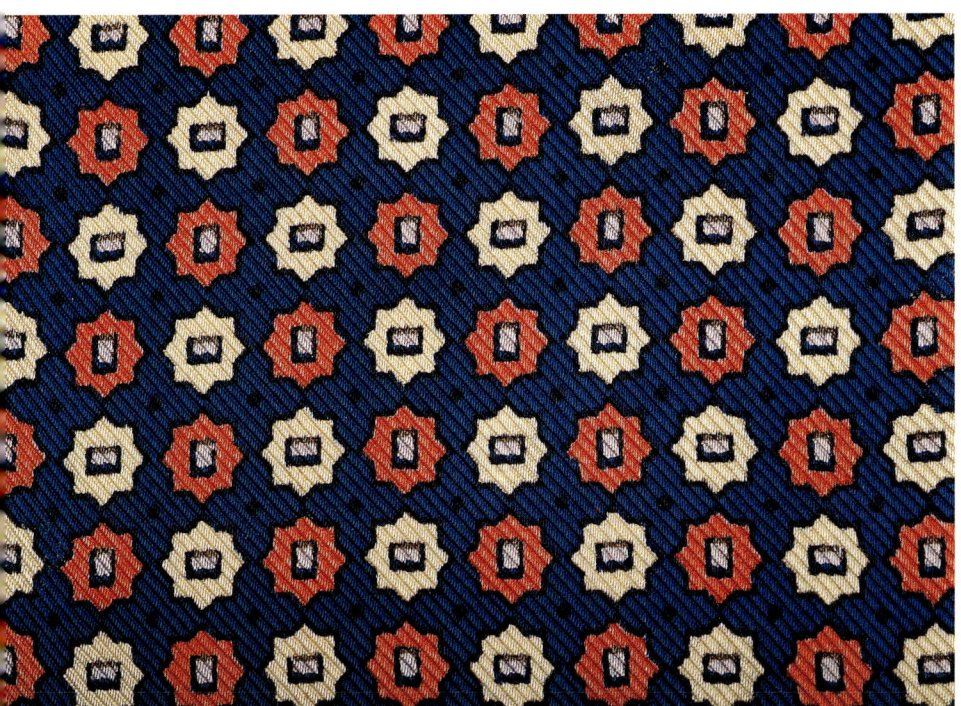
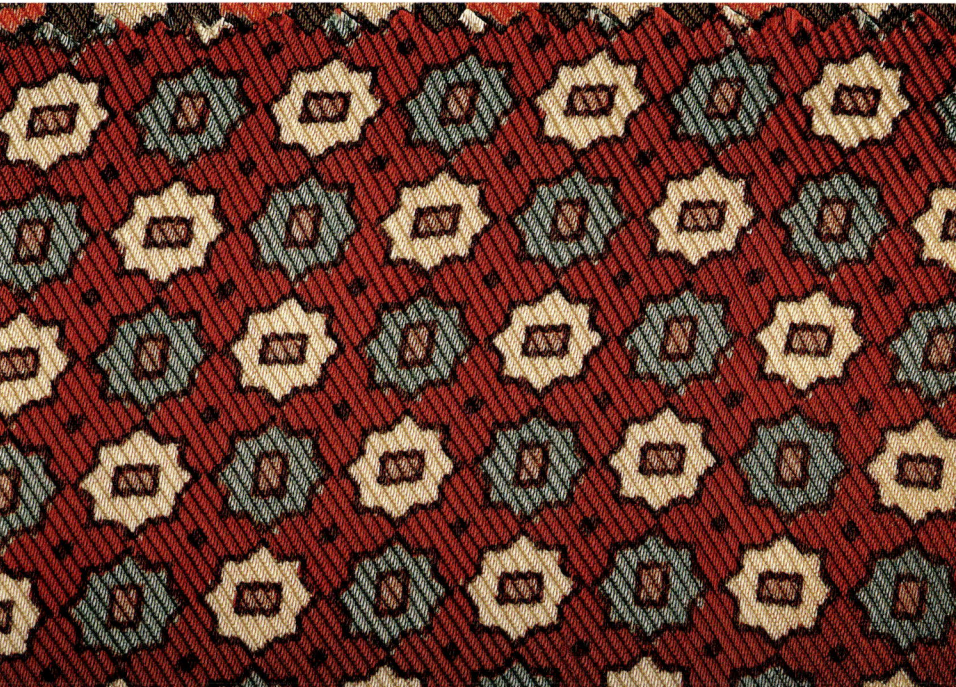
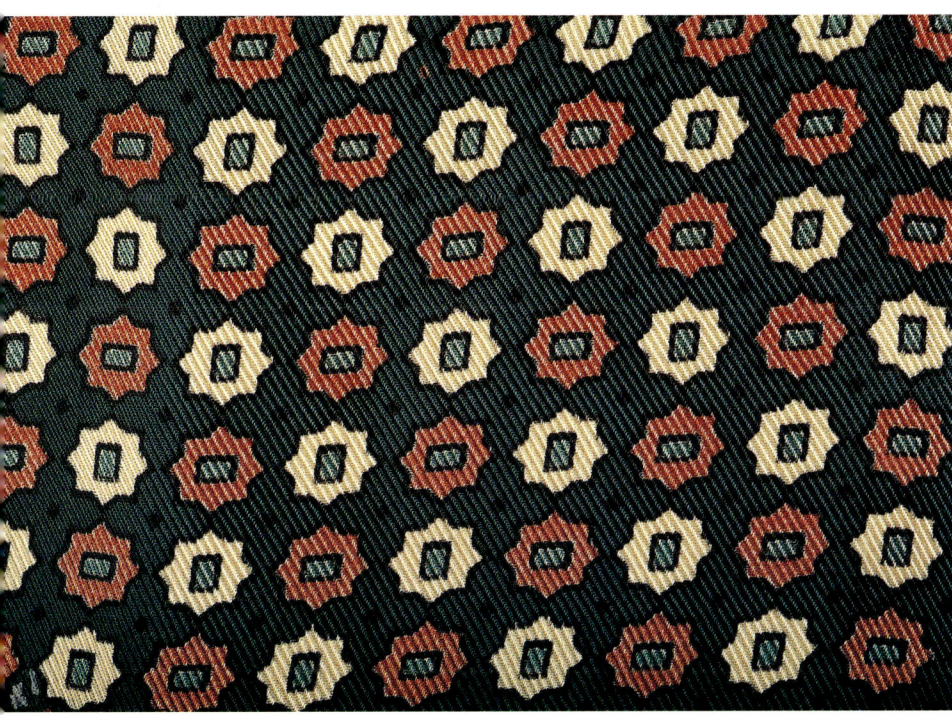
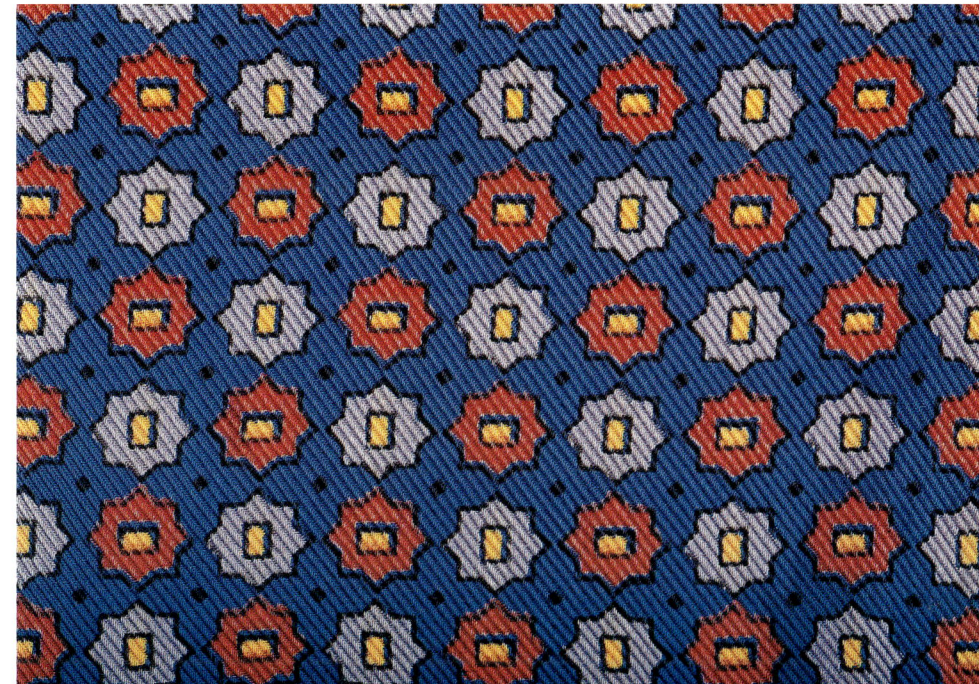

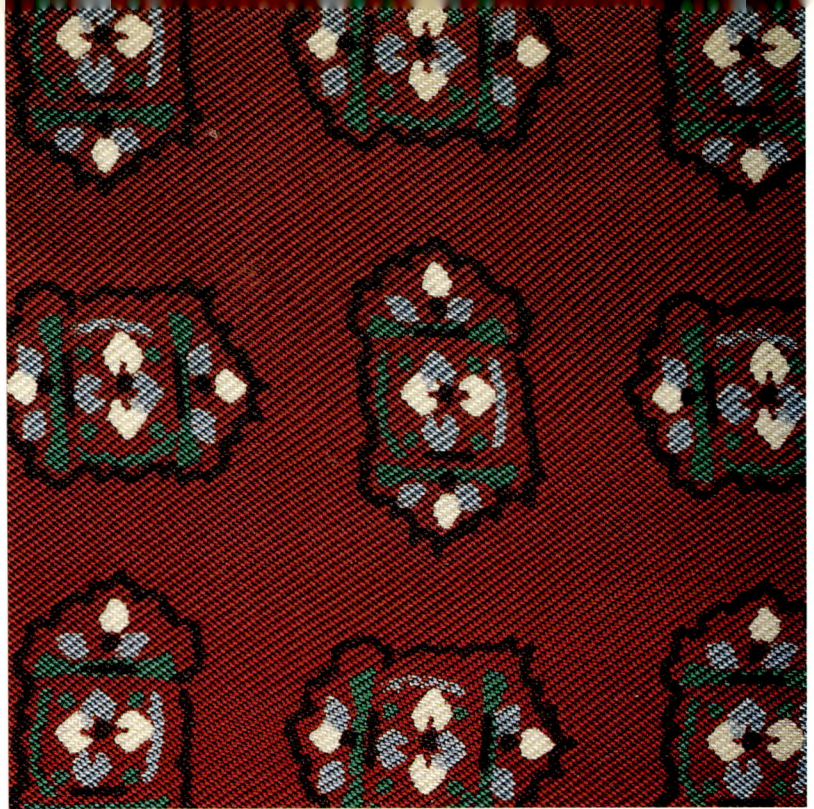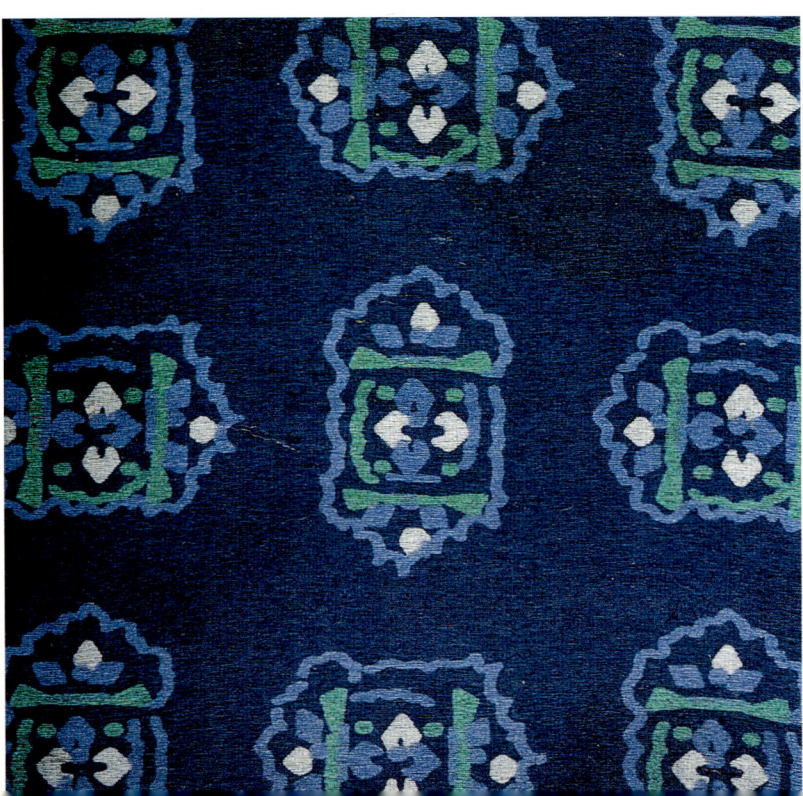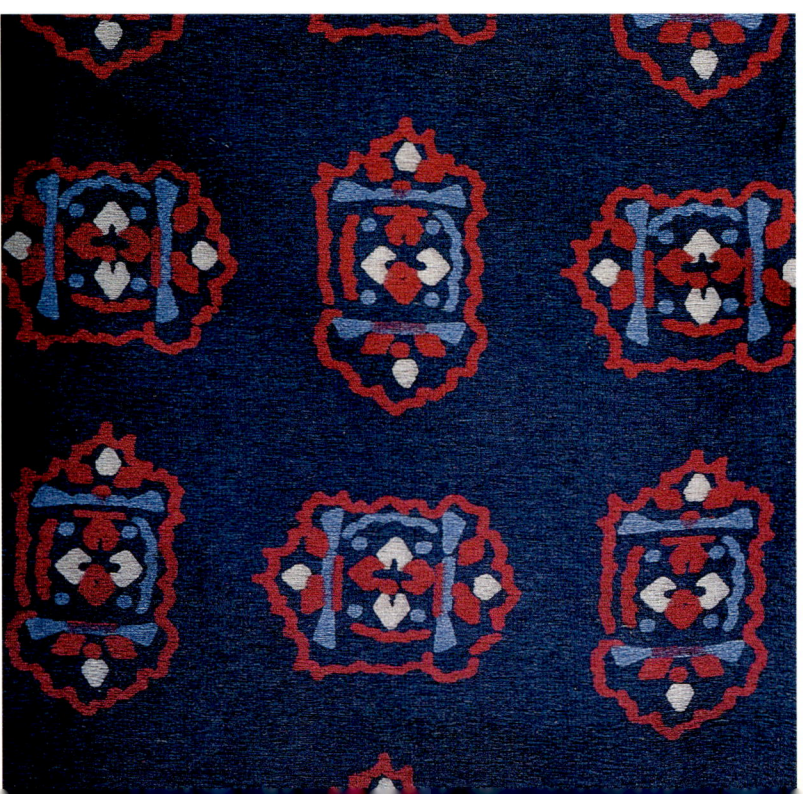

42

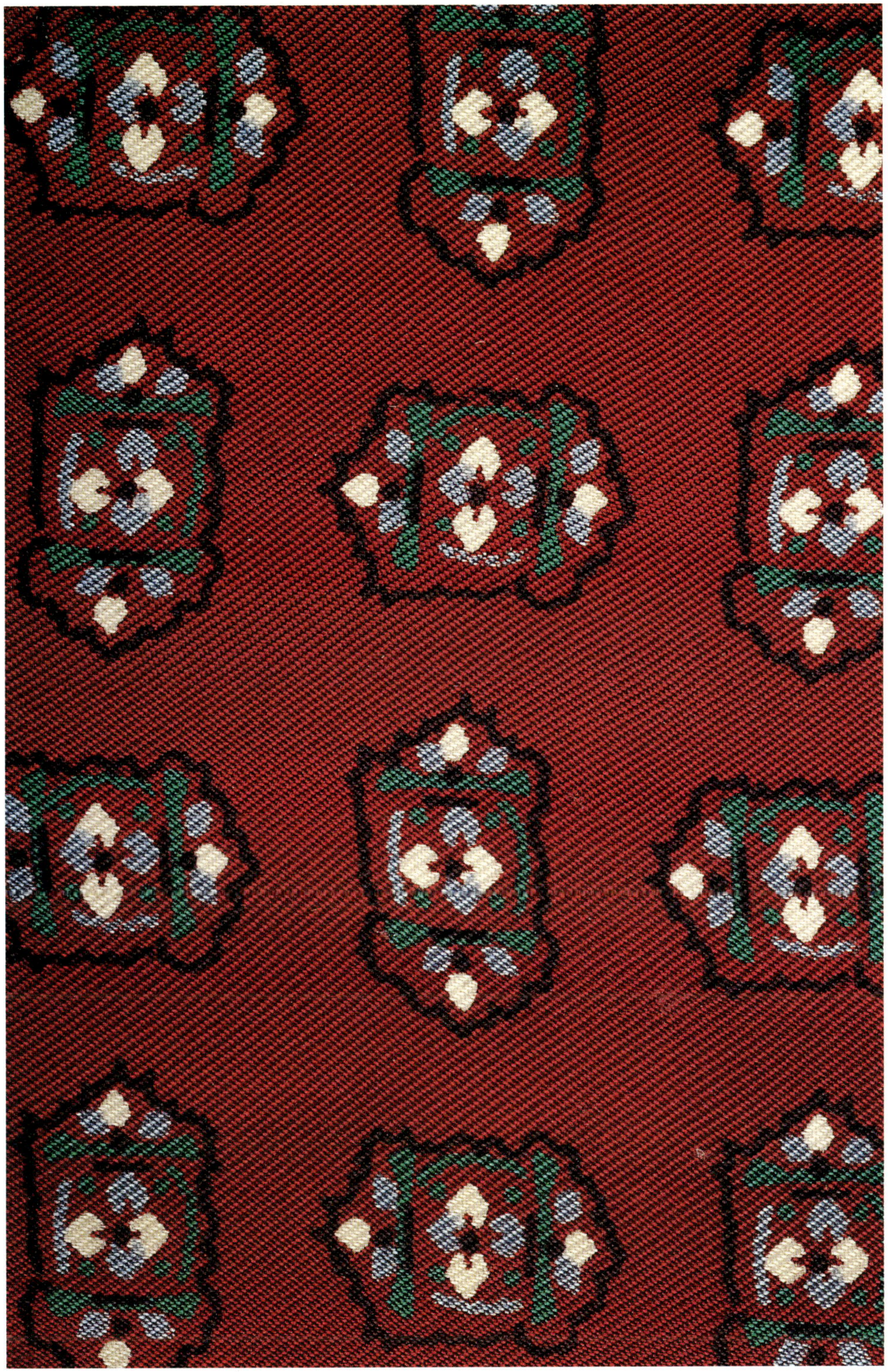

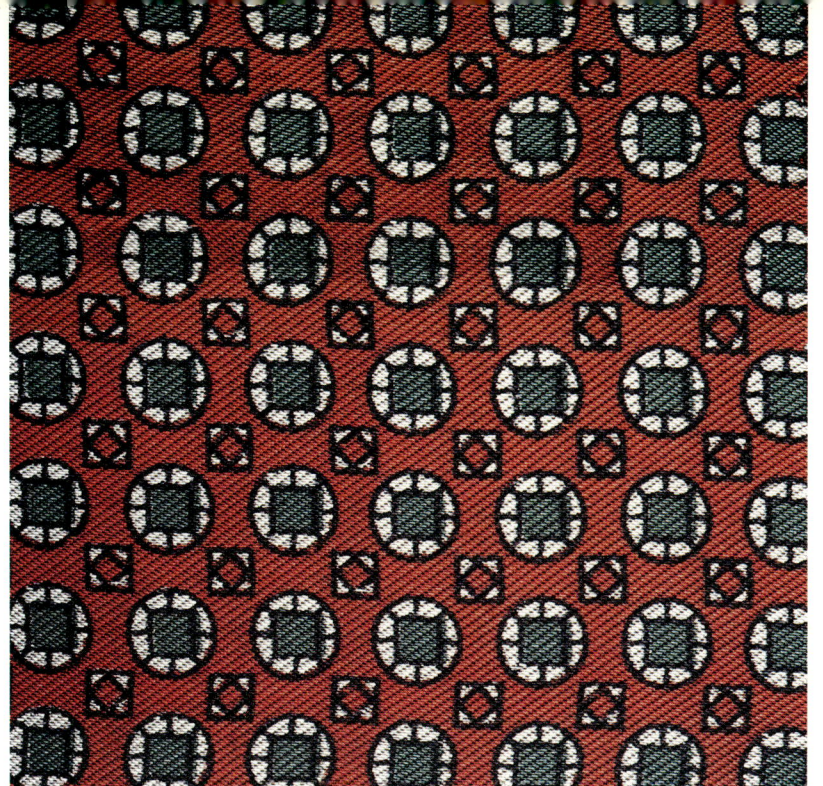

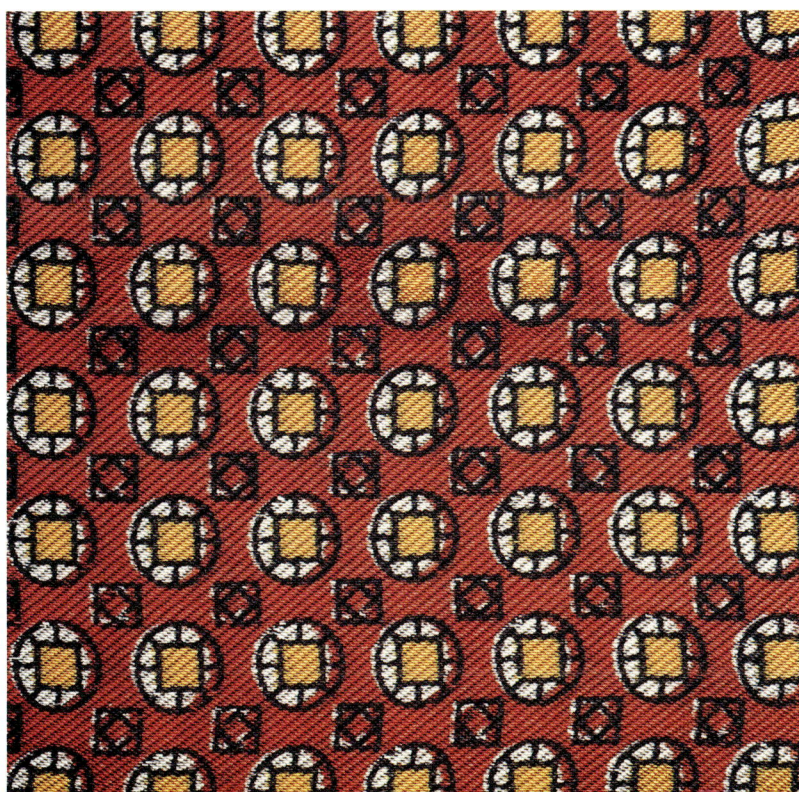

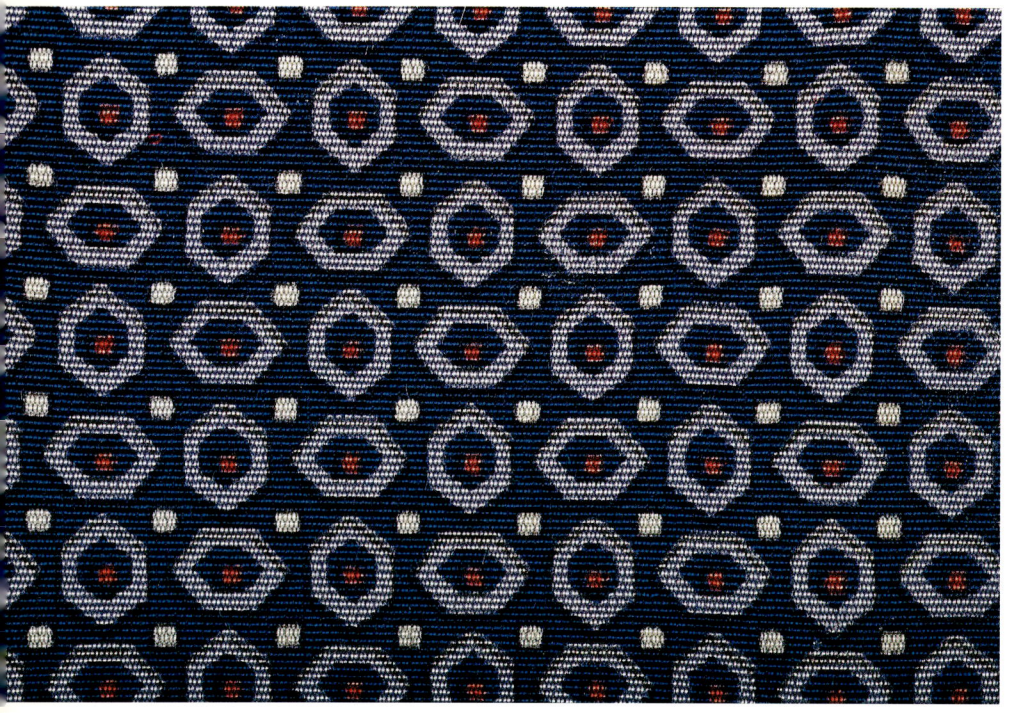
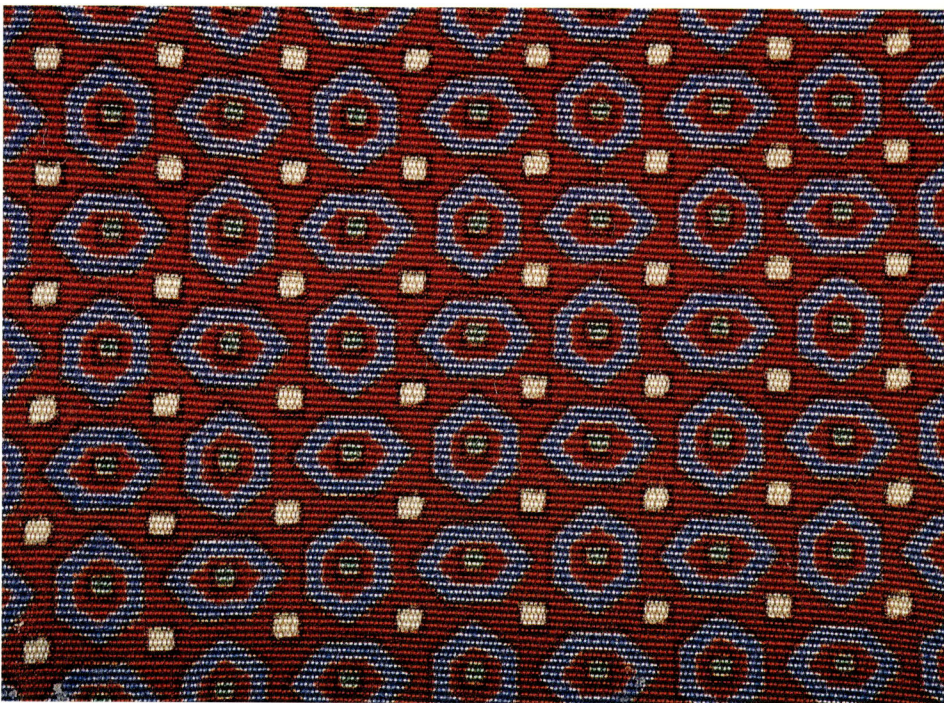
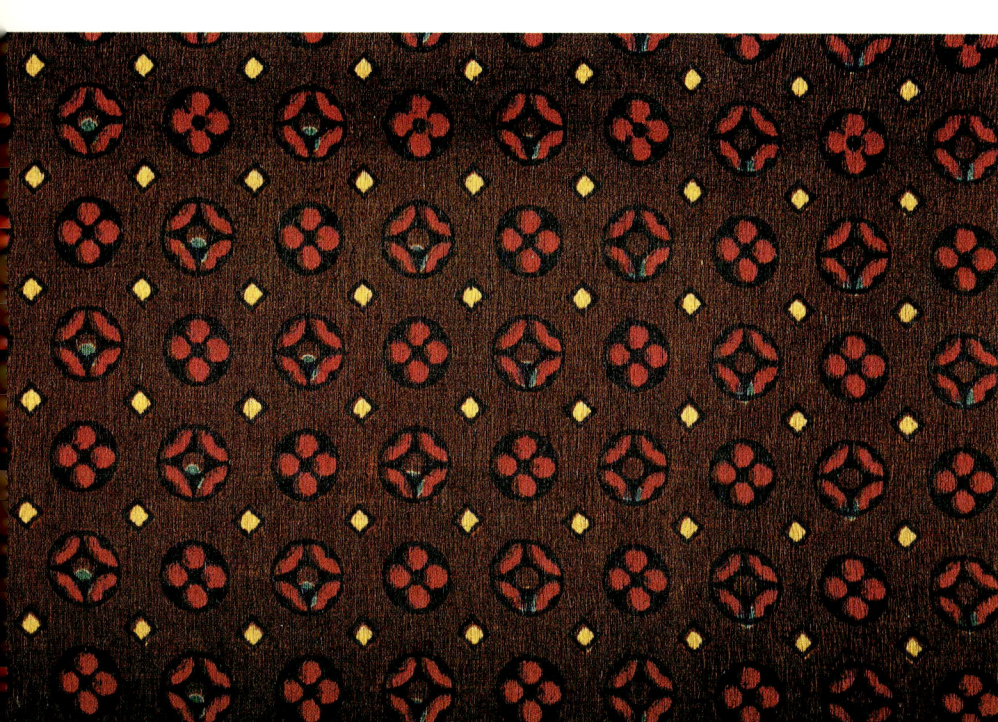
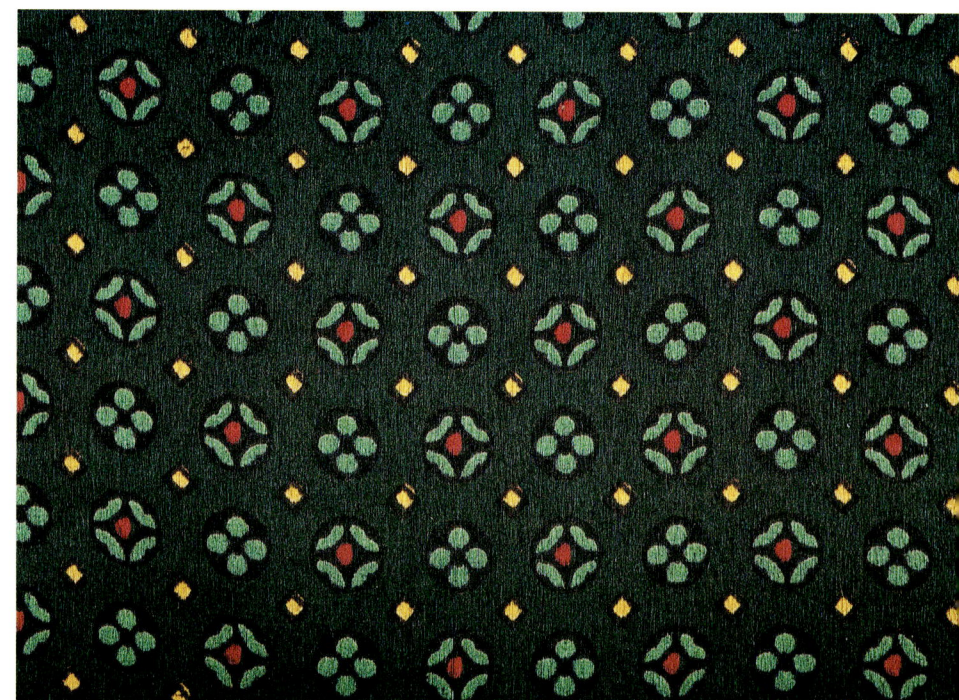

44

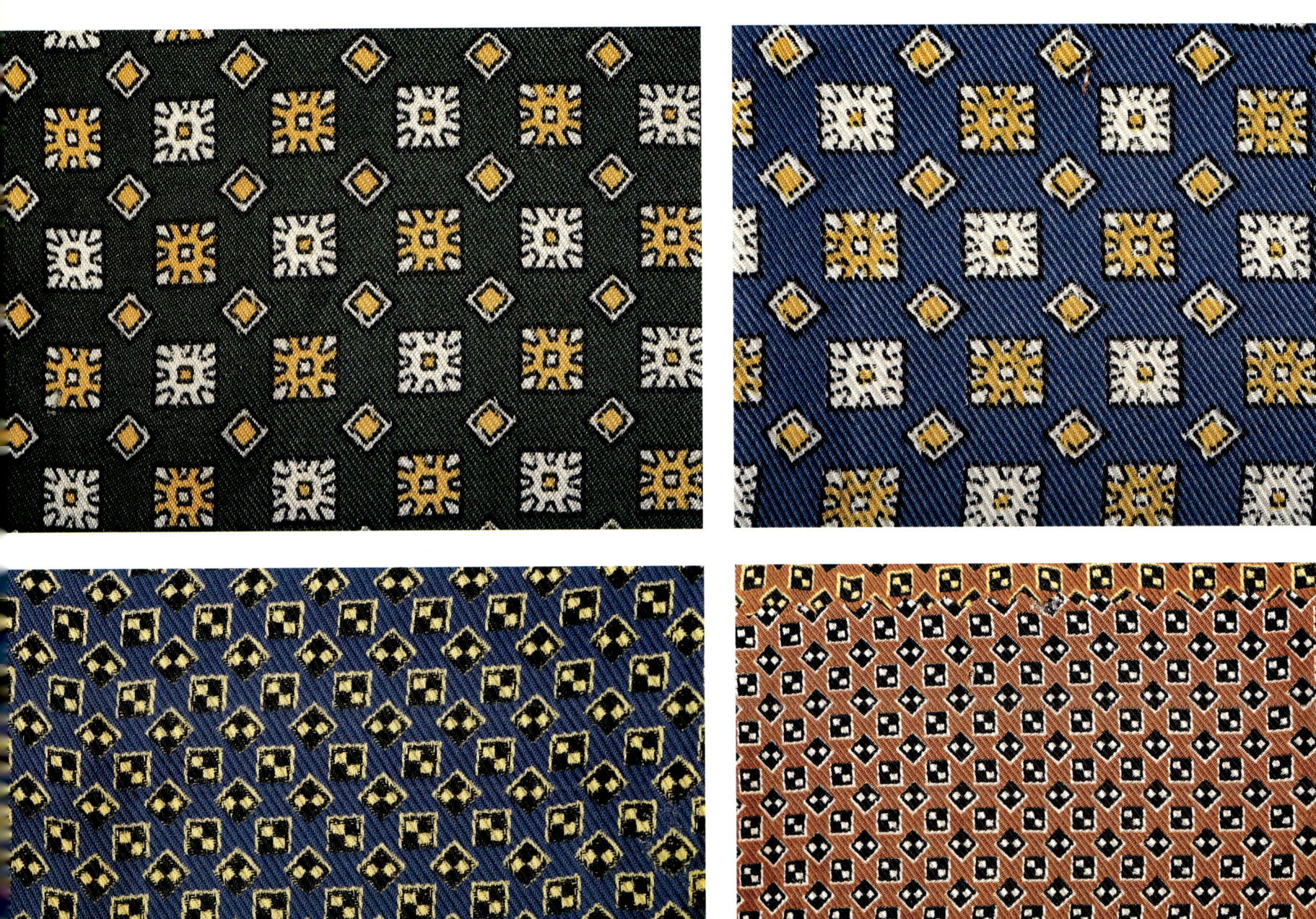

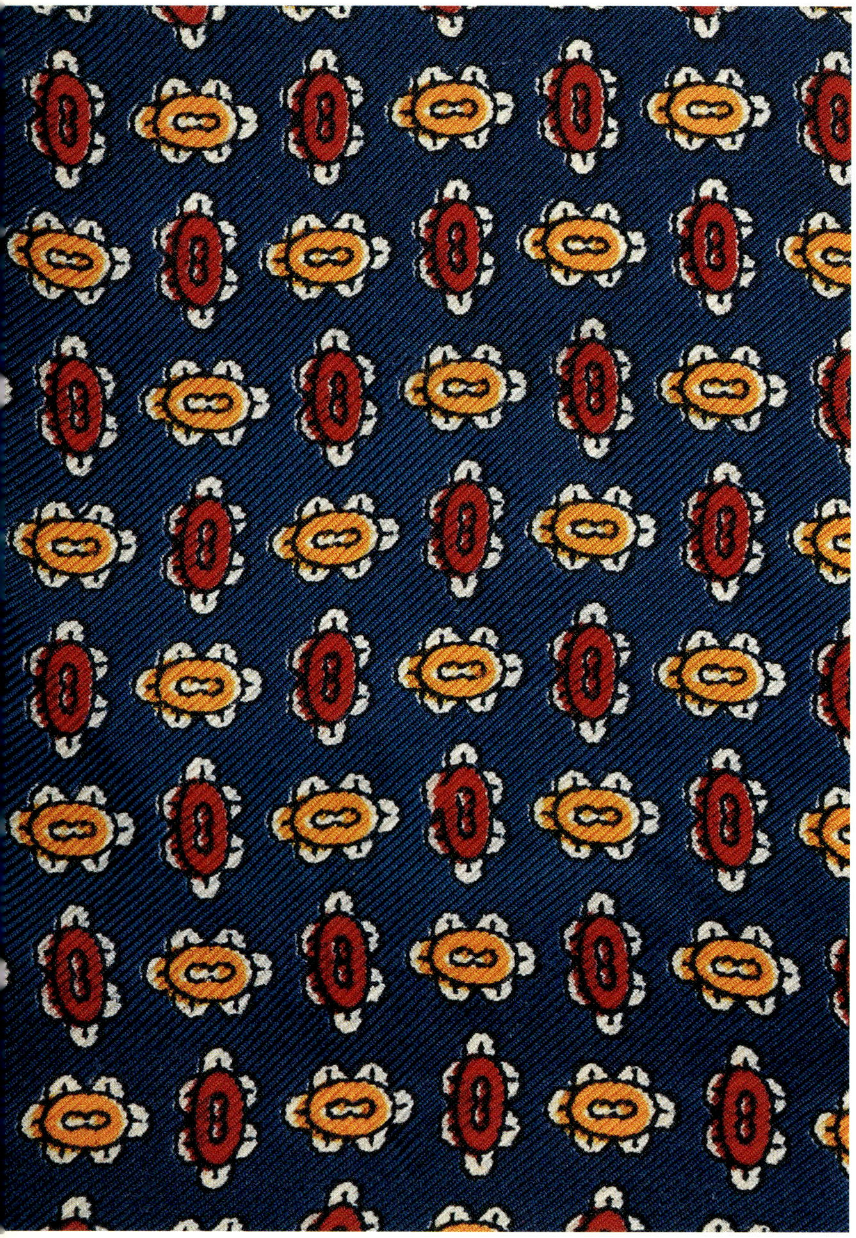

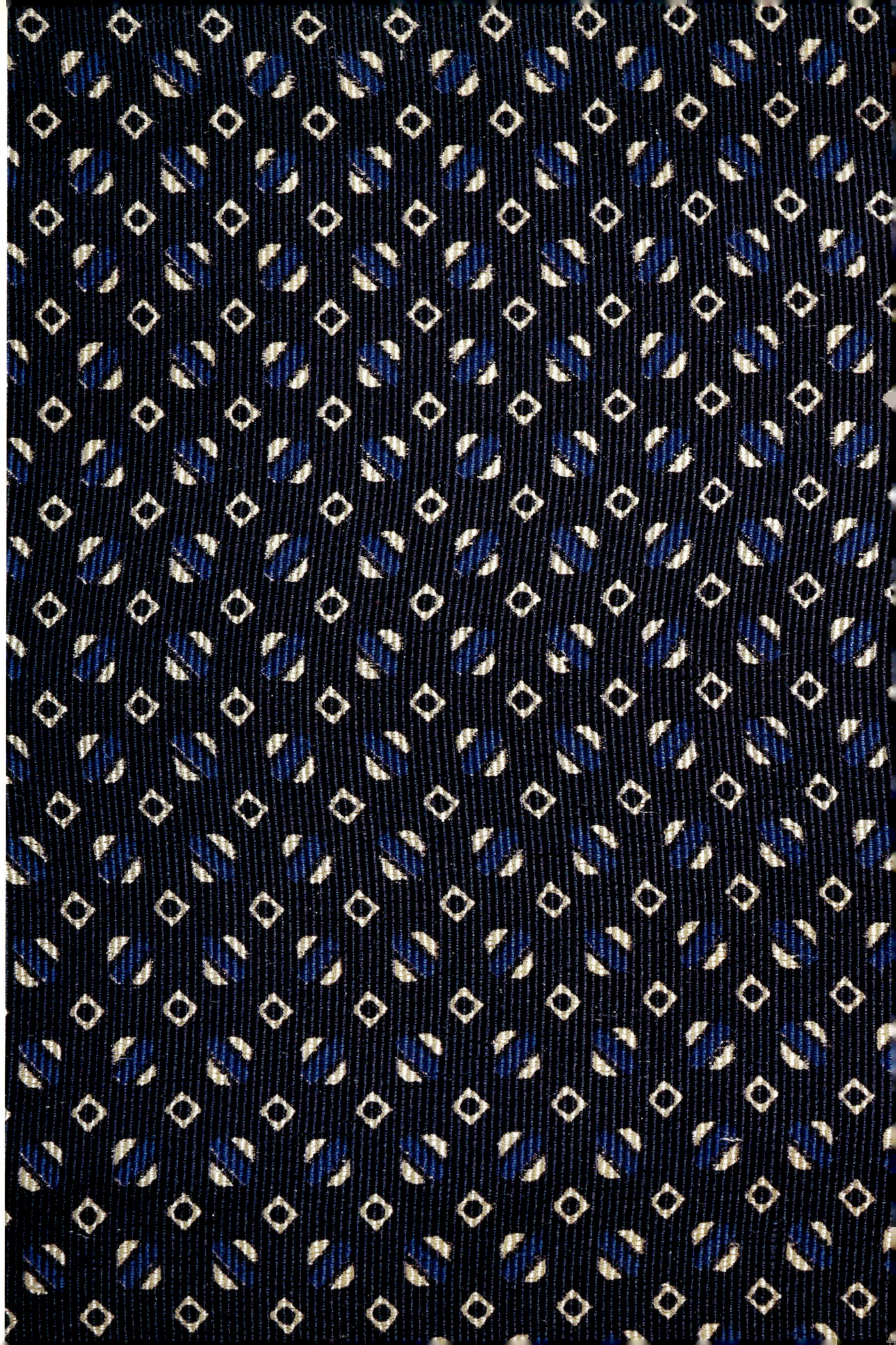

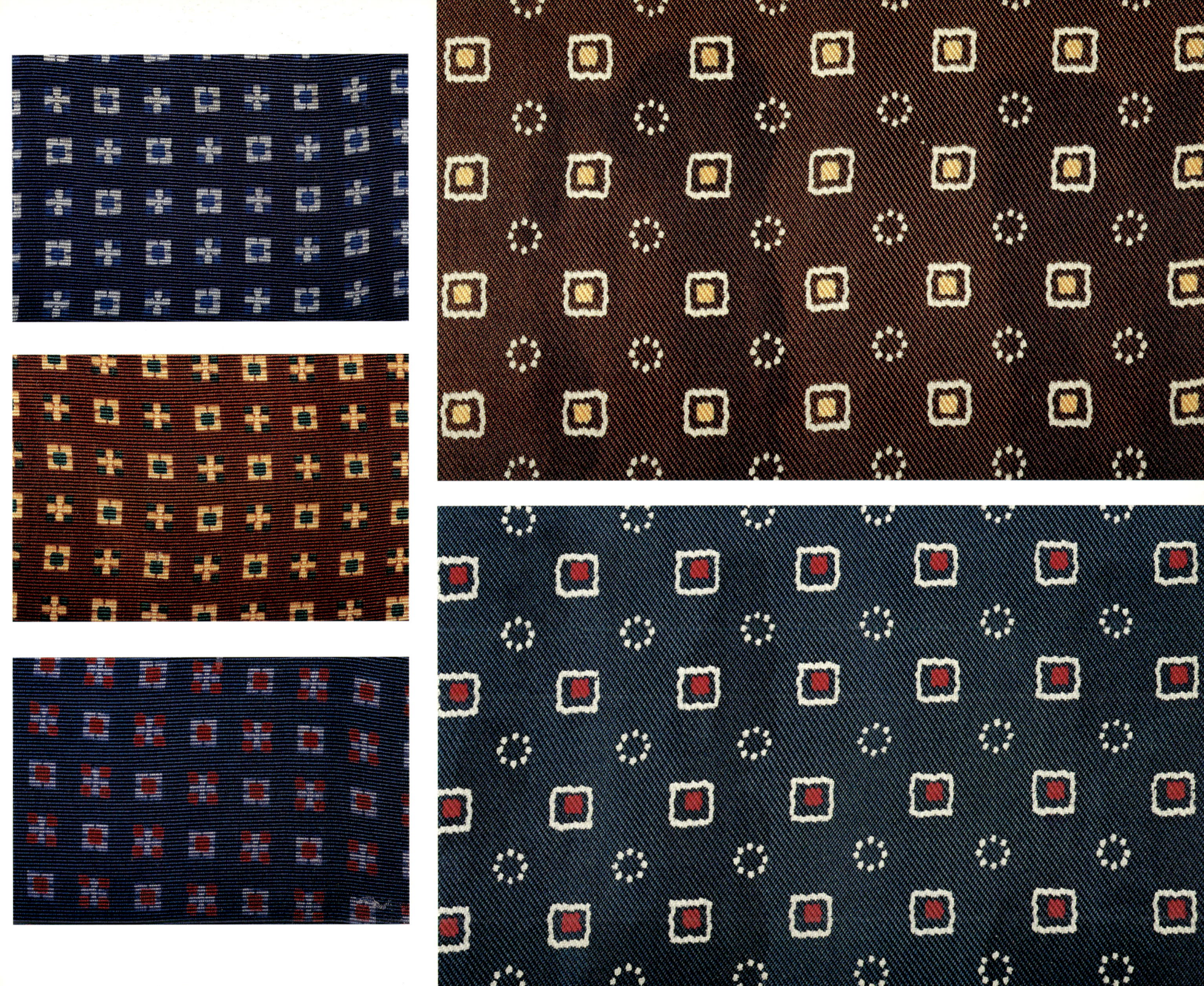

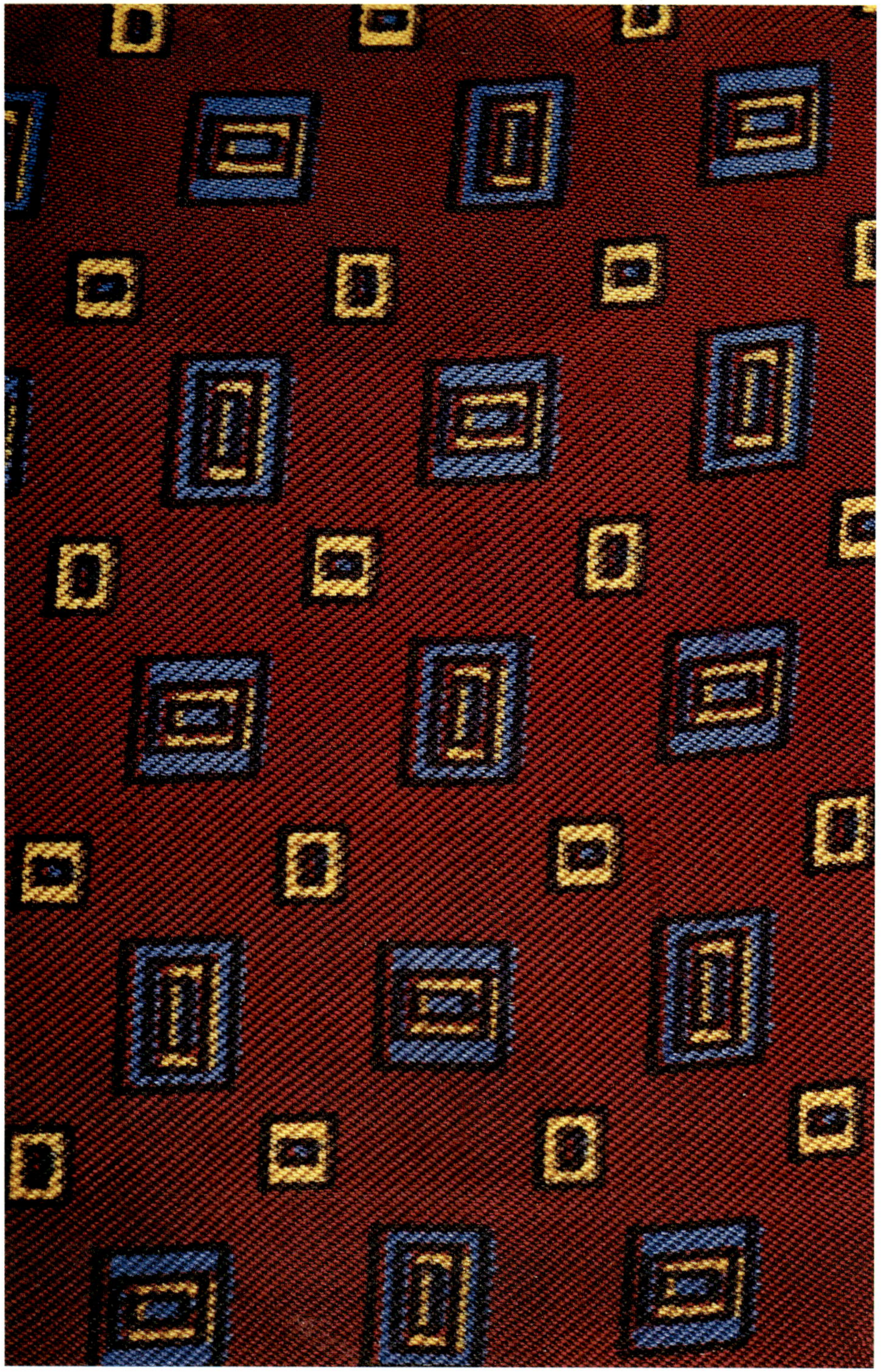

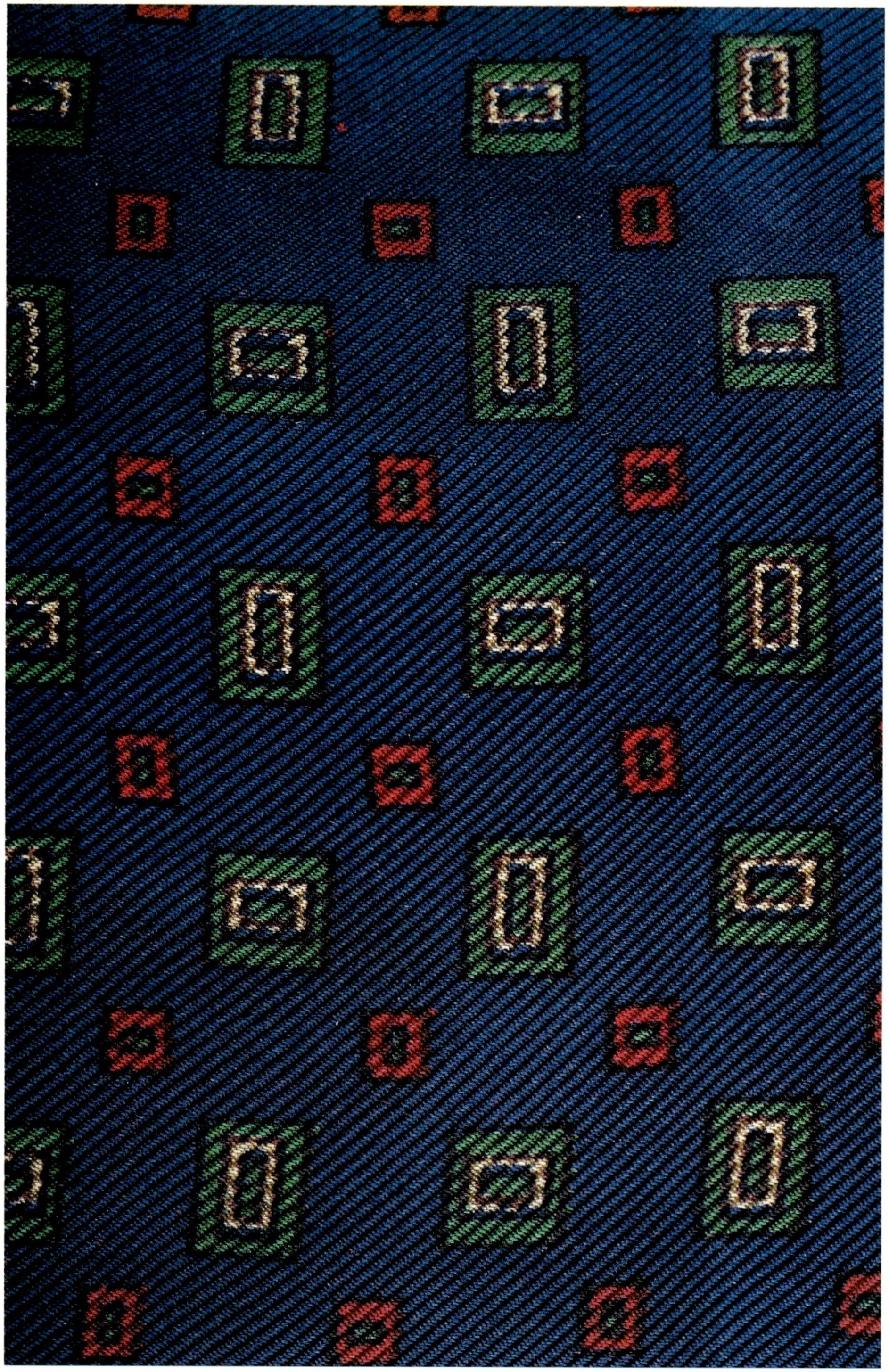

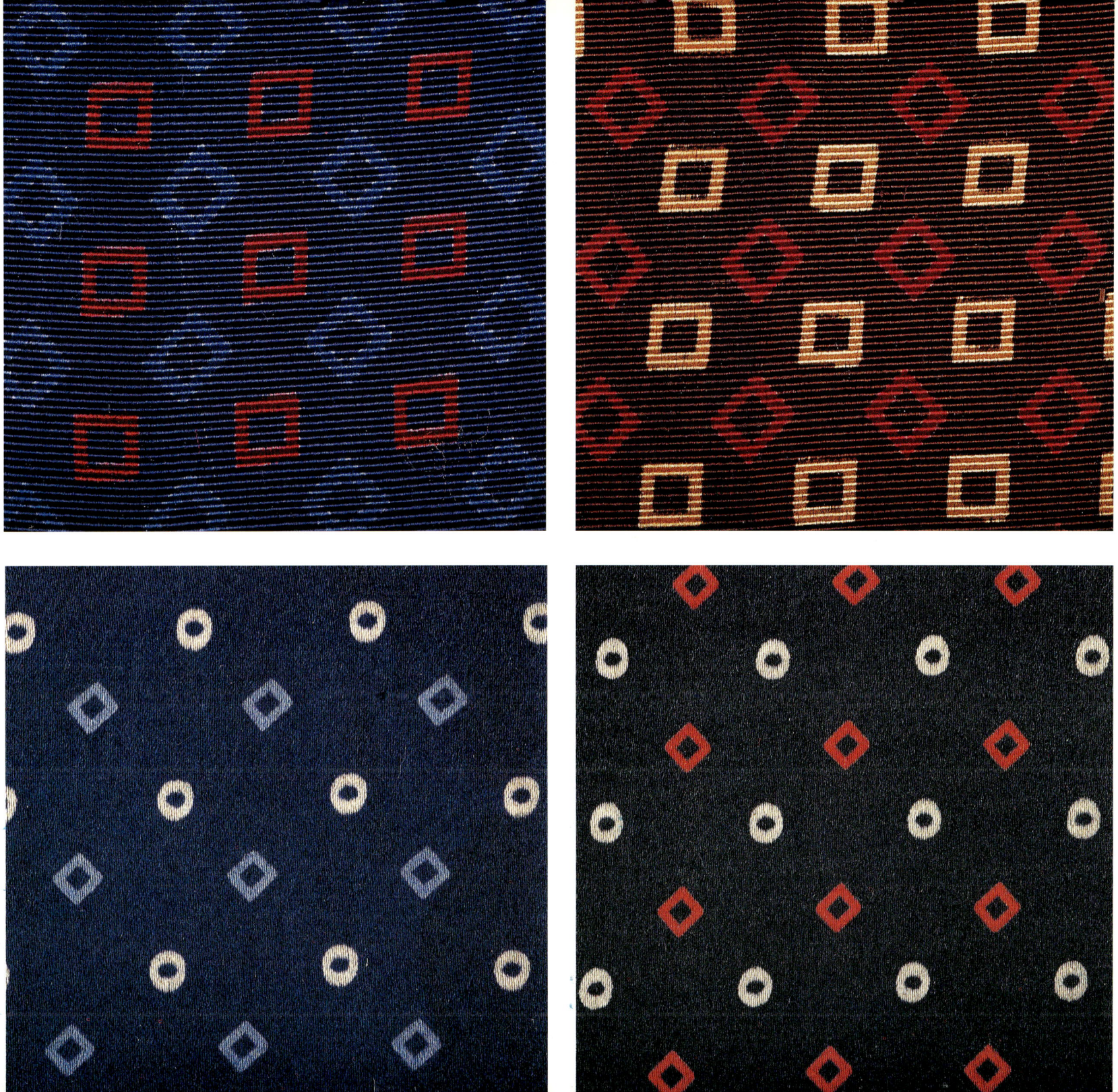

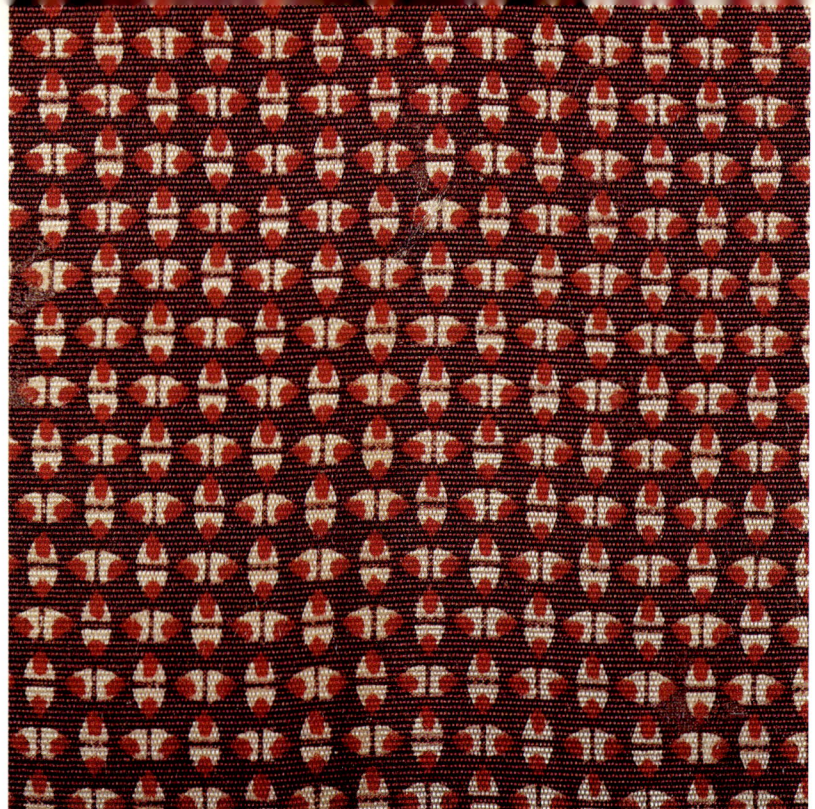
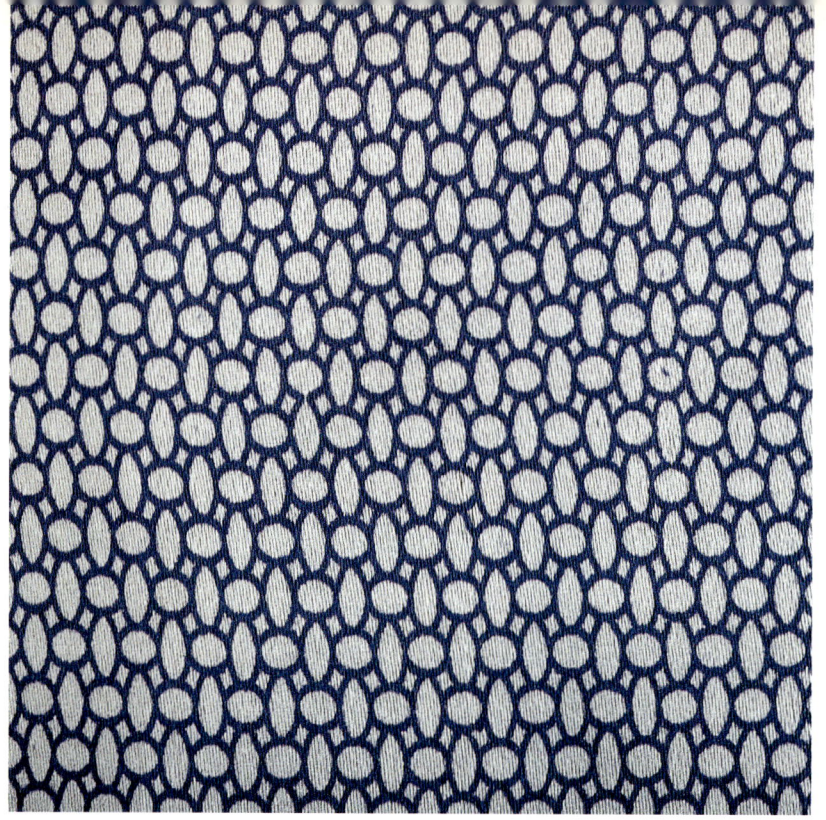
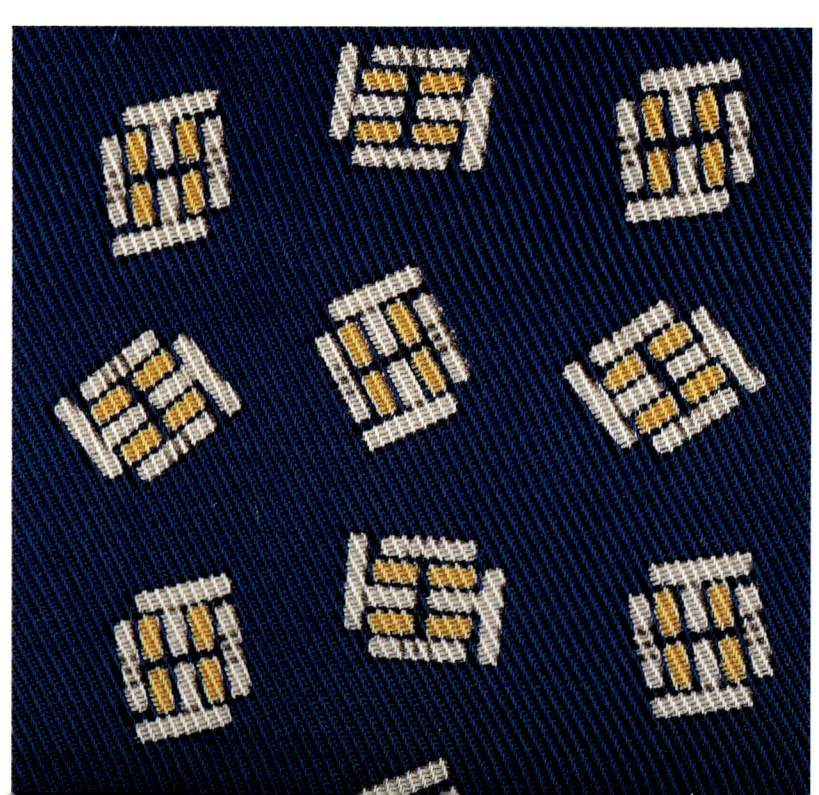
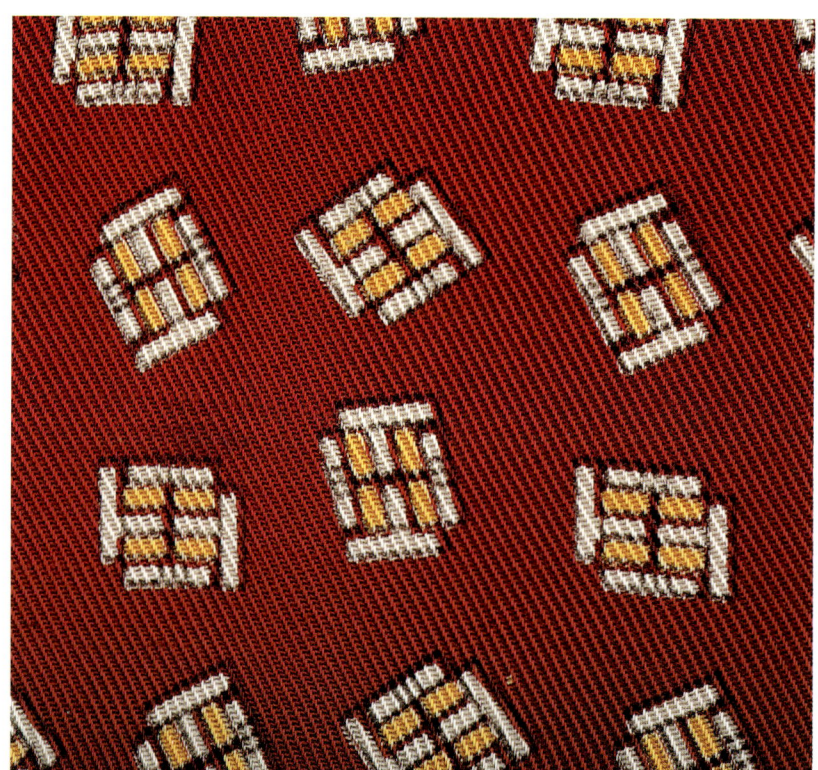

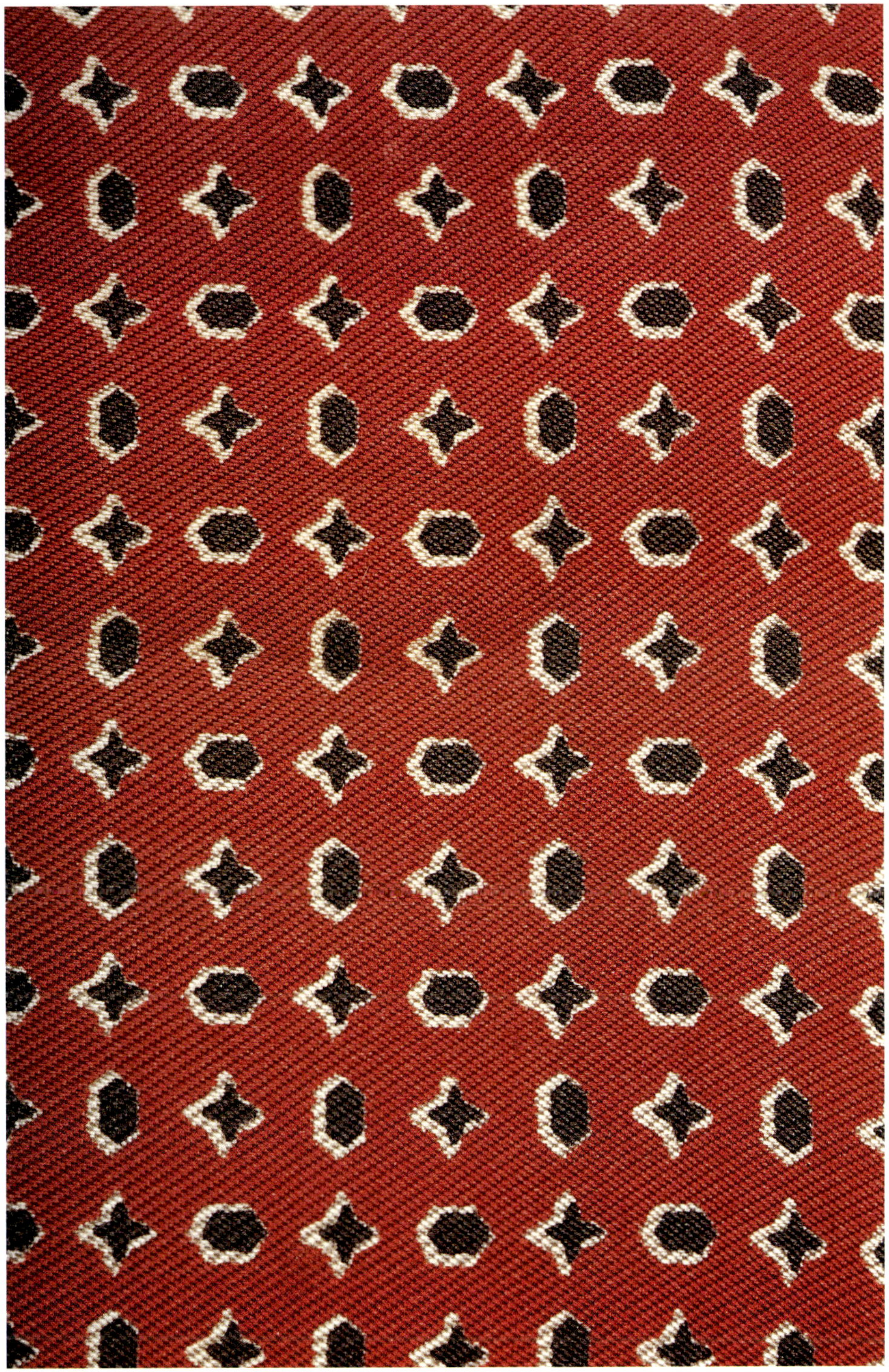

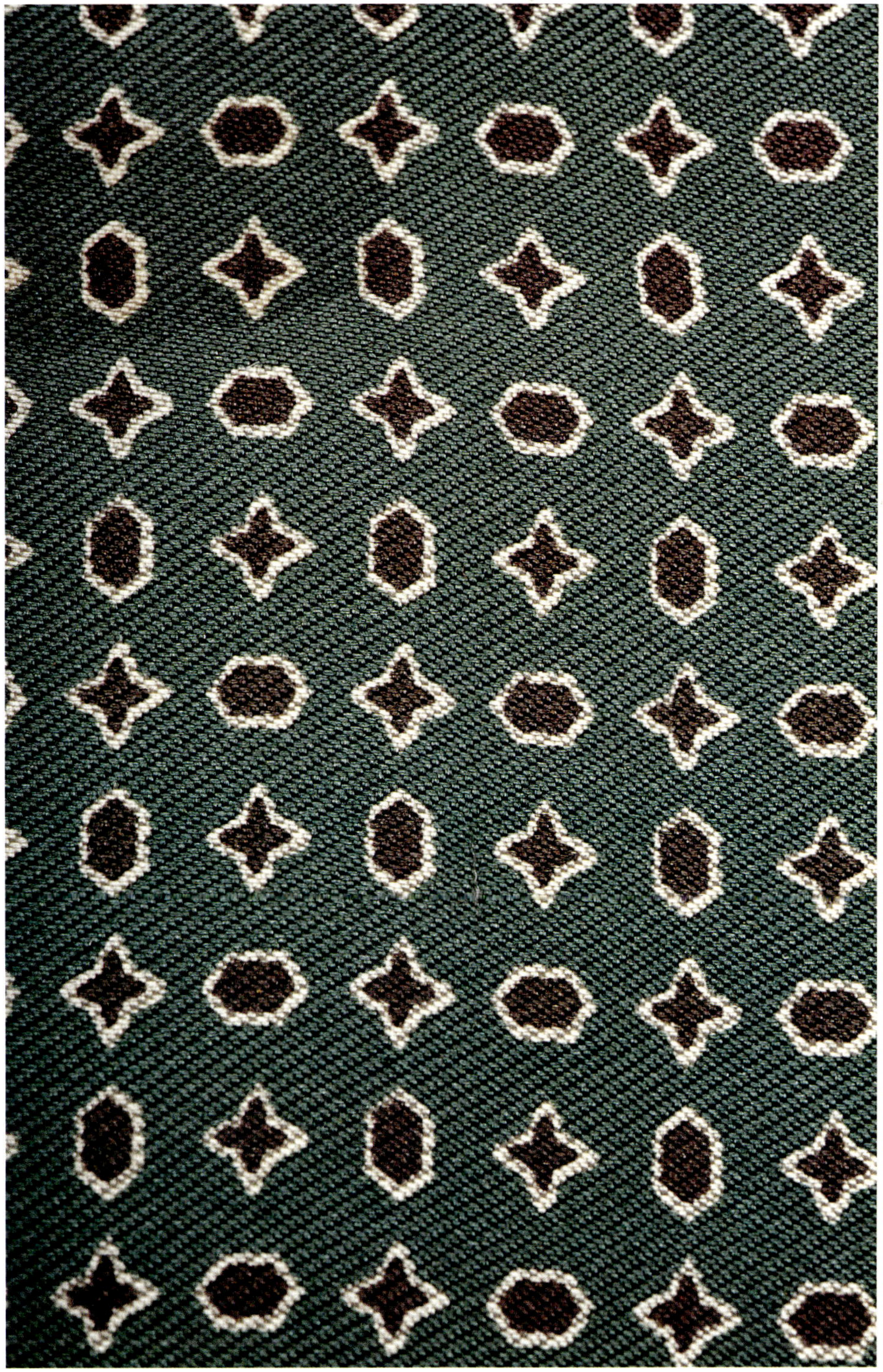

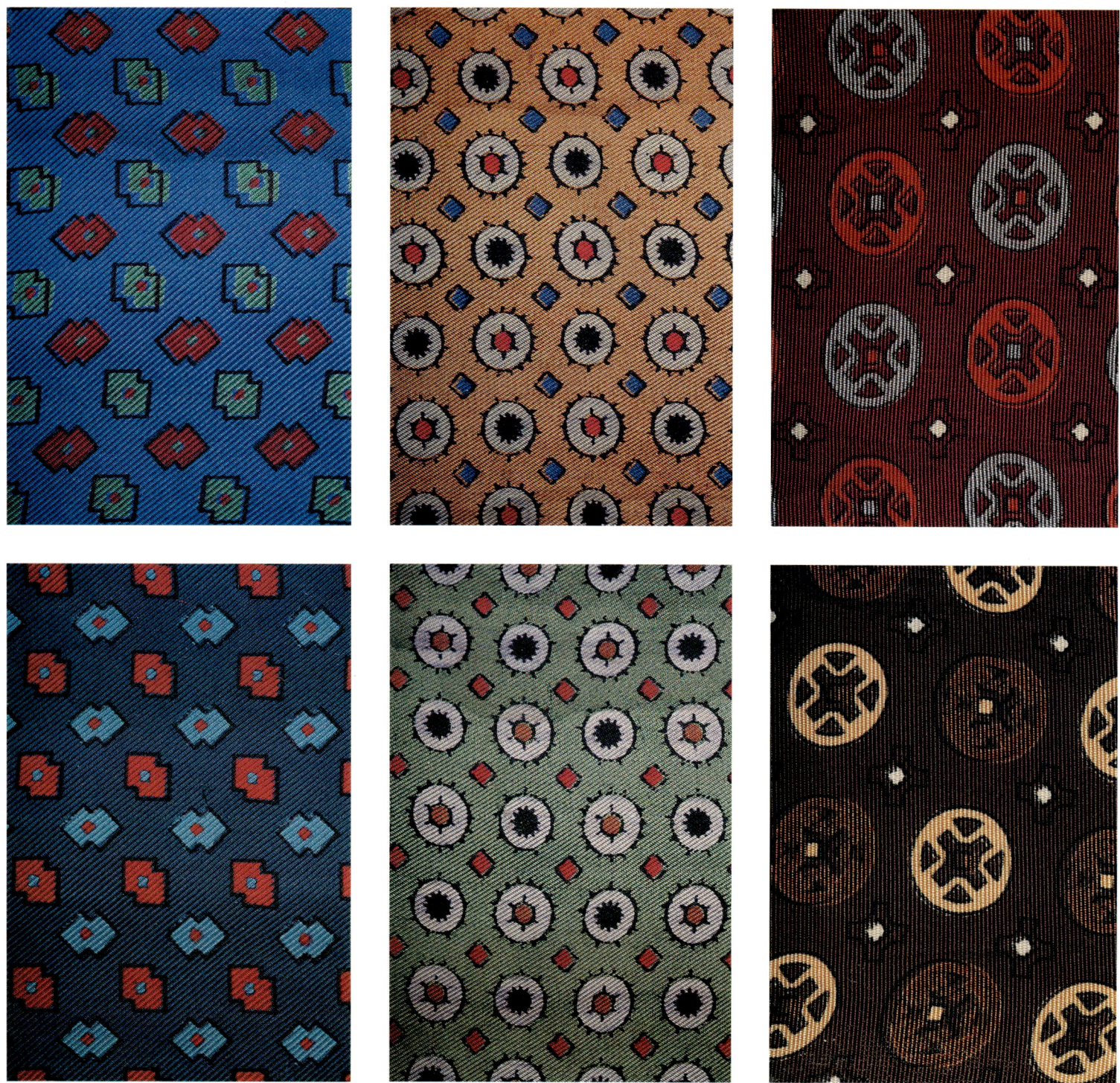

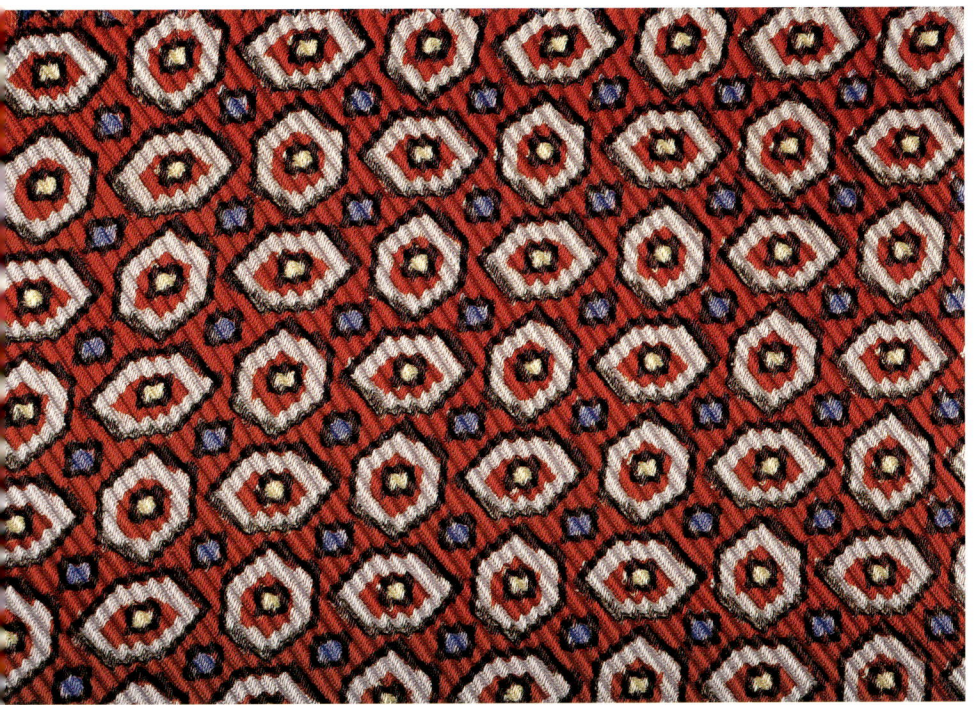
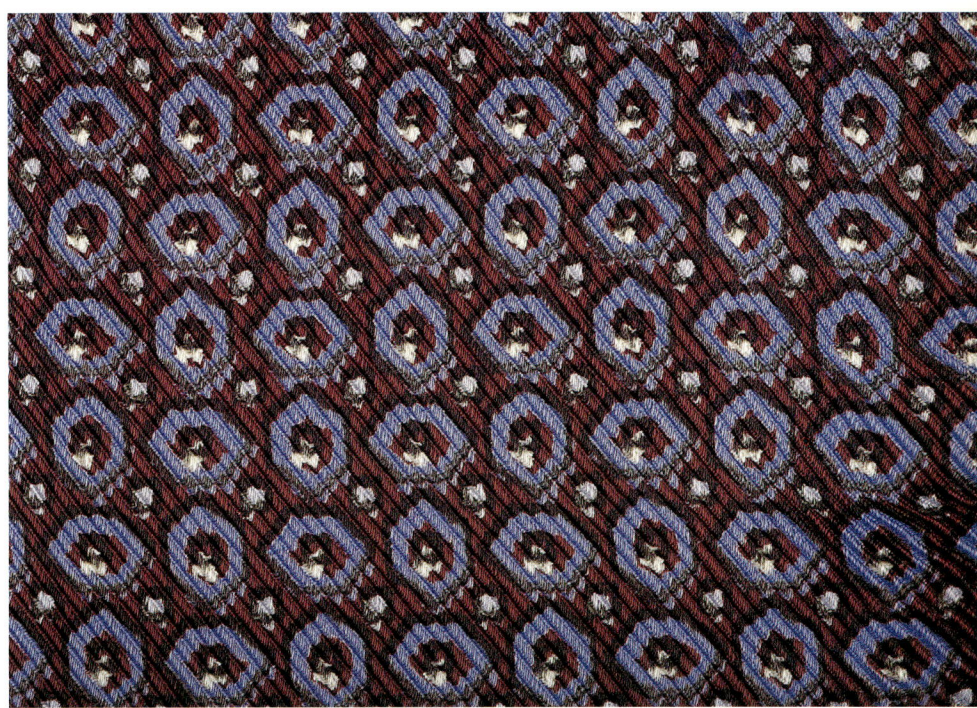
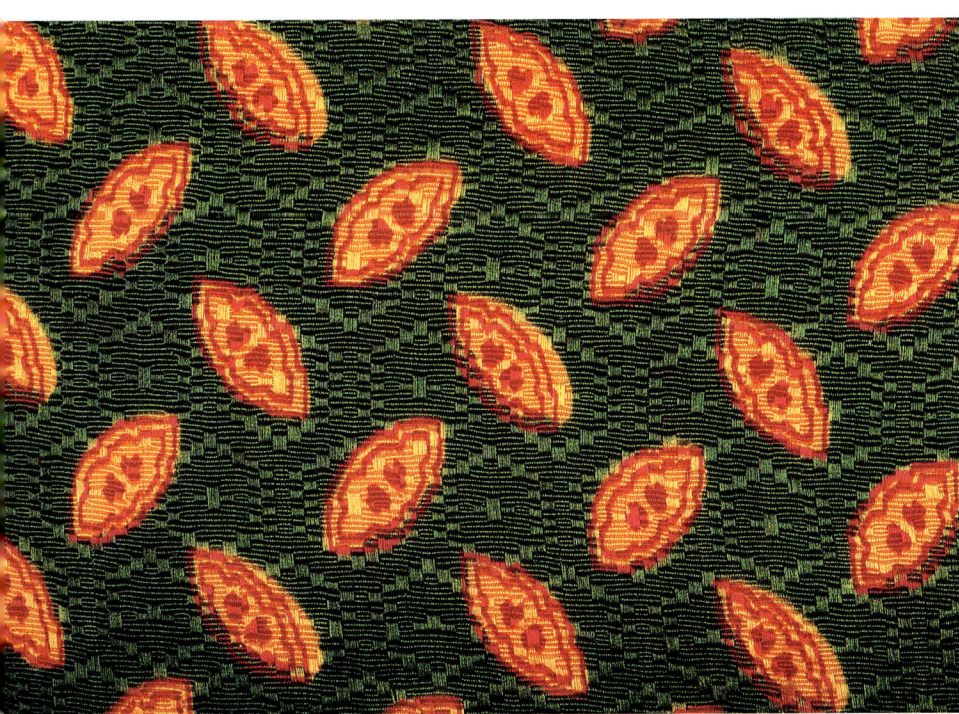
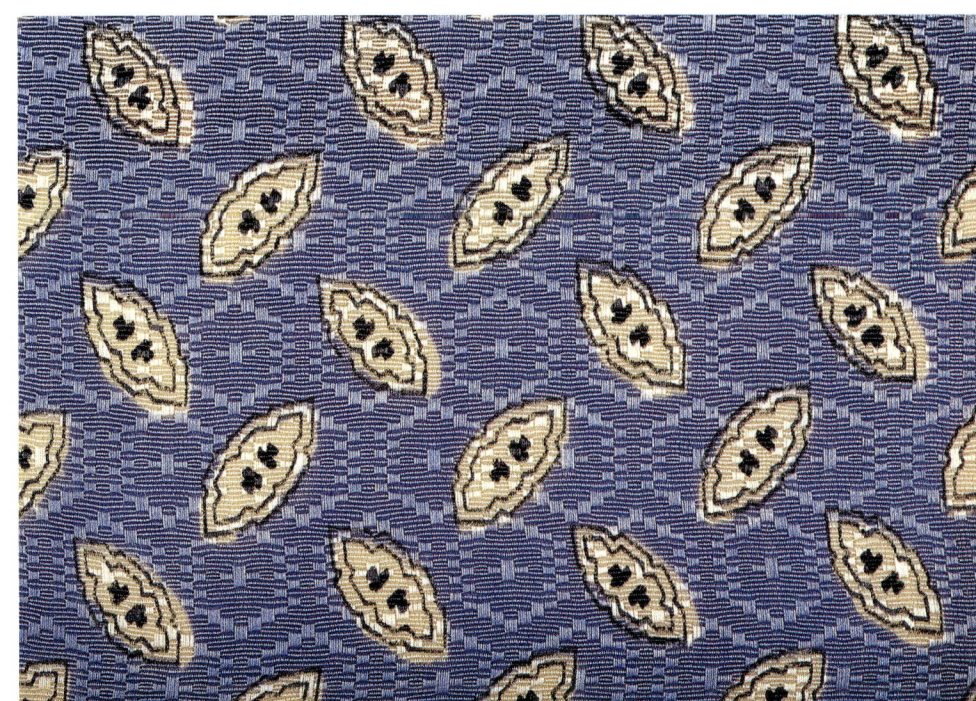

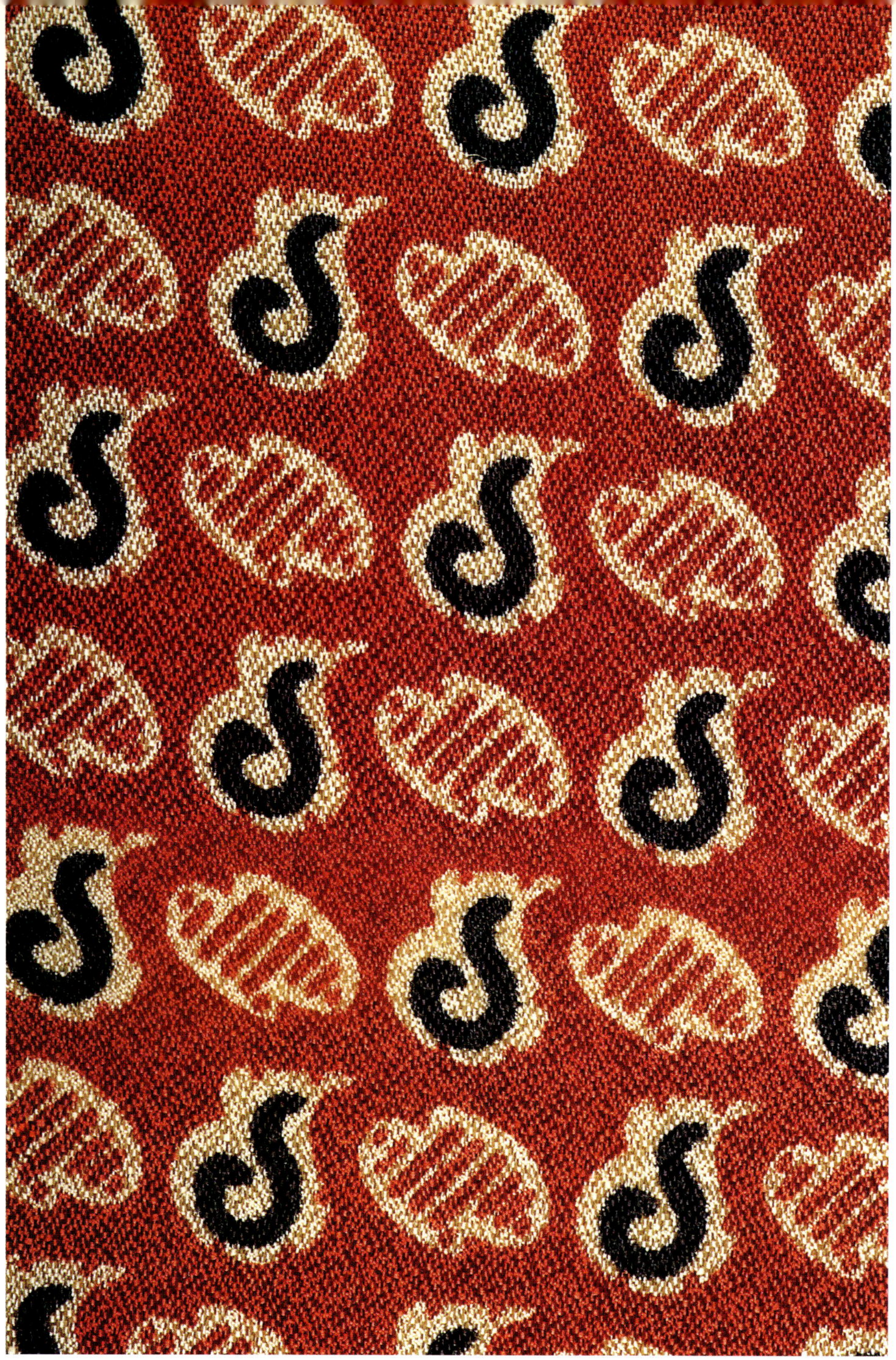

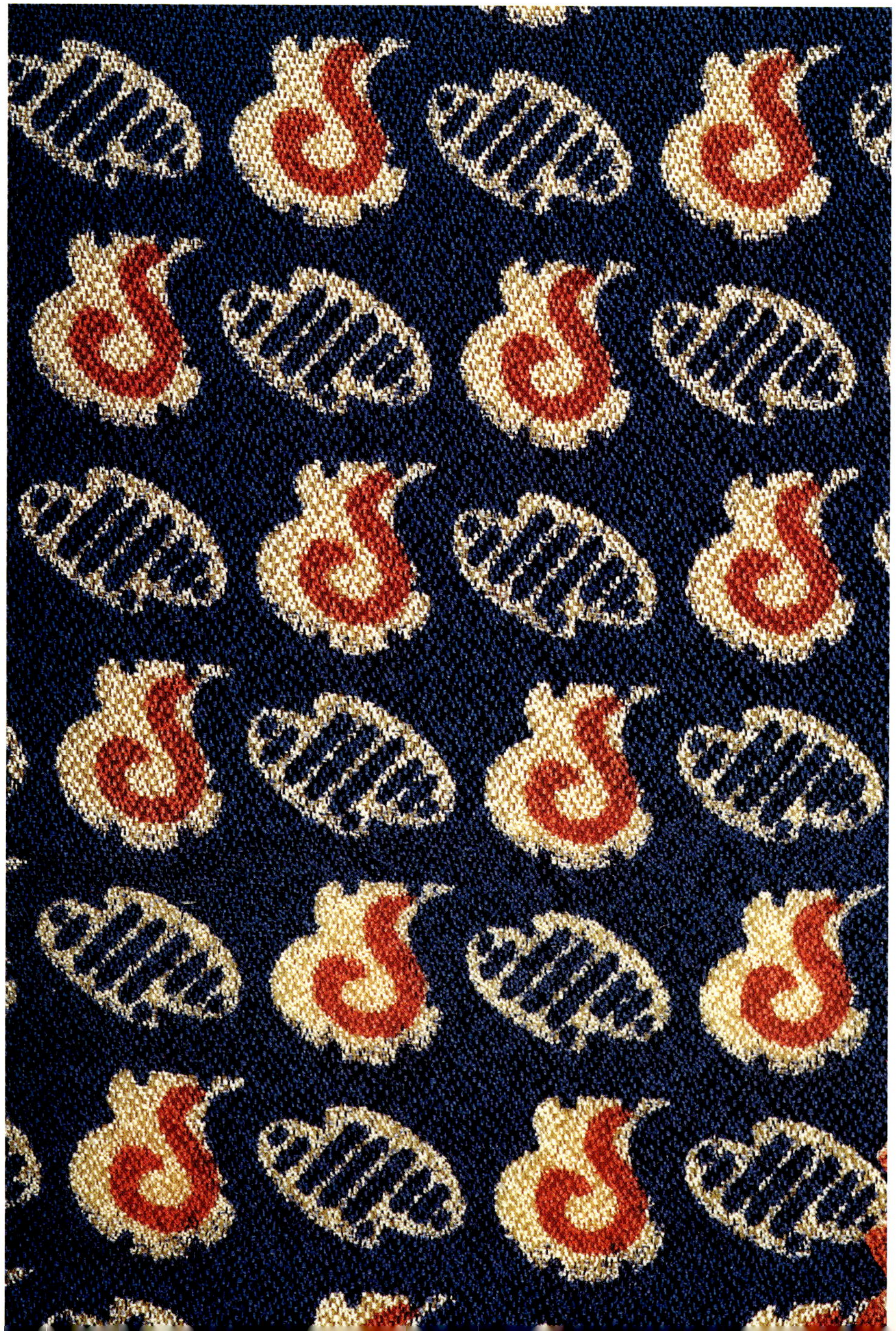

54

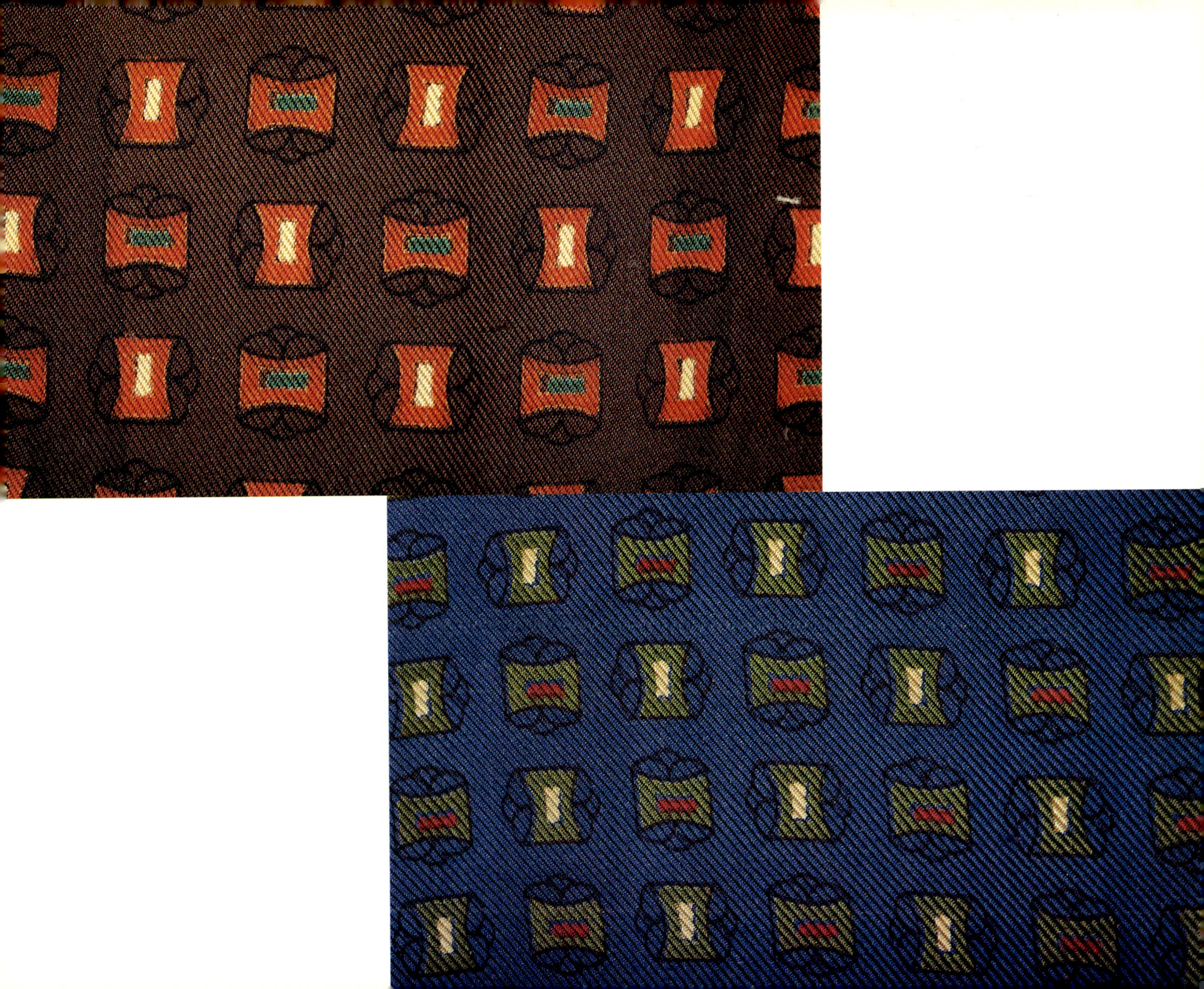

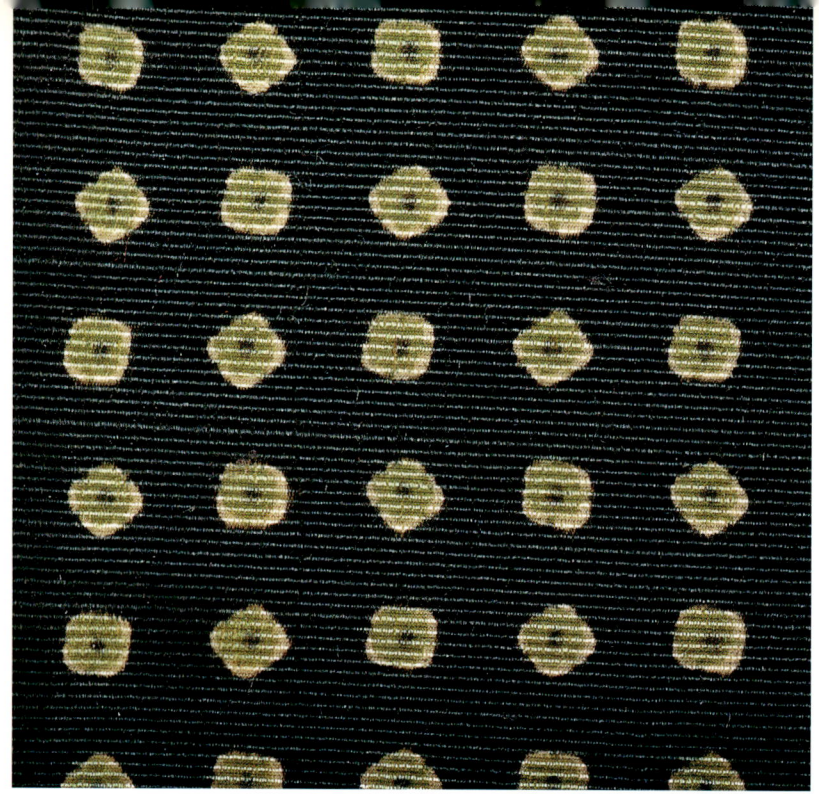
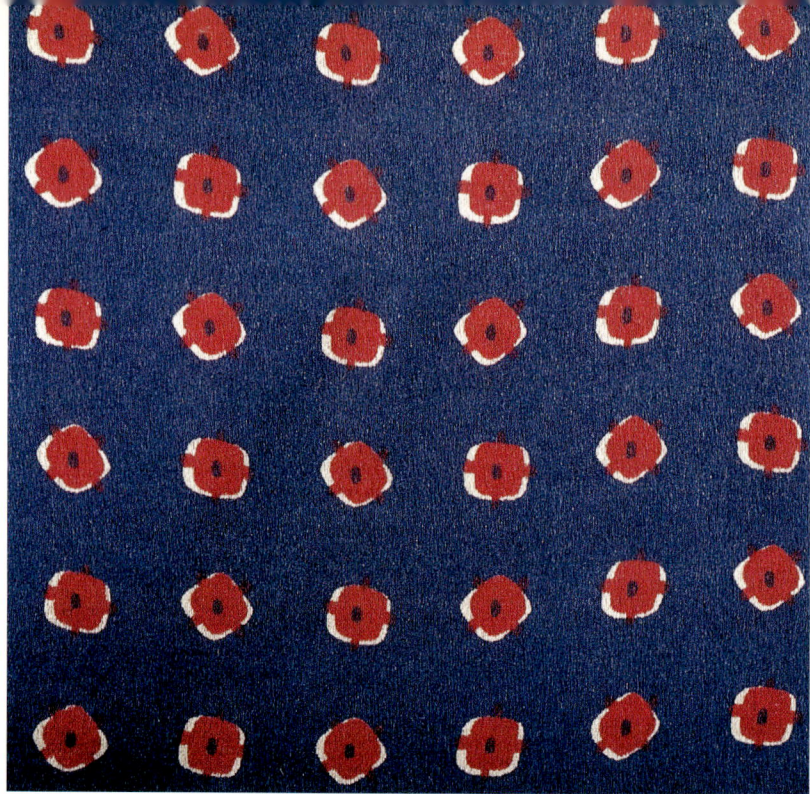
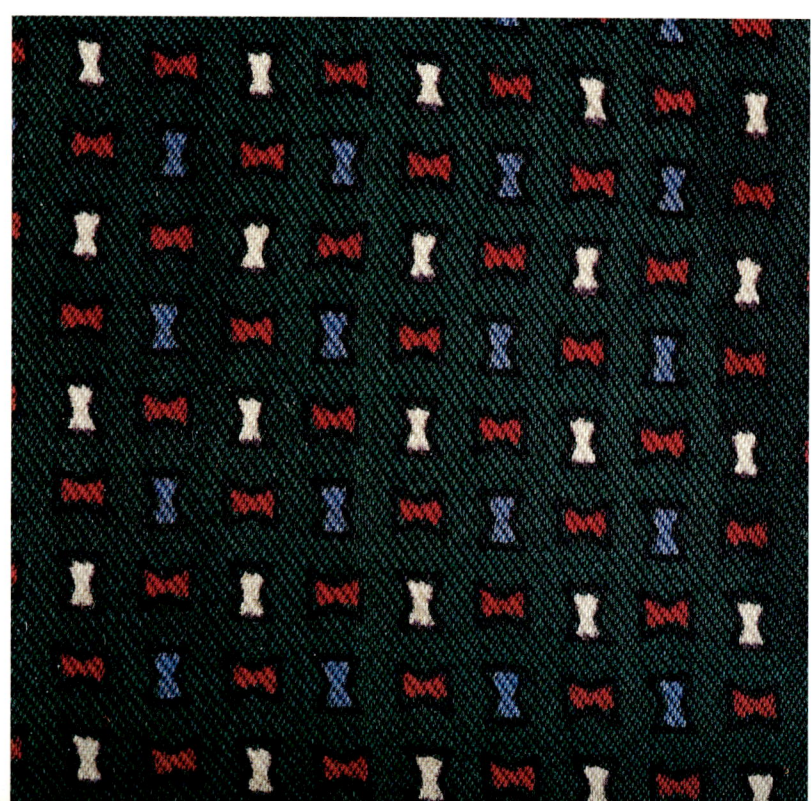
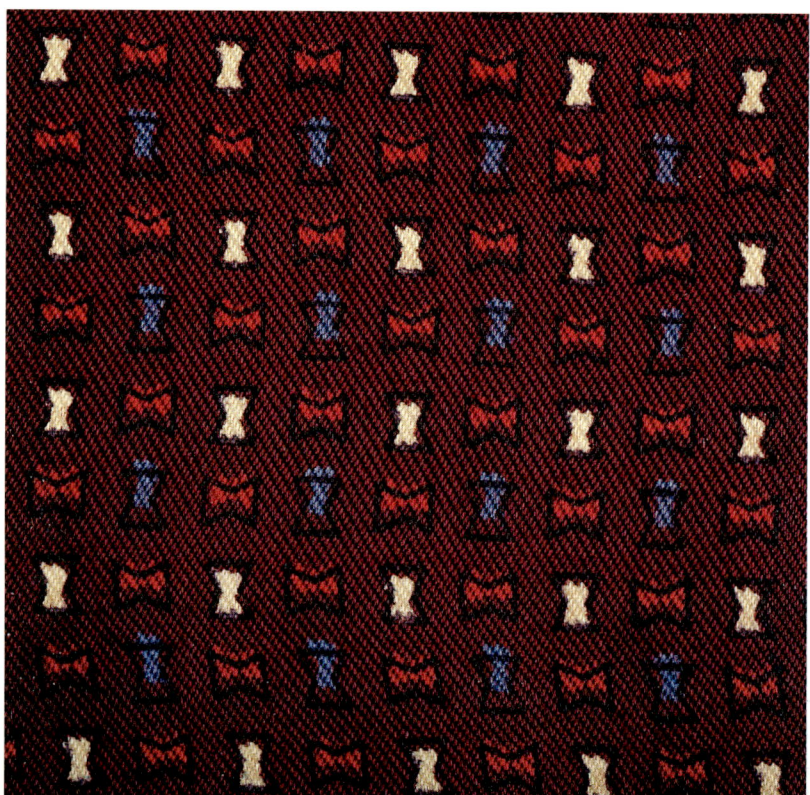

56

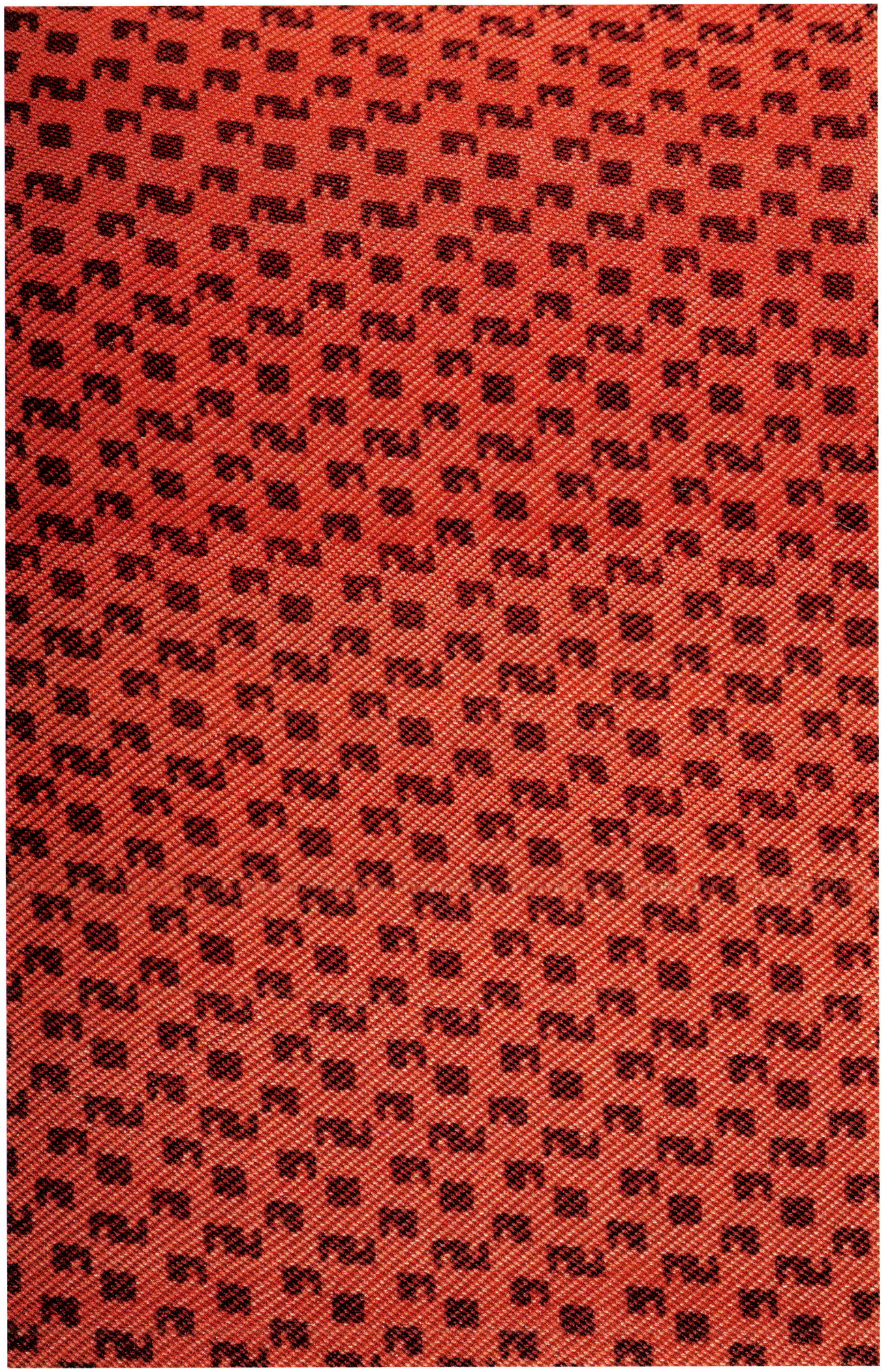

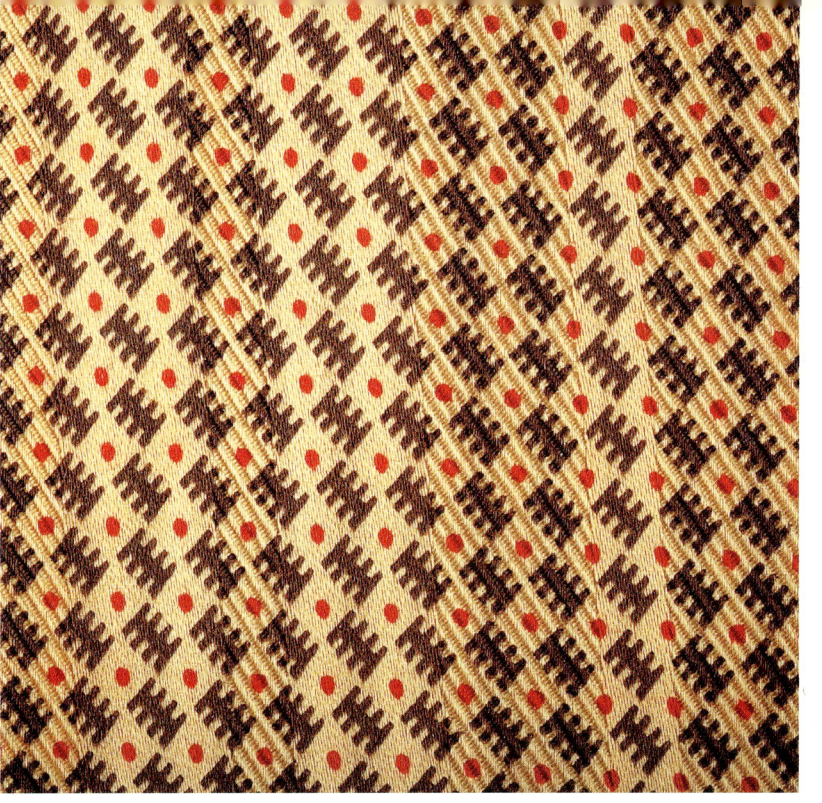

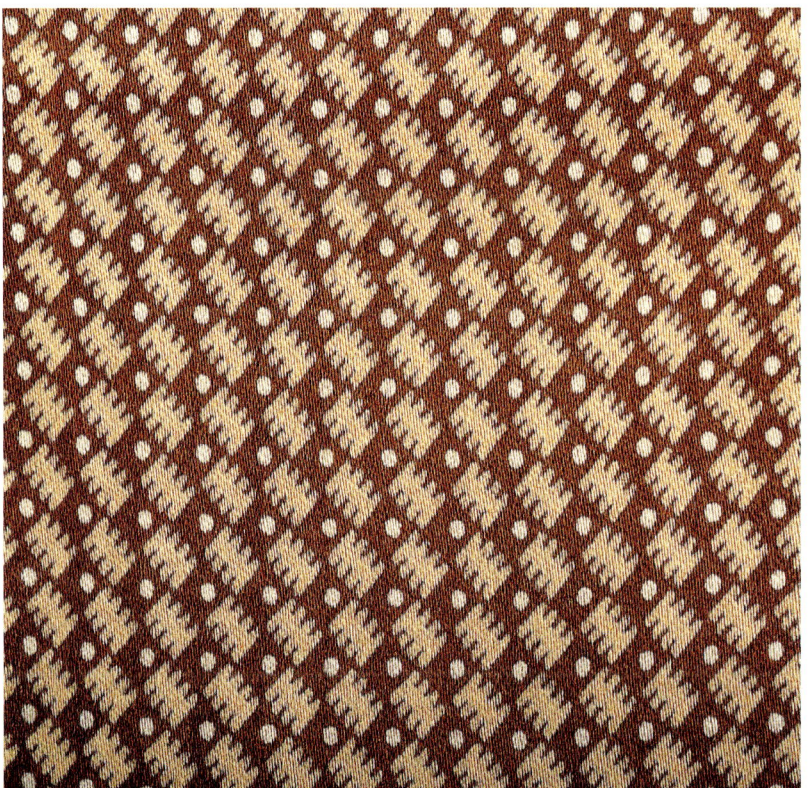

57

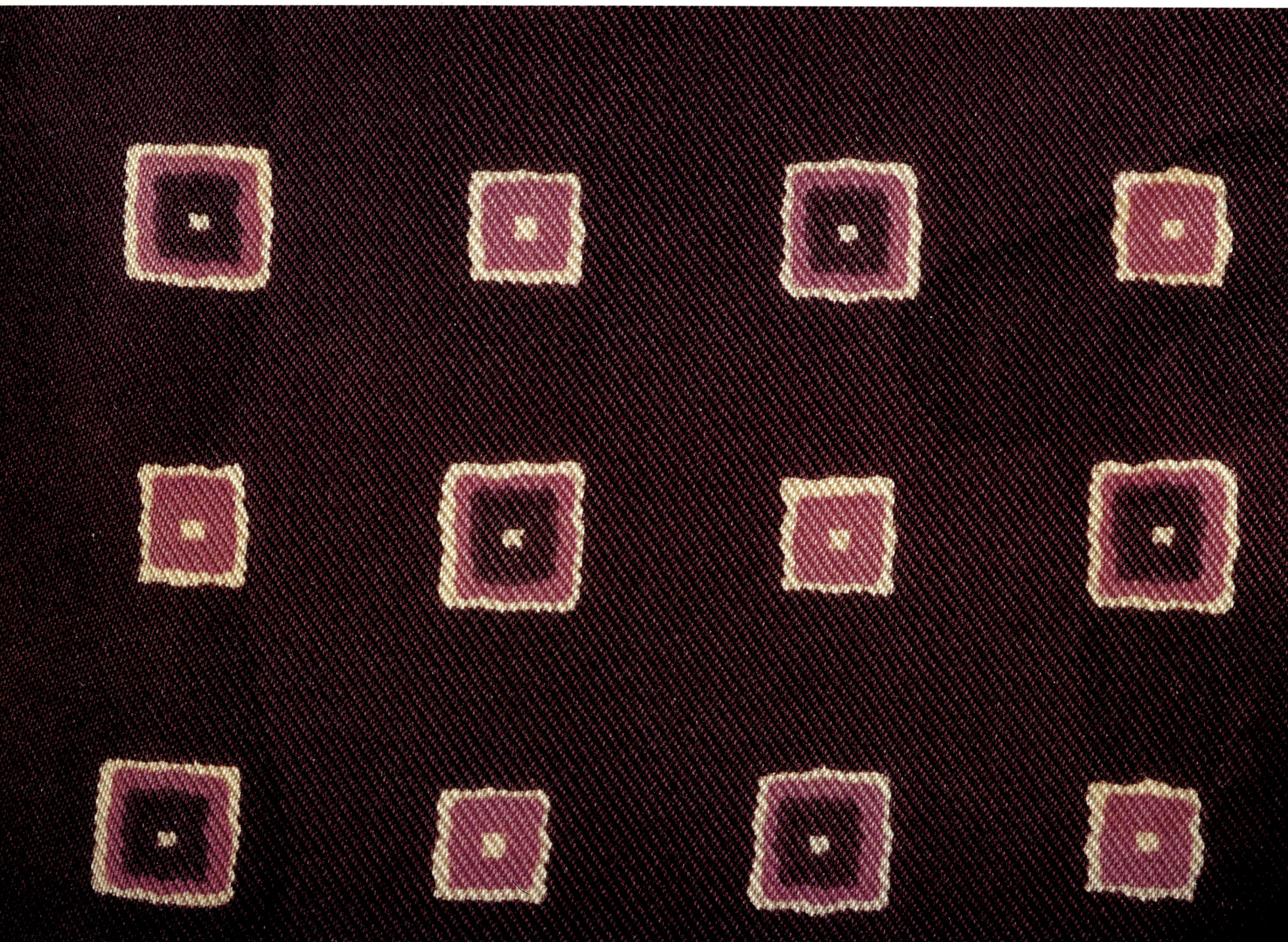

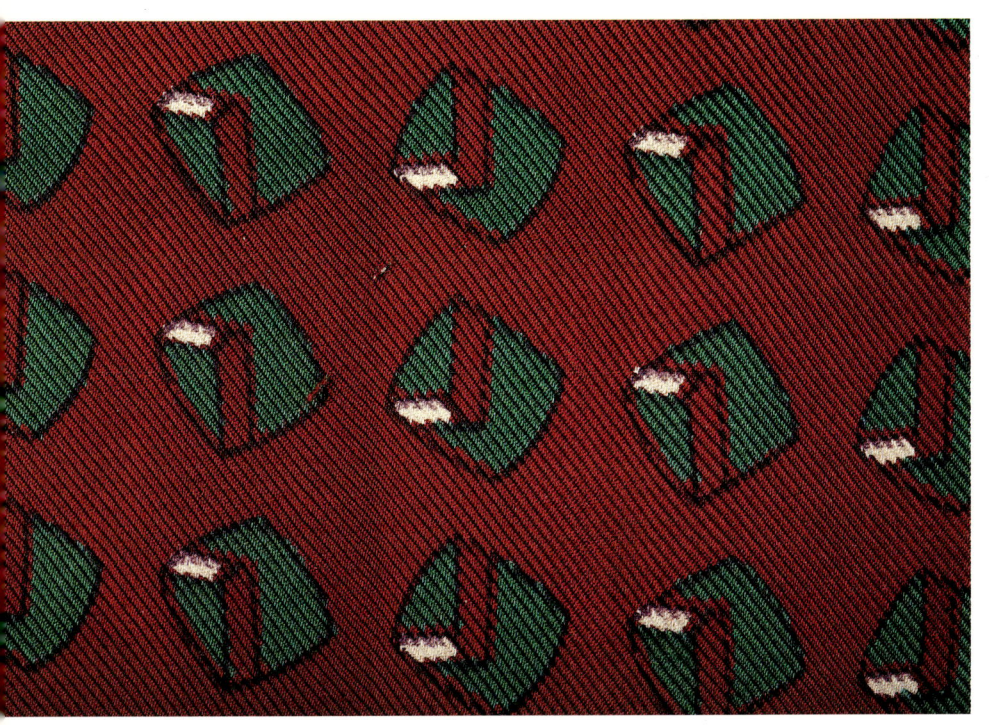
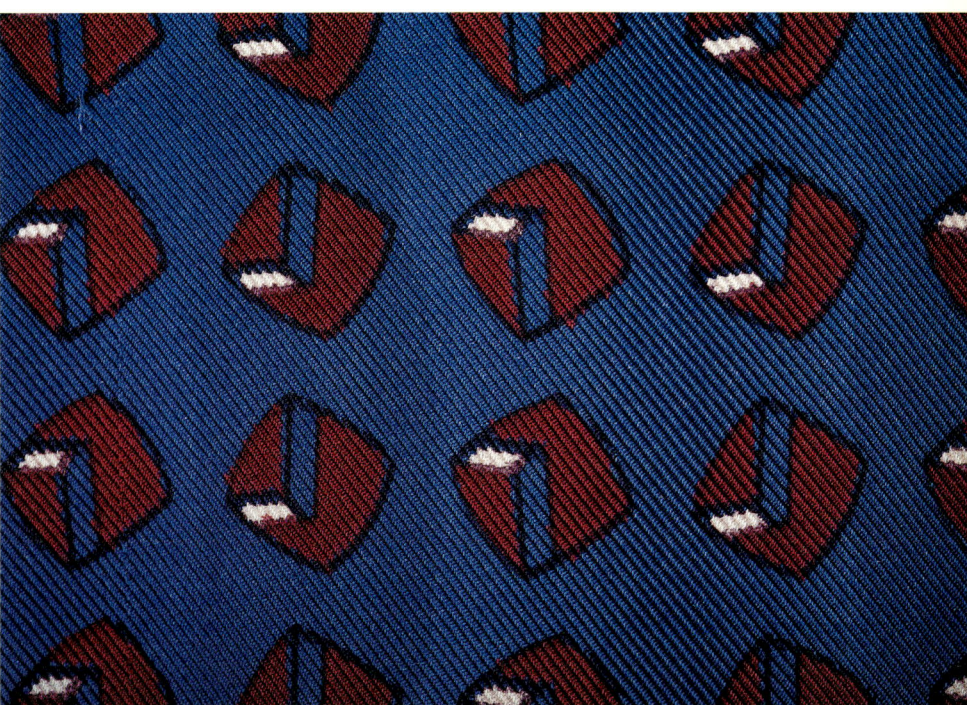
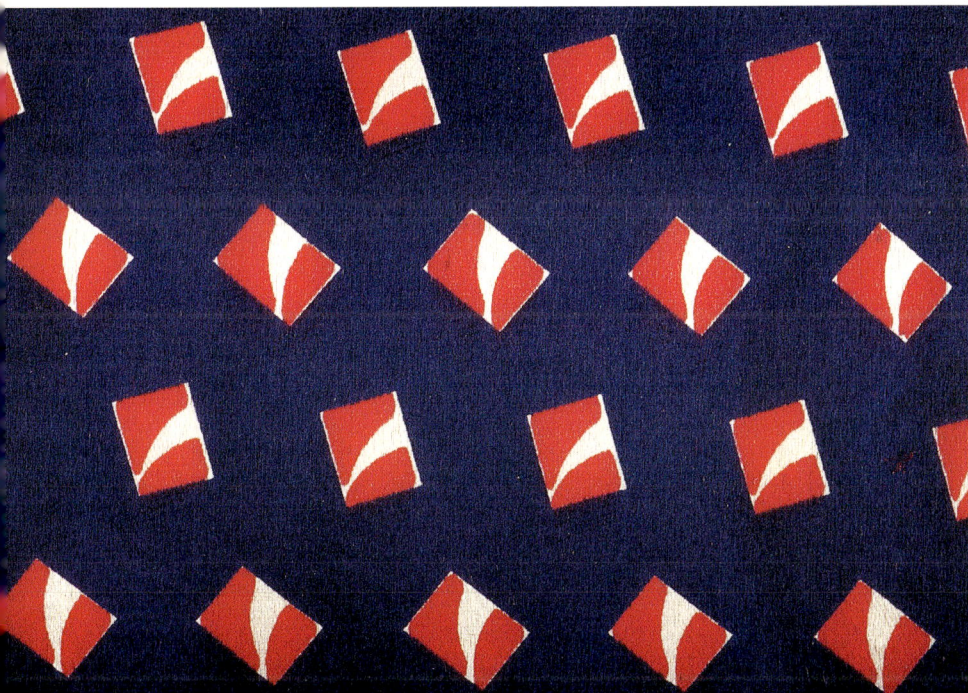

59

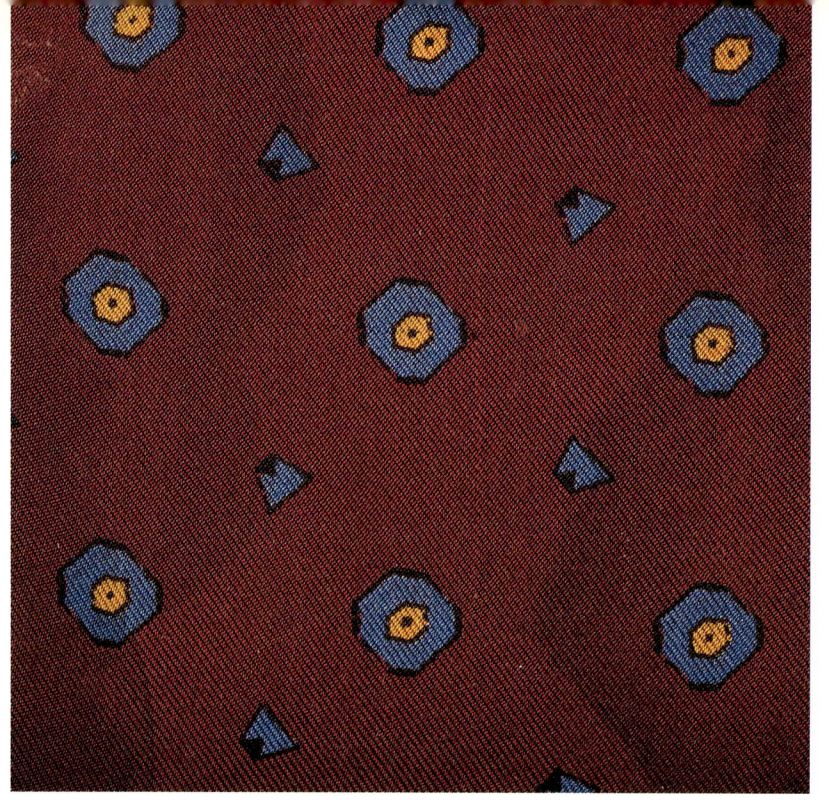
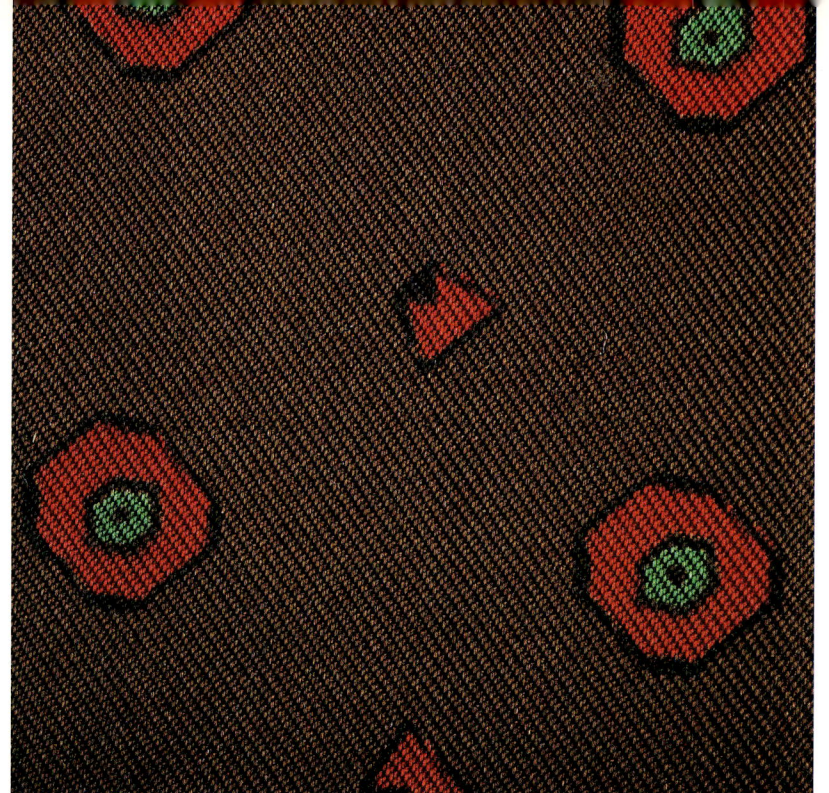
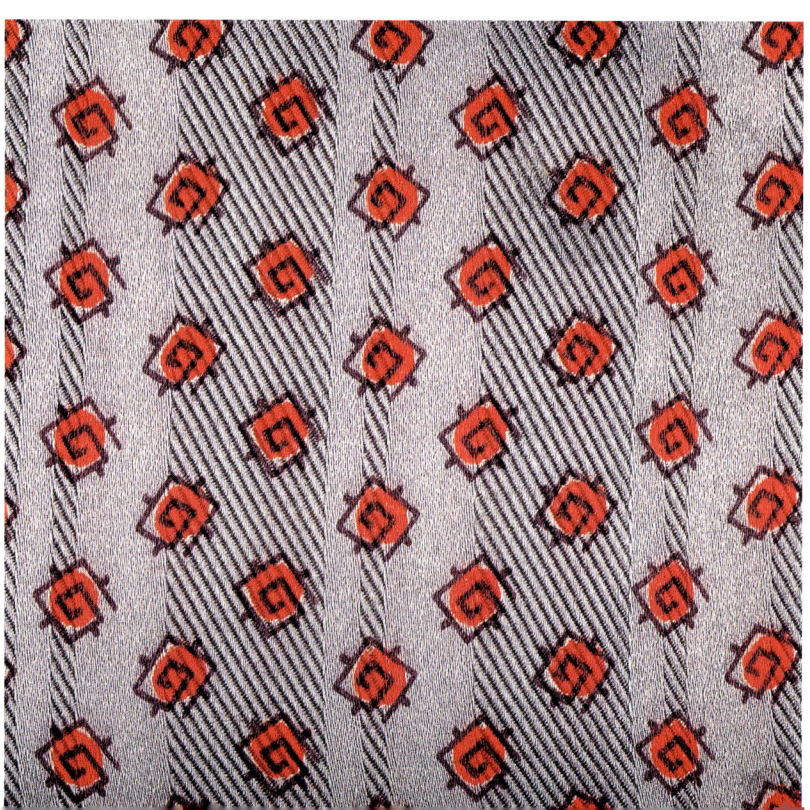
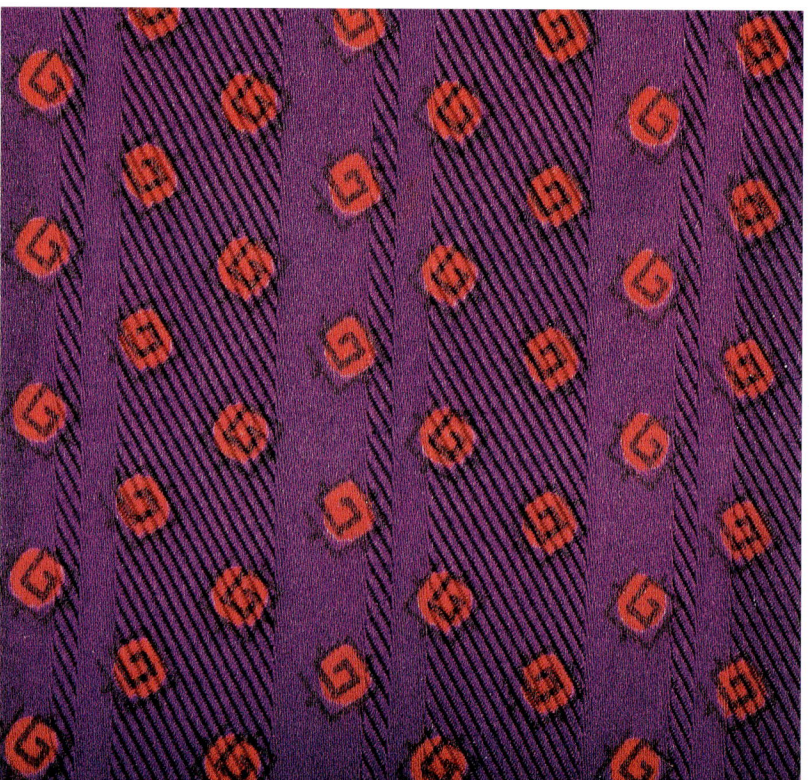

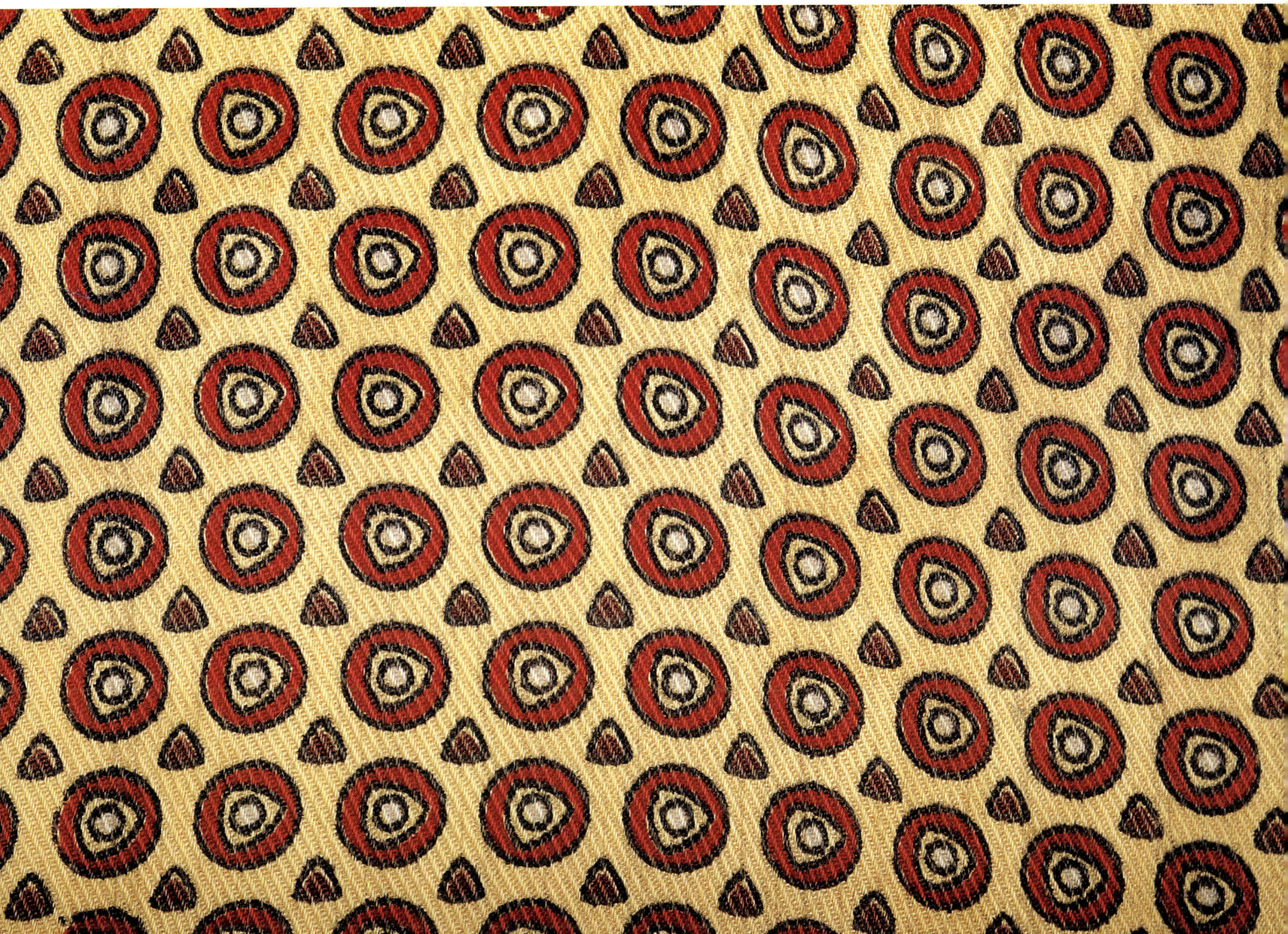

Floral & Fancy

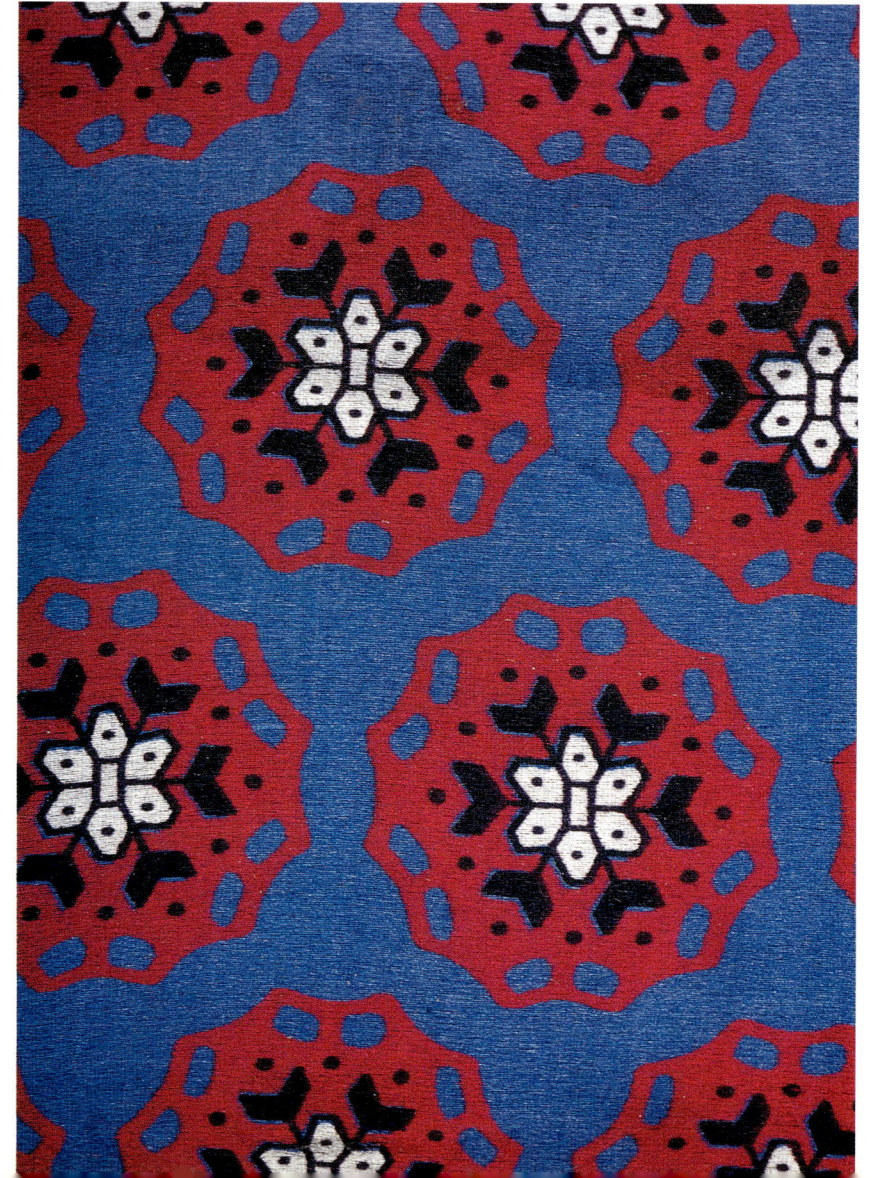

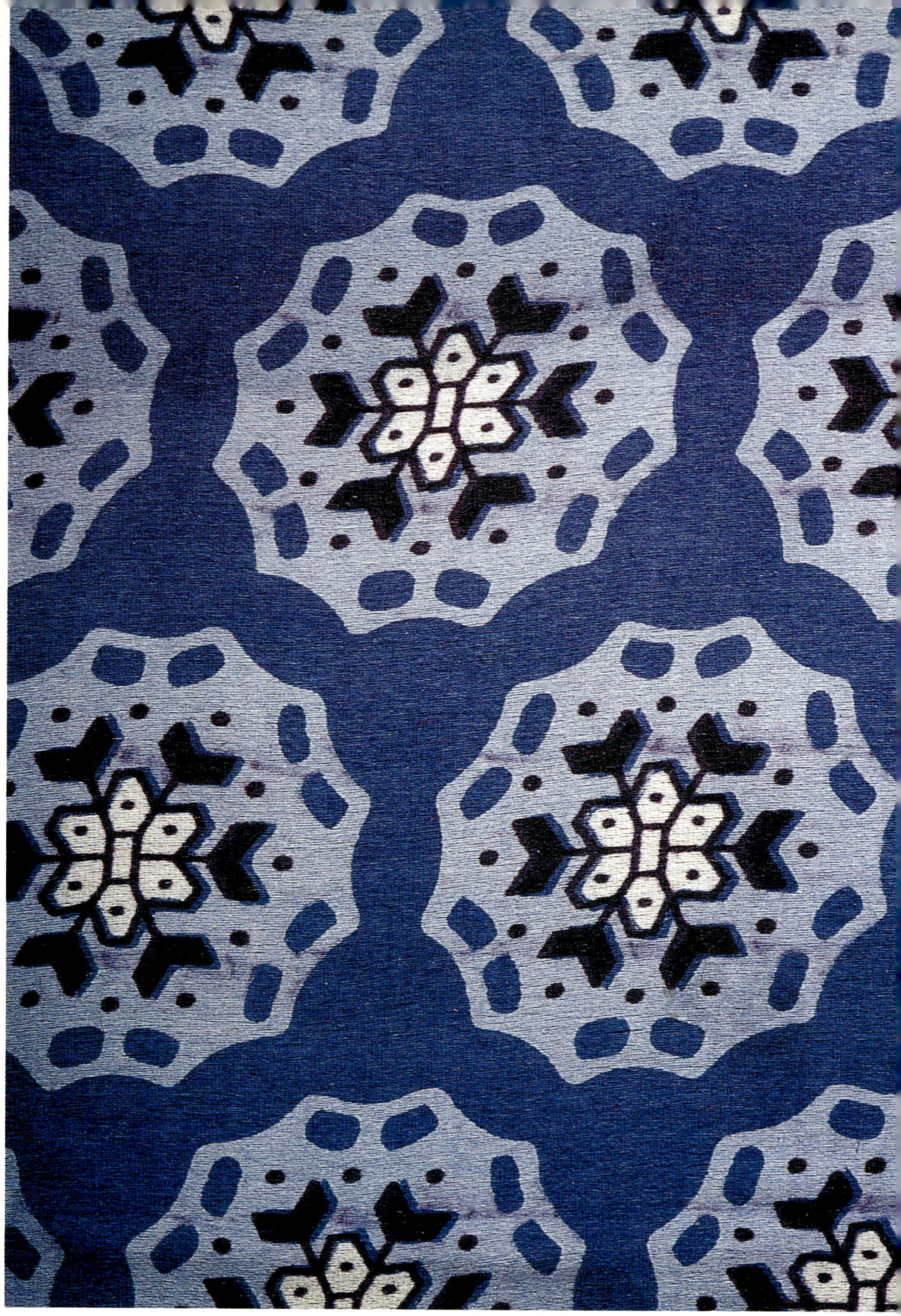

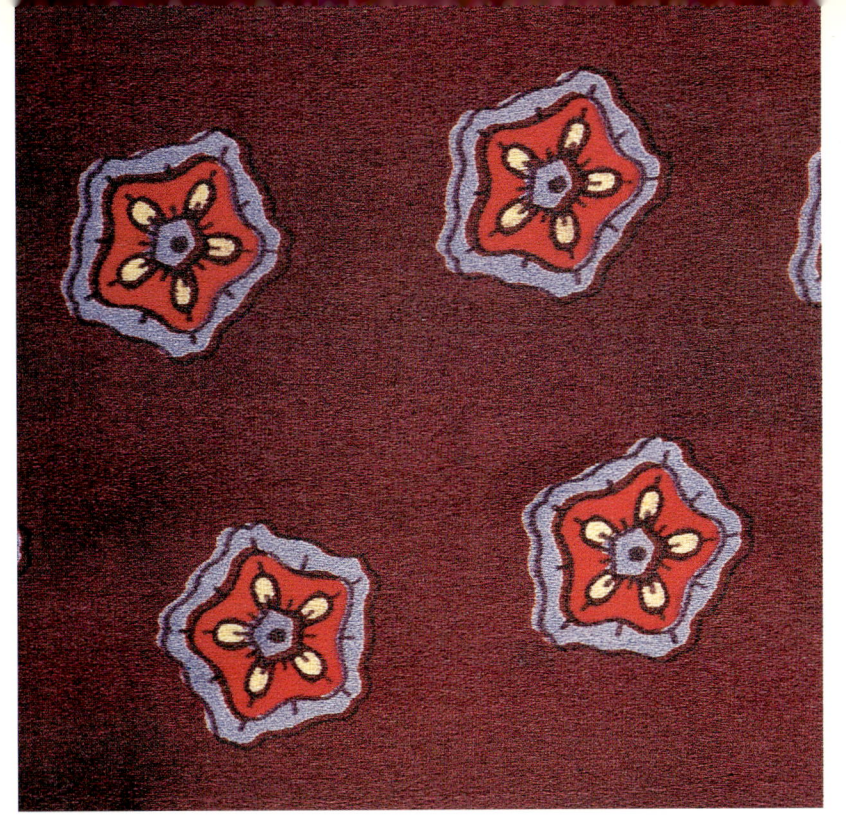
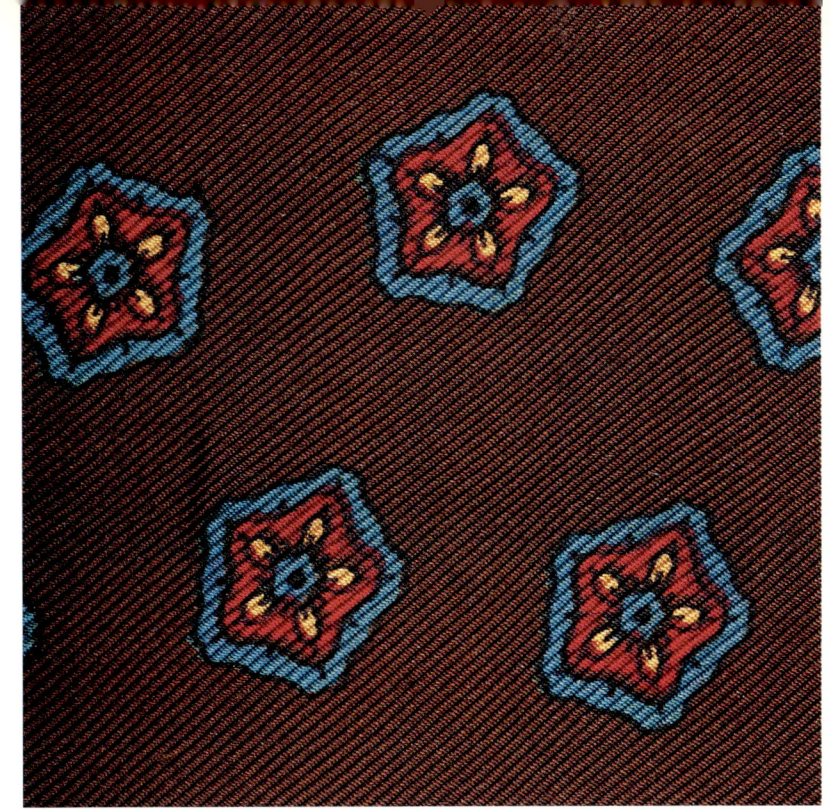
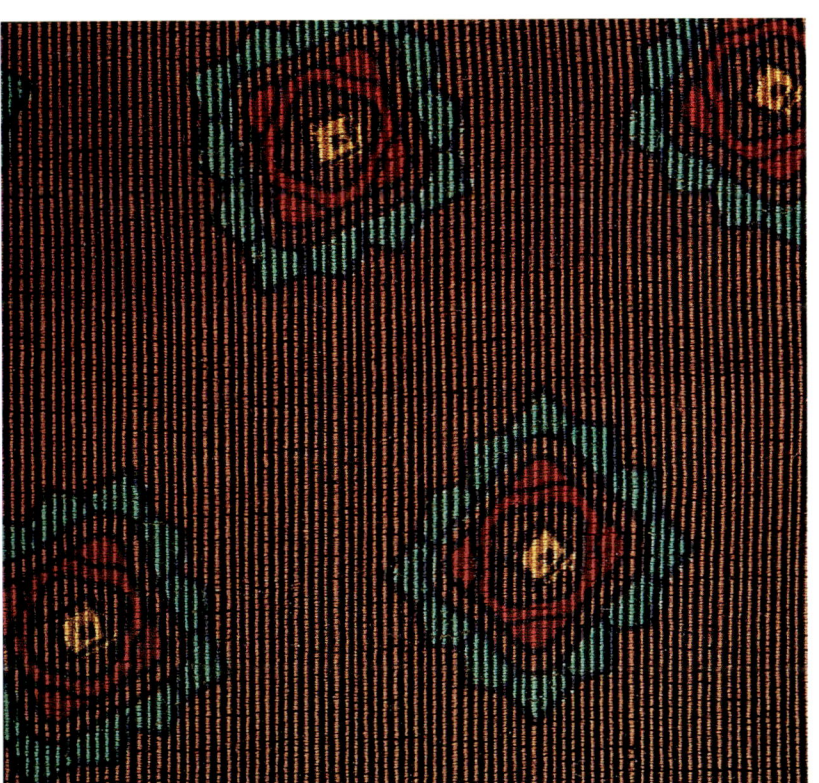
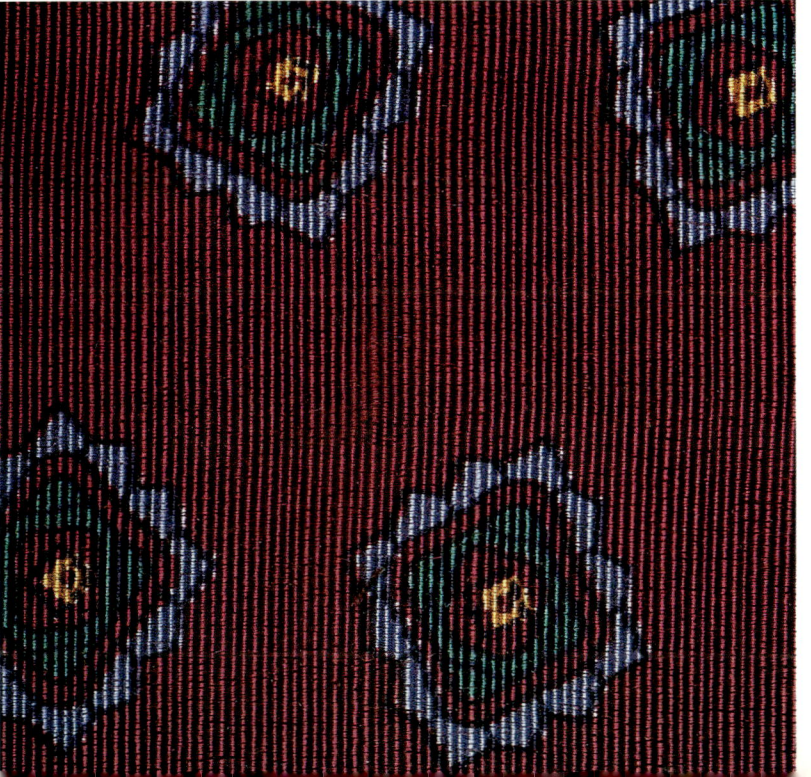

63

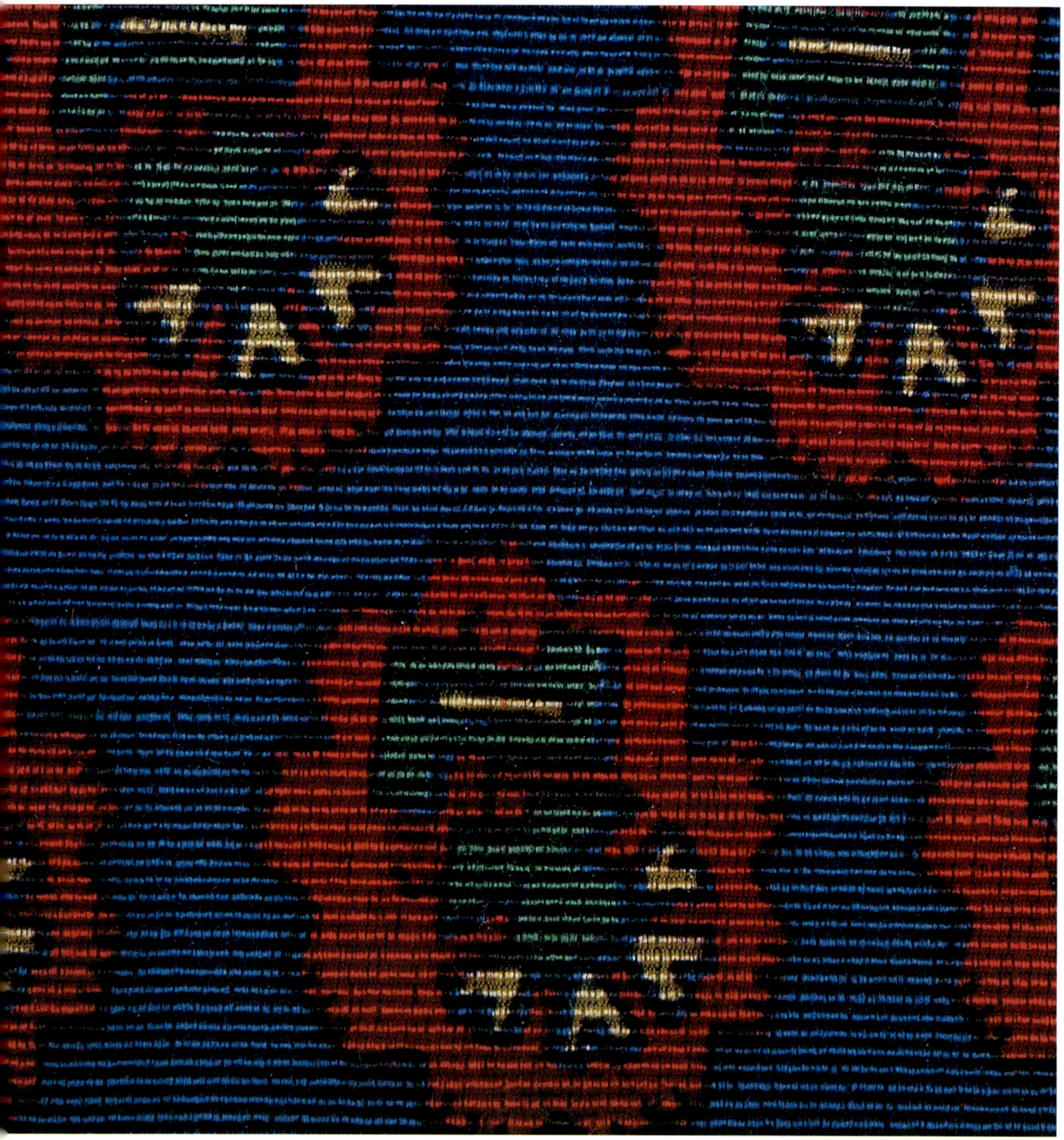

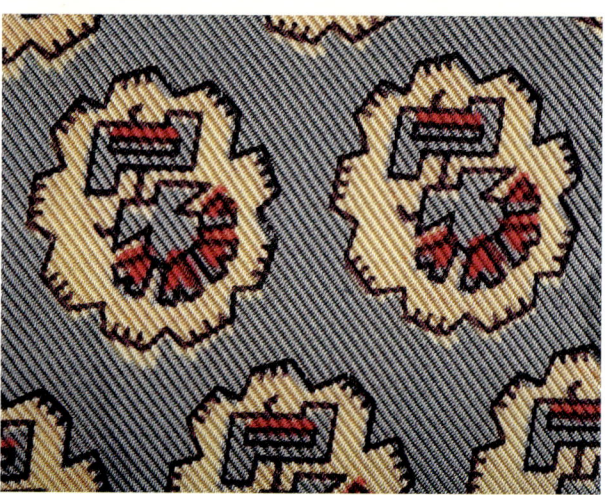

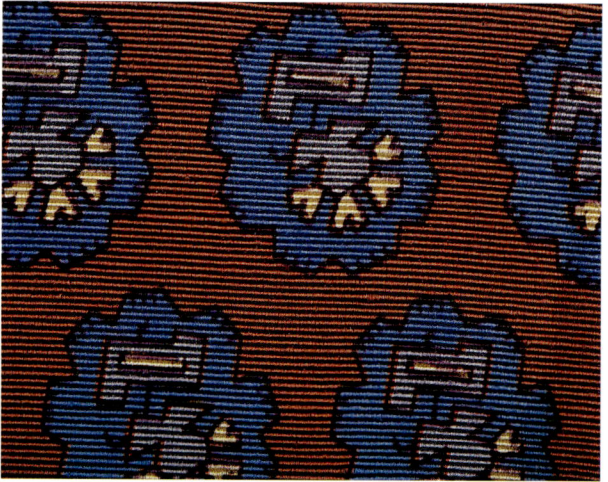

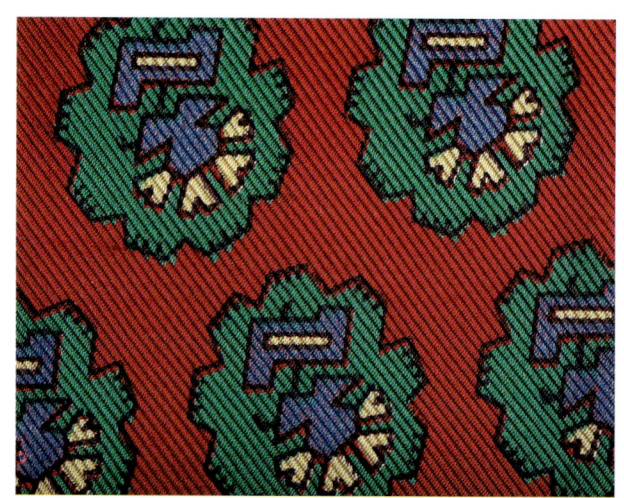

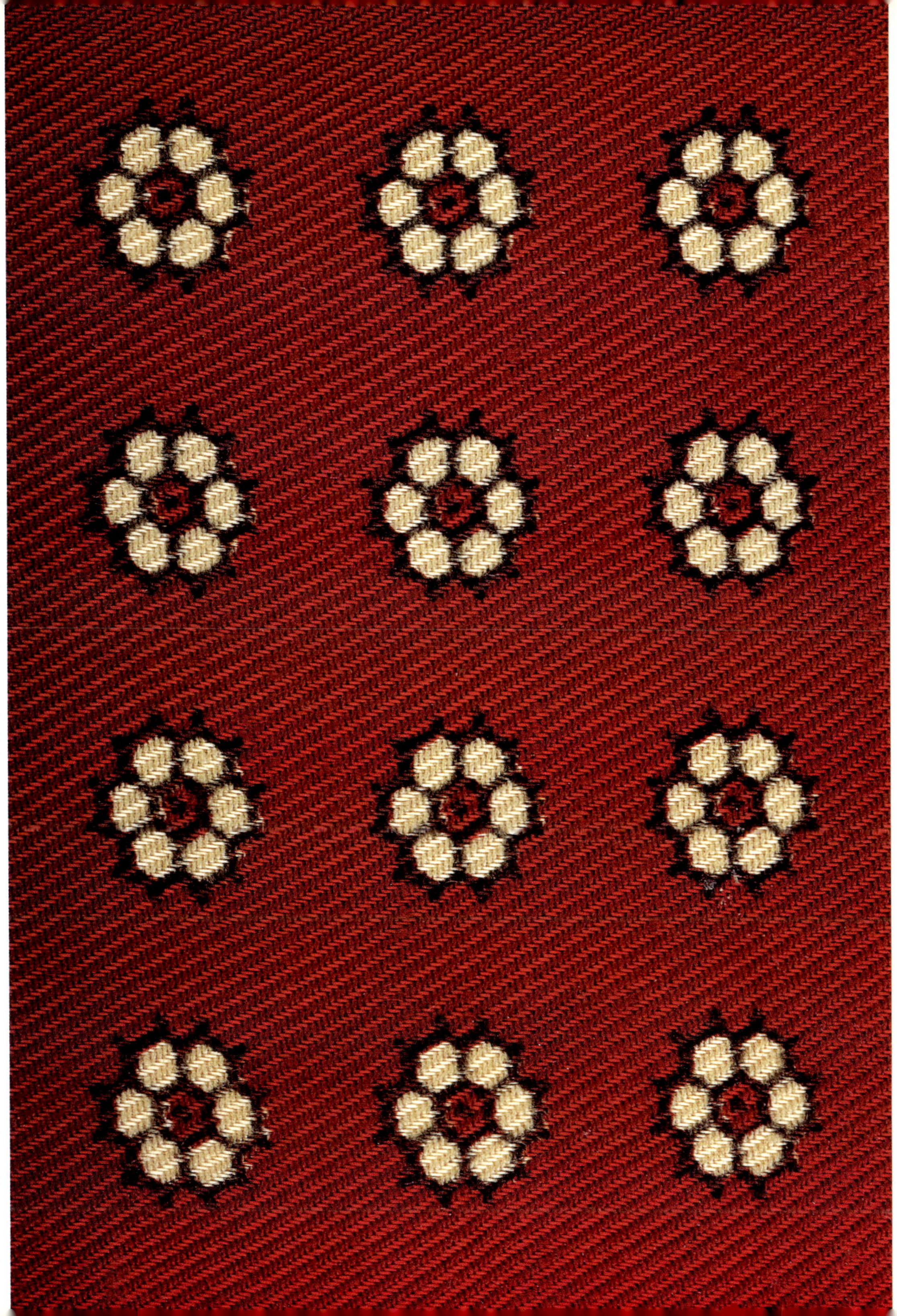

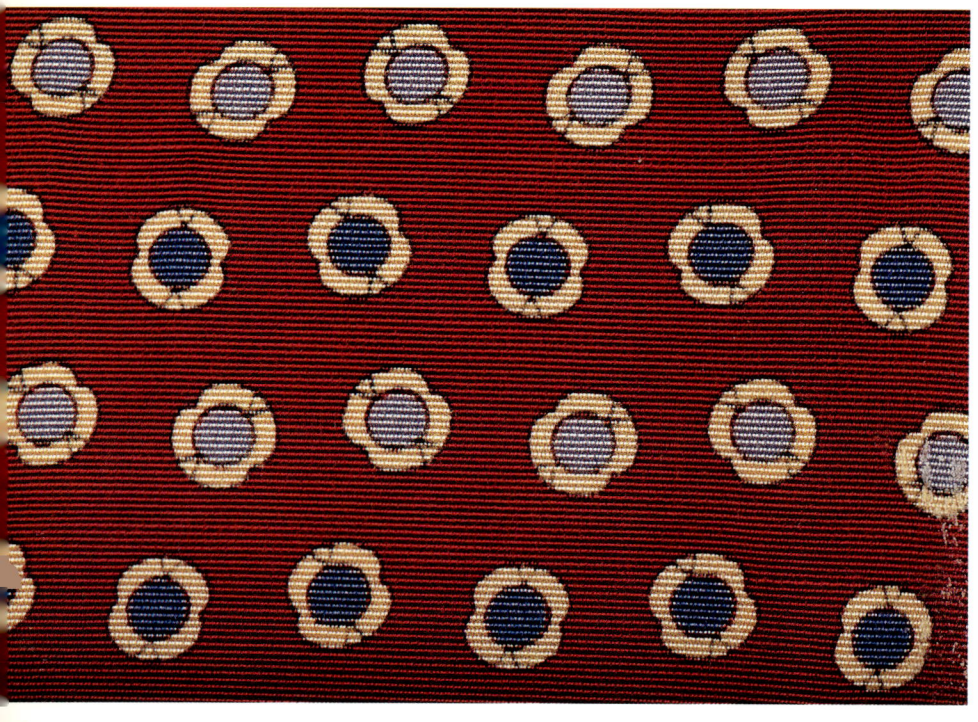

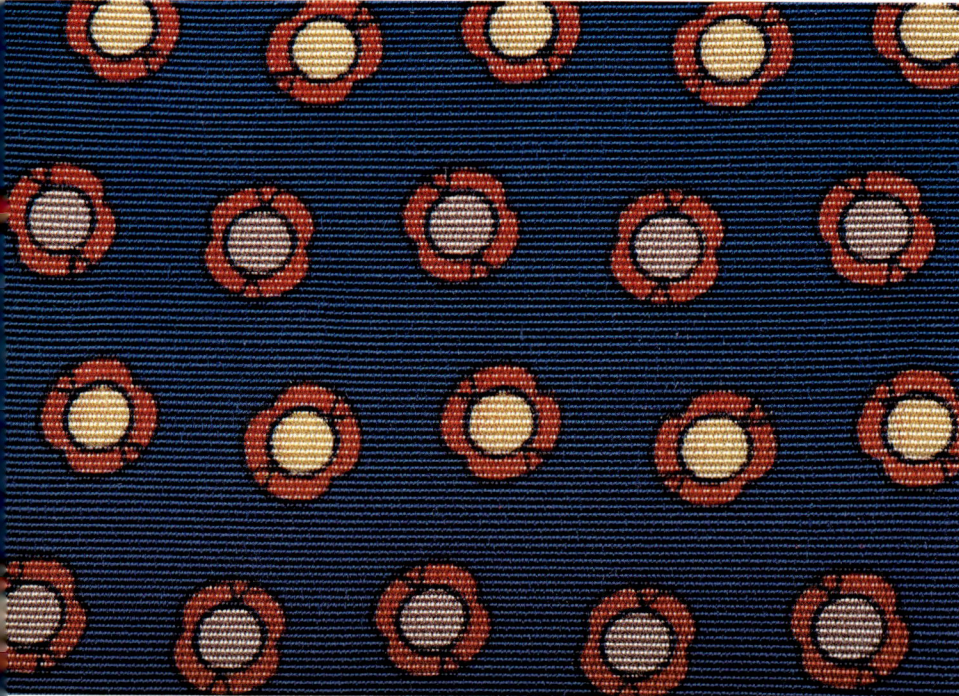

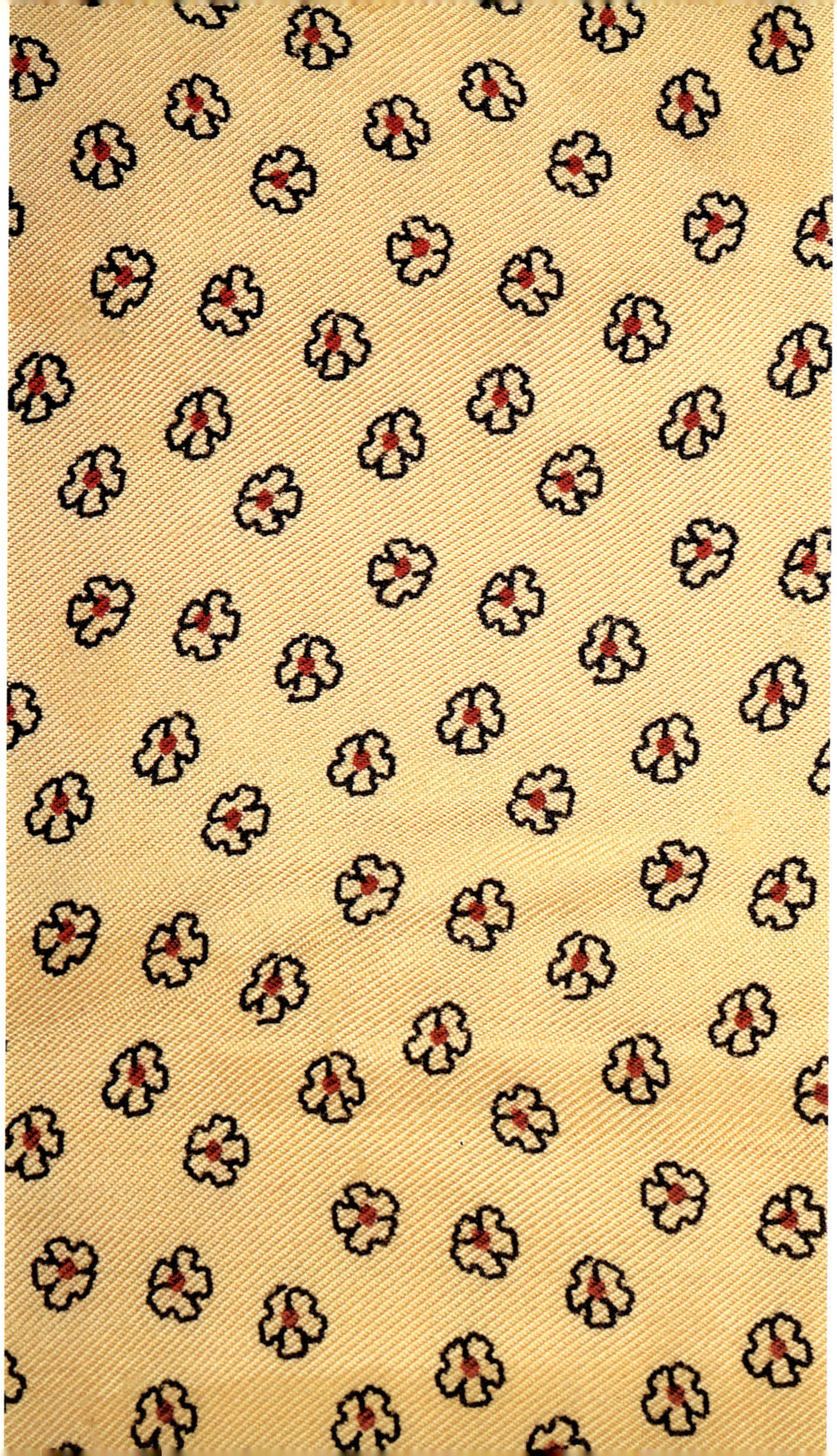

66

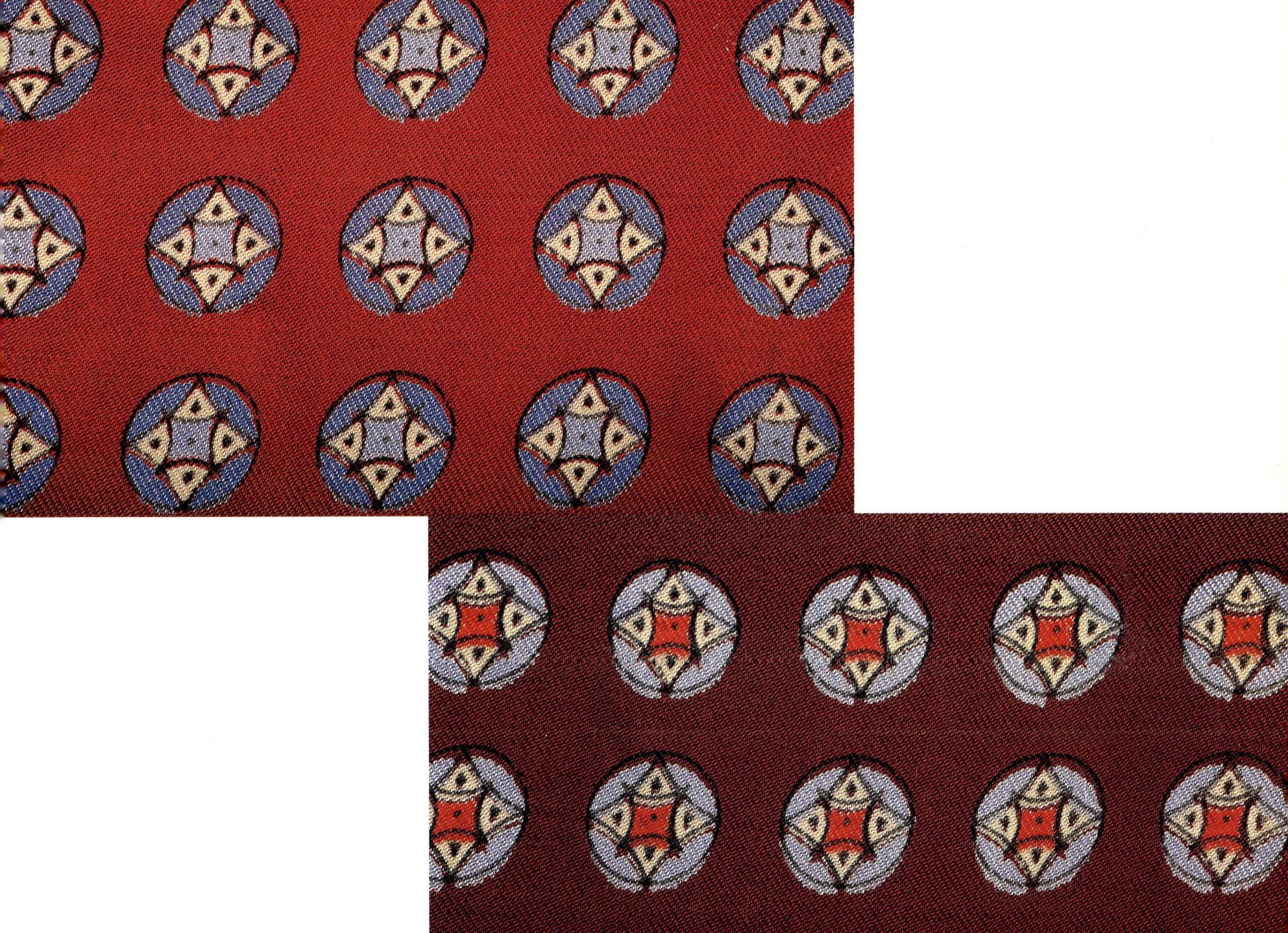

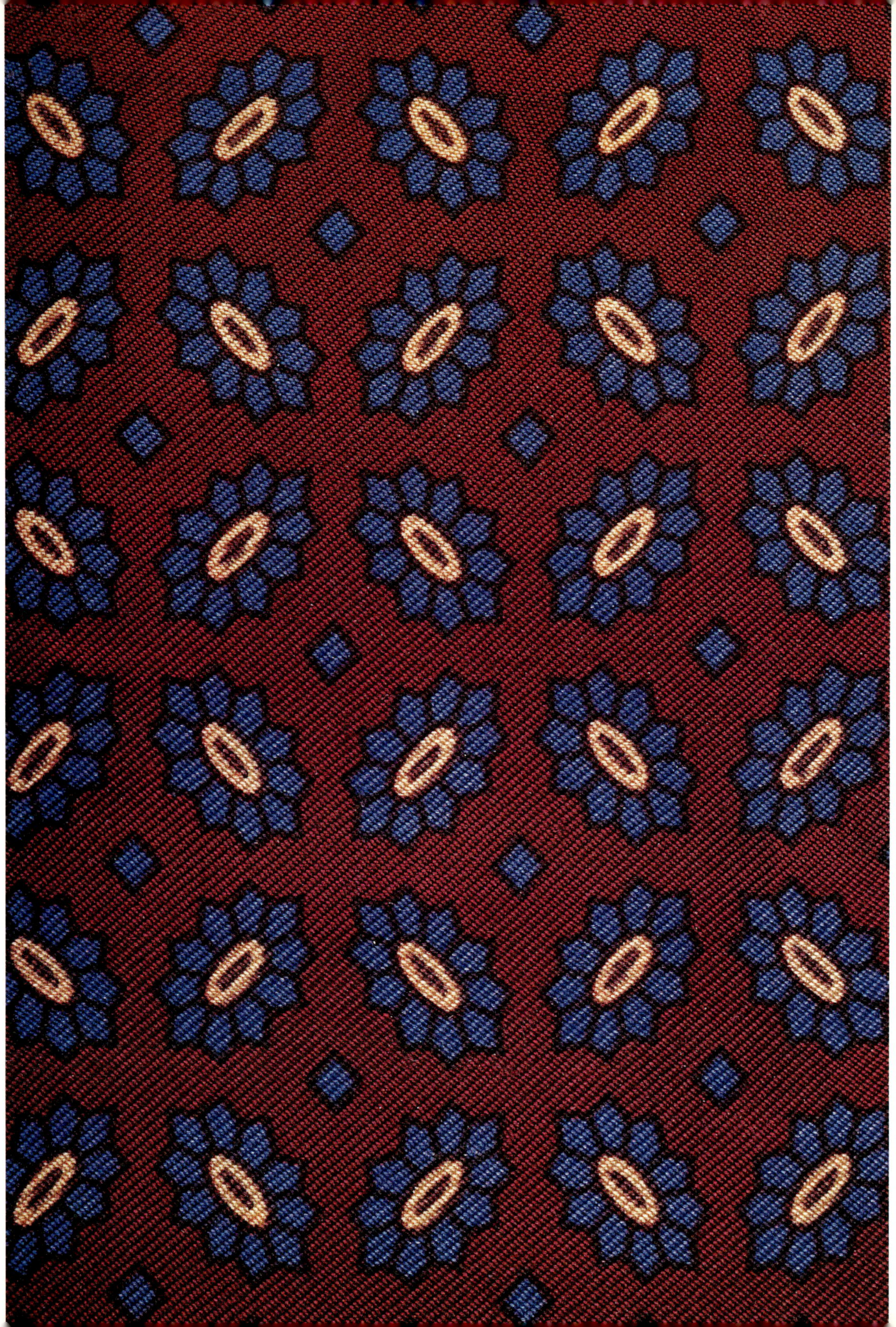

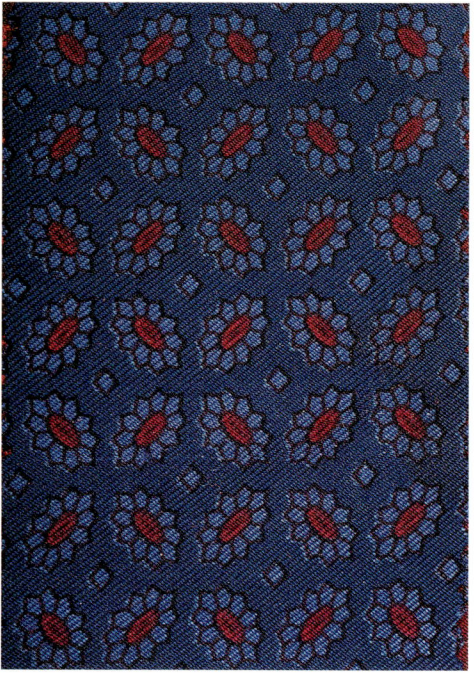

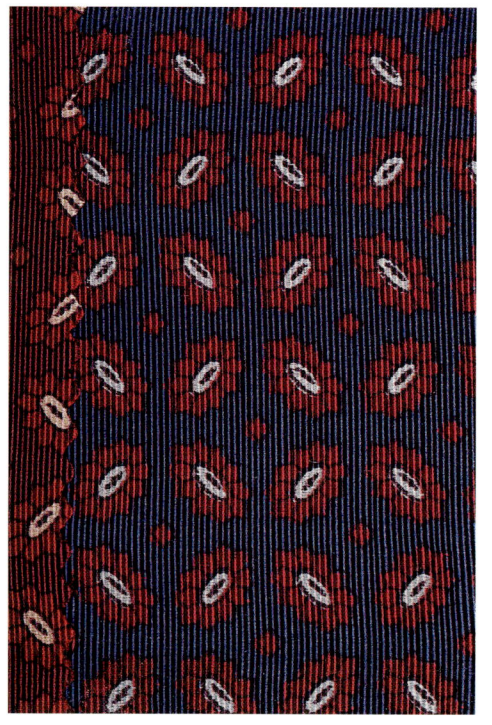

68

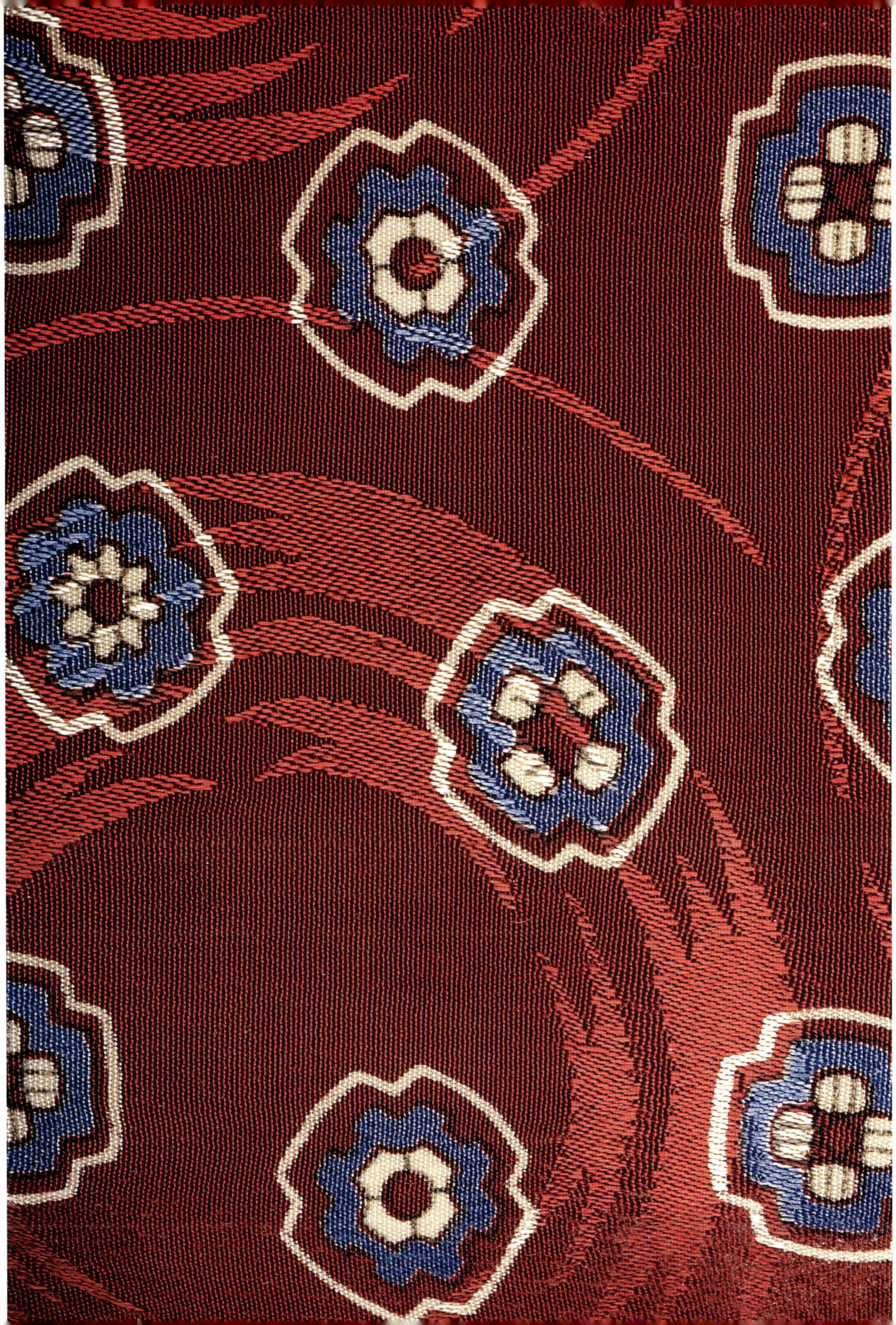

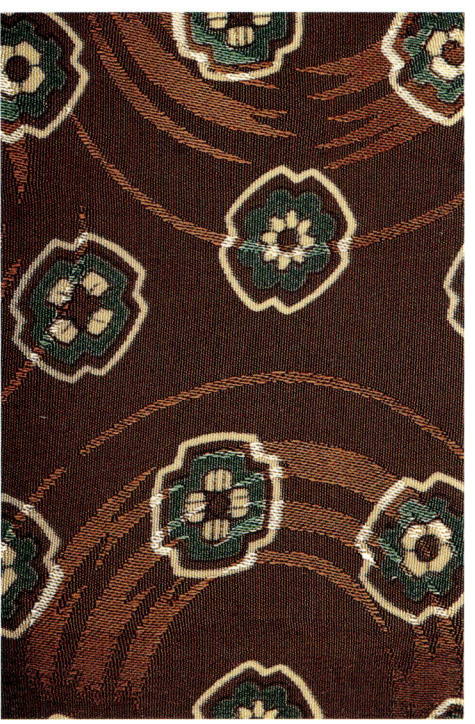

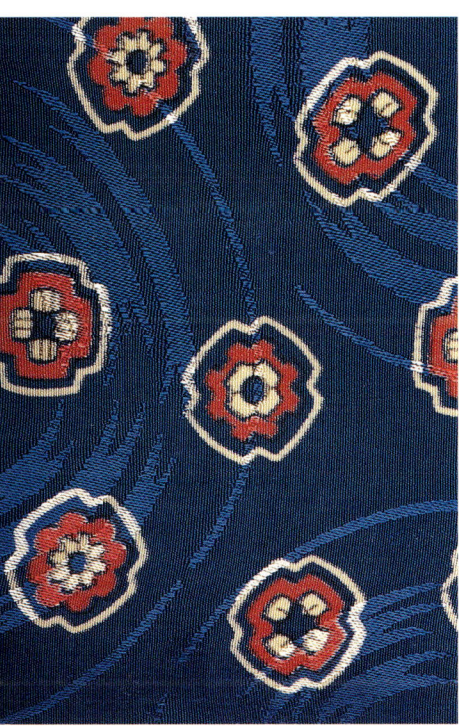

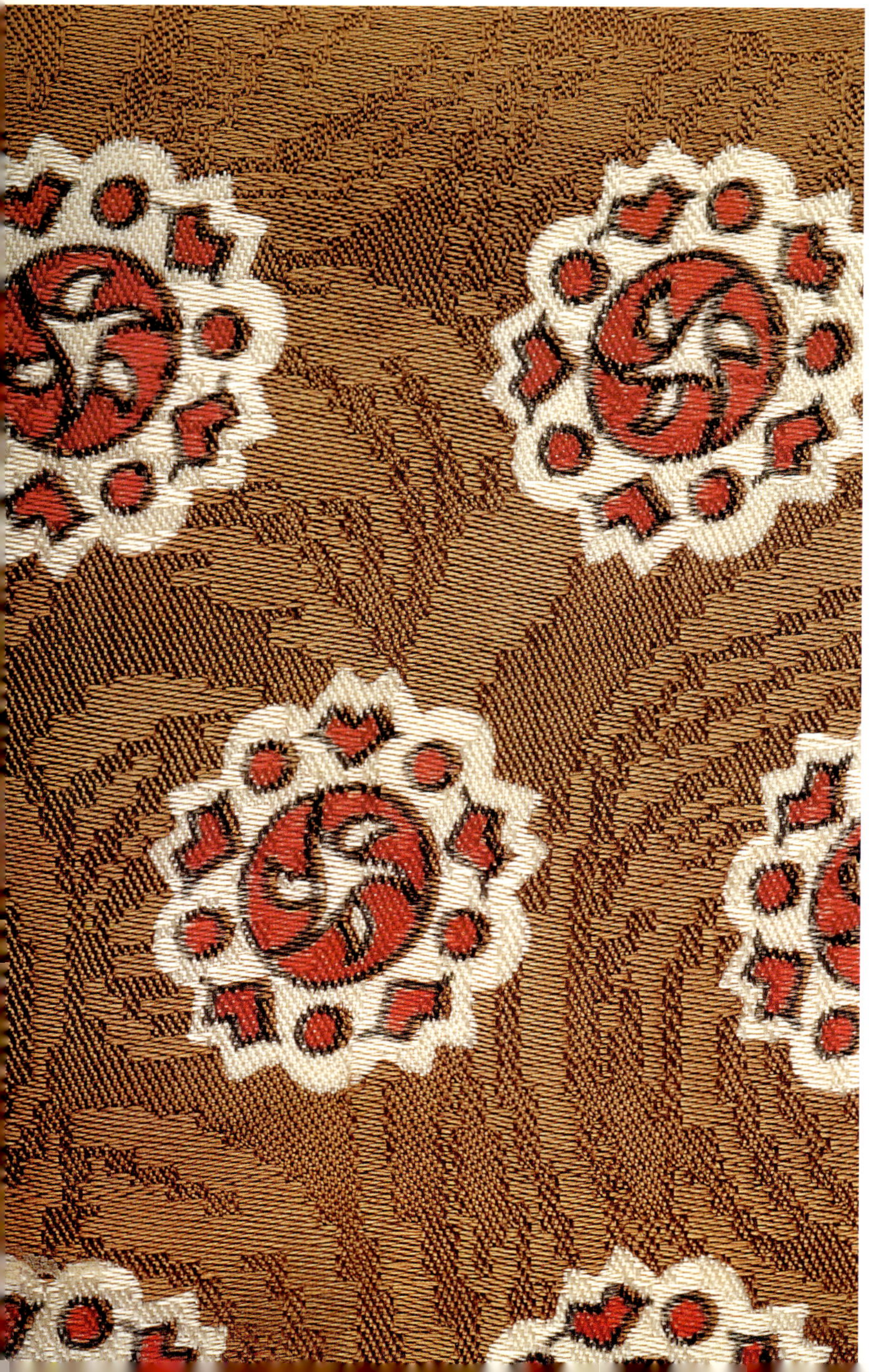

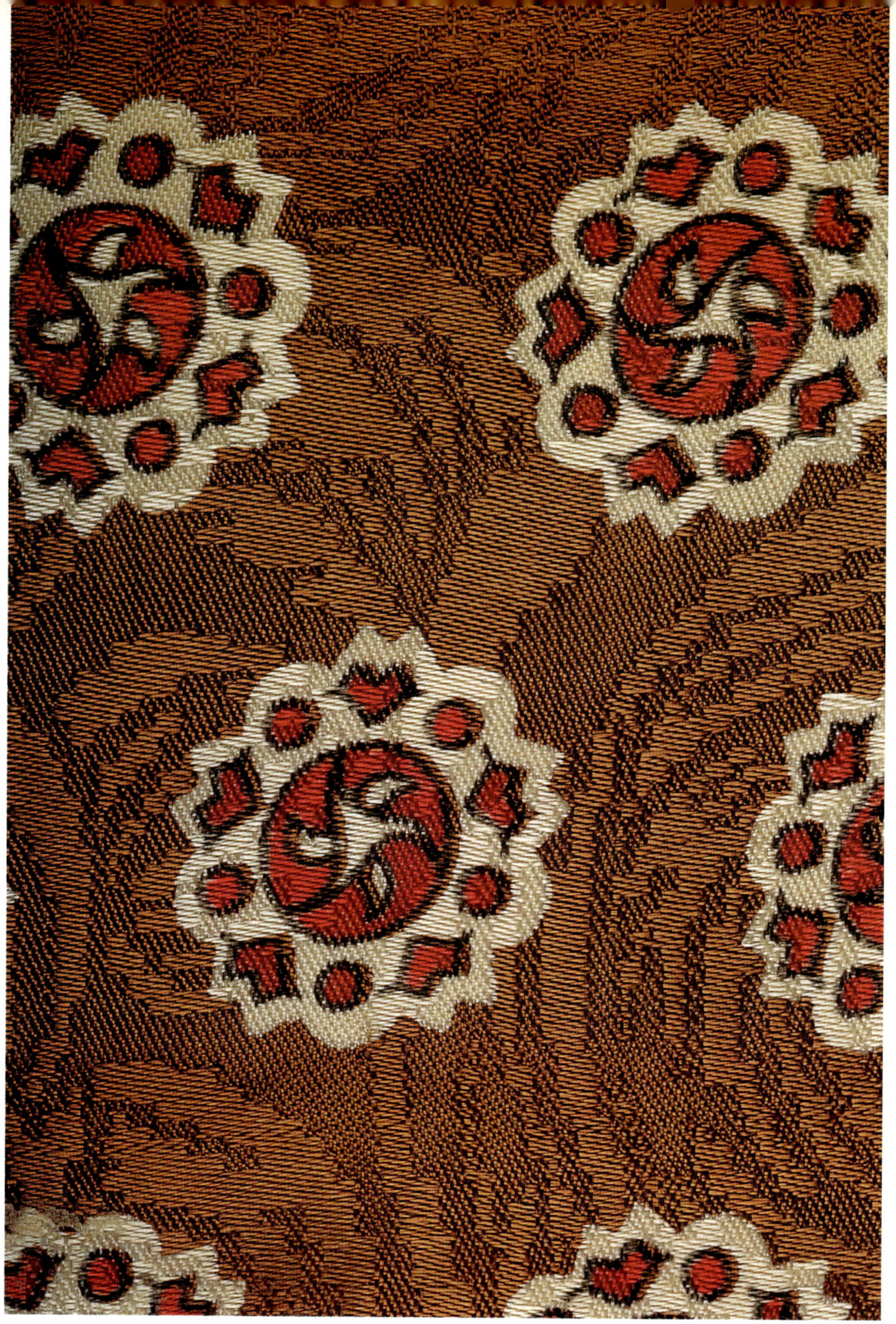

Rounded Edges

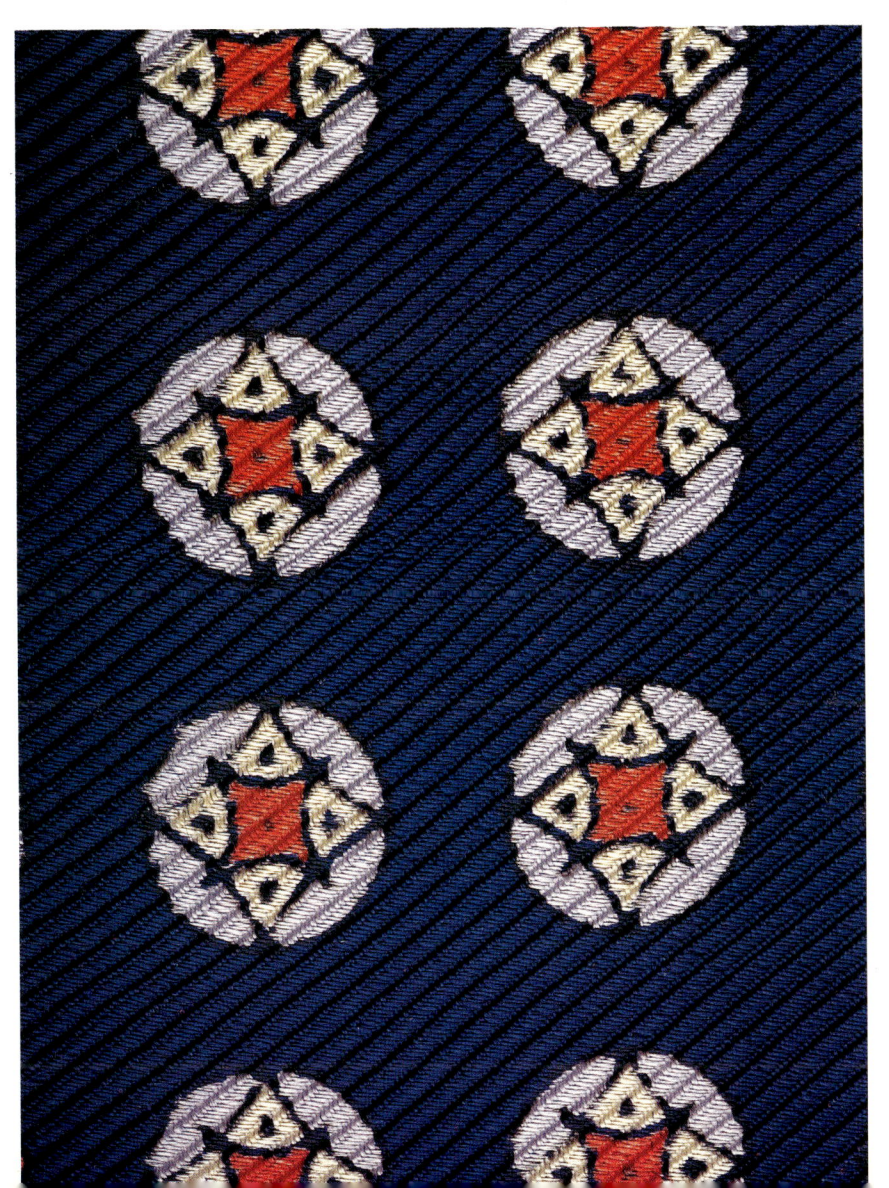

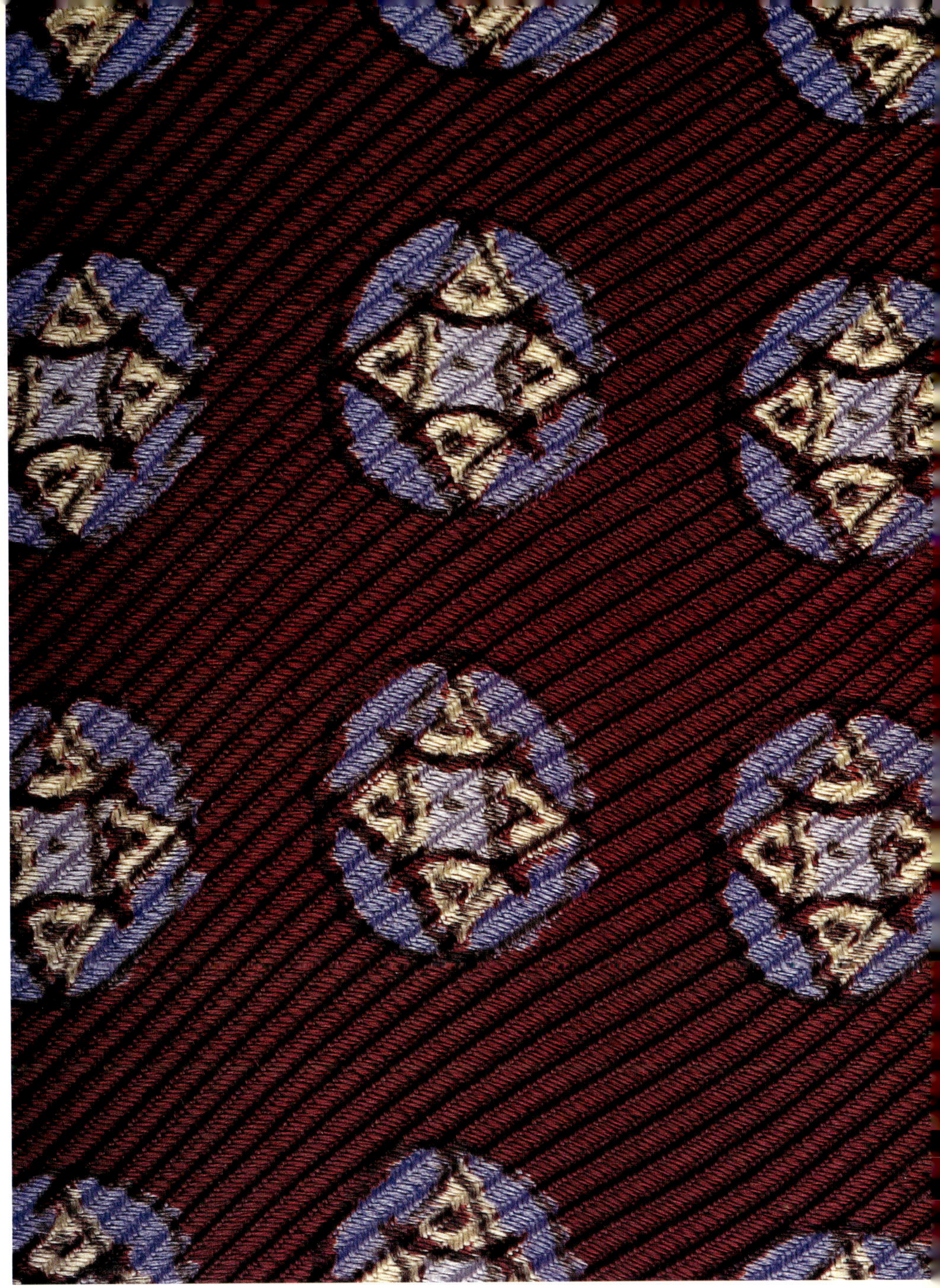

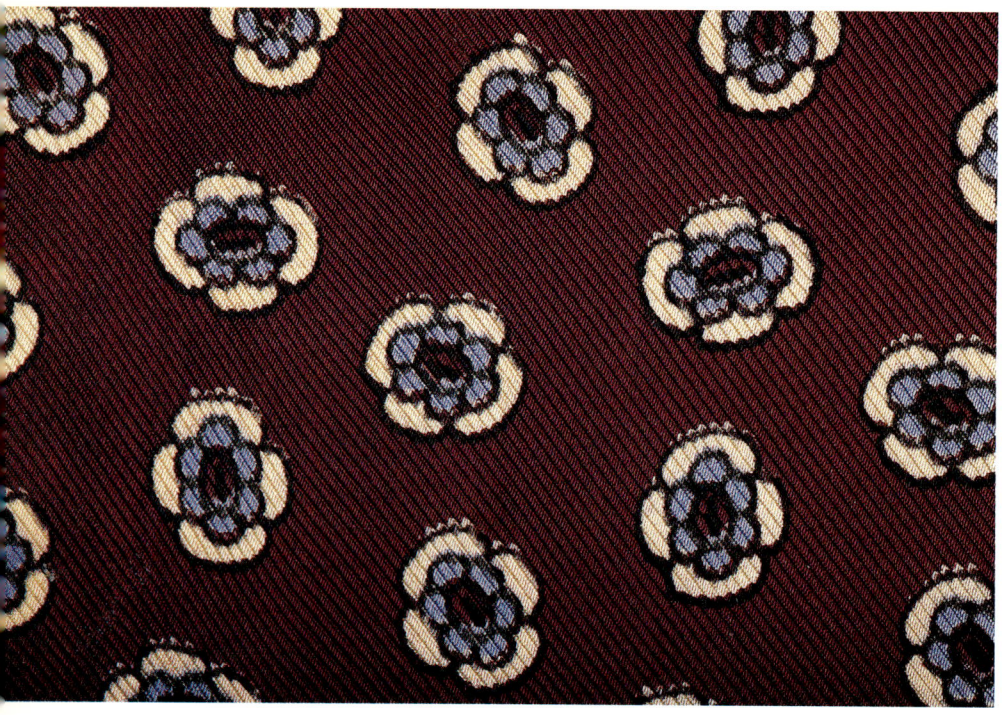
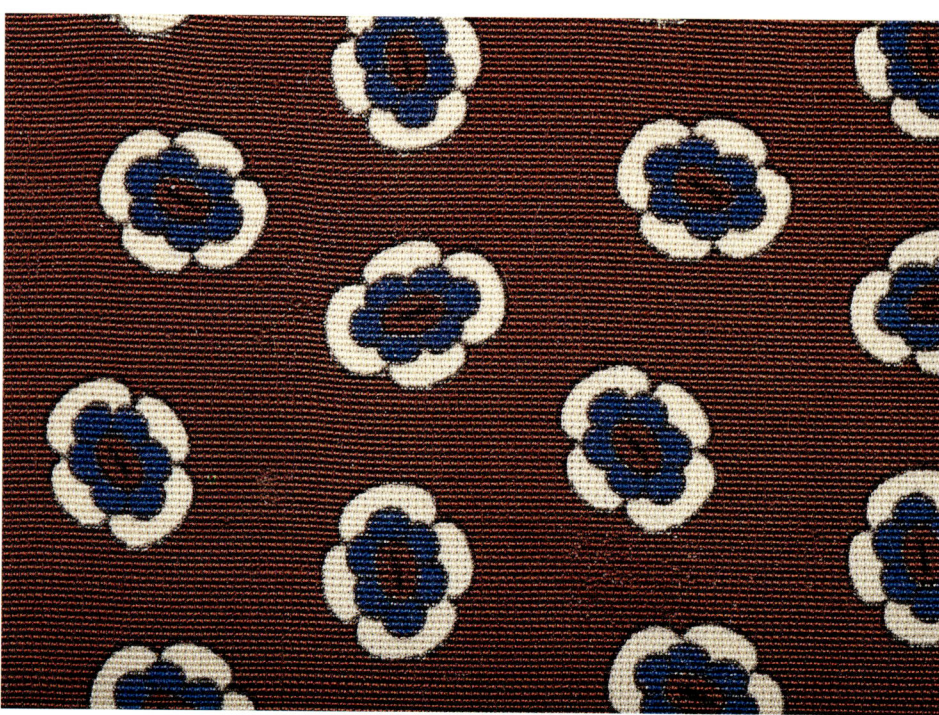
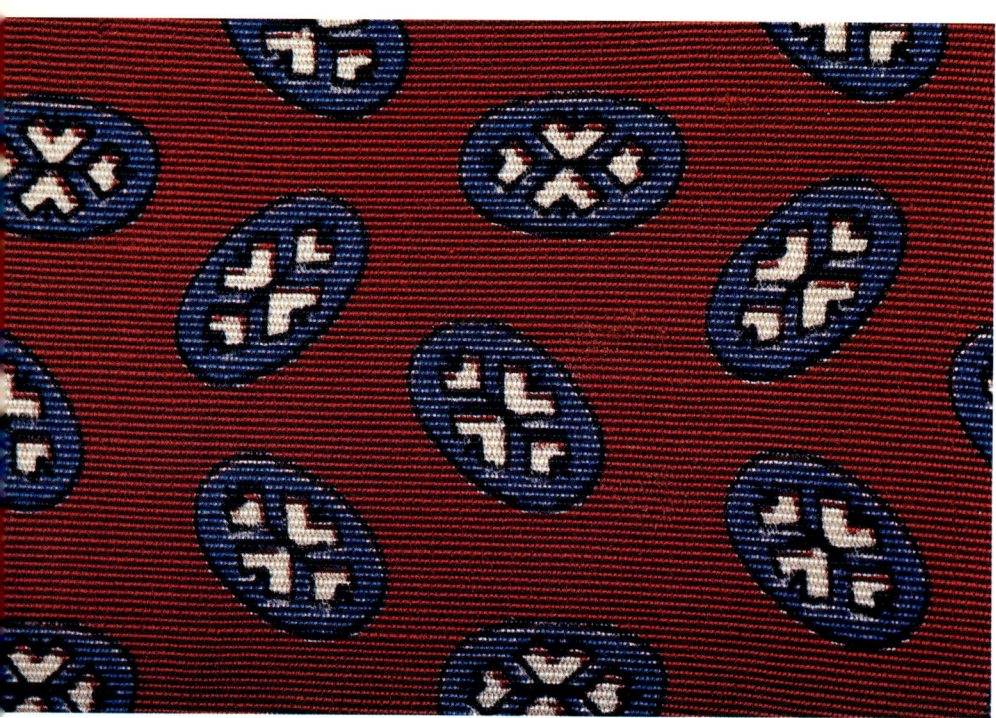
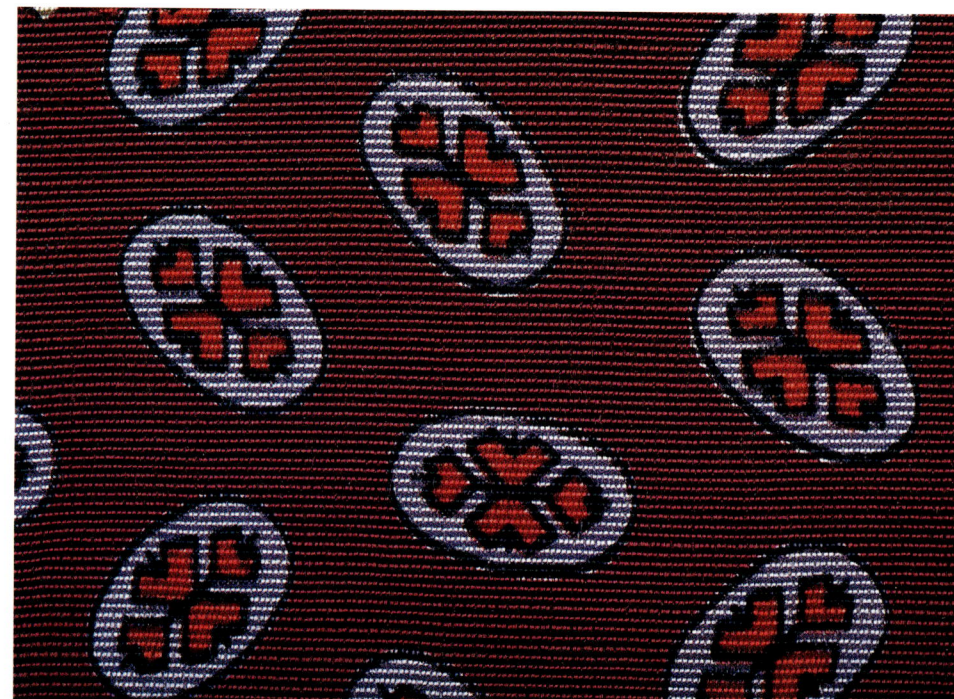

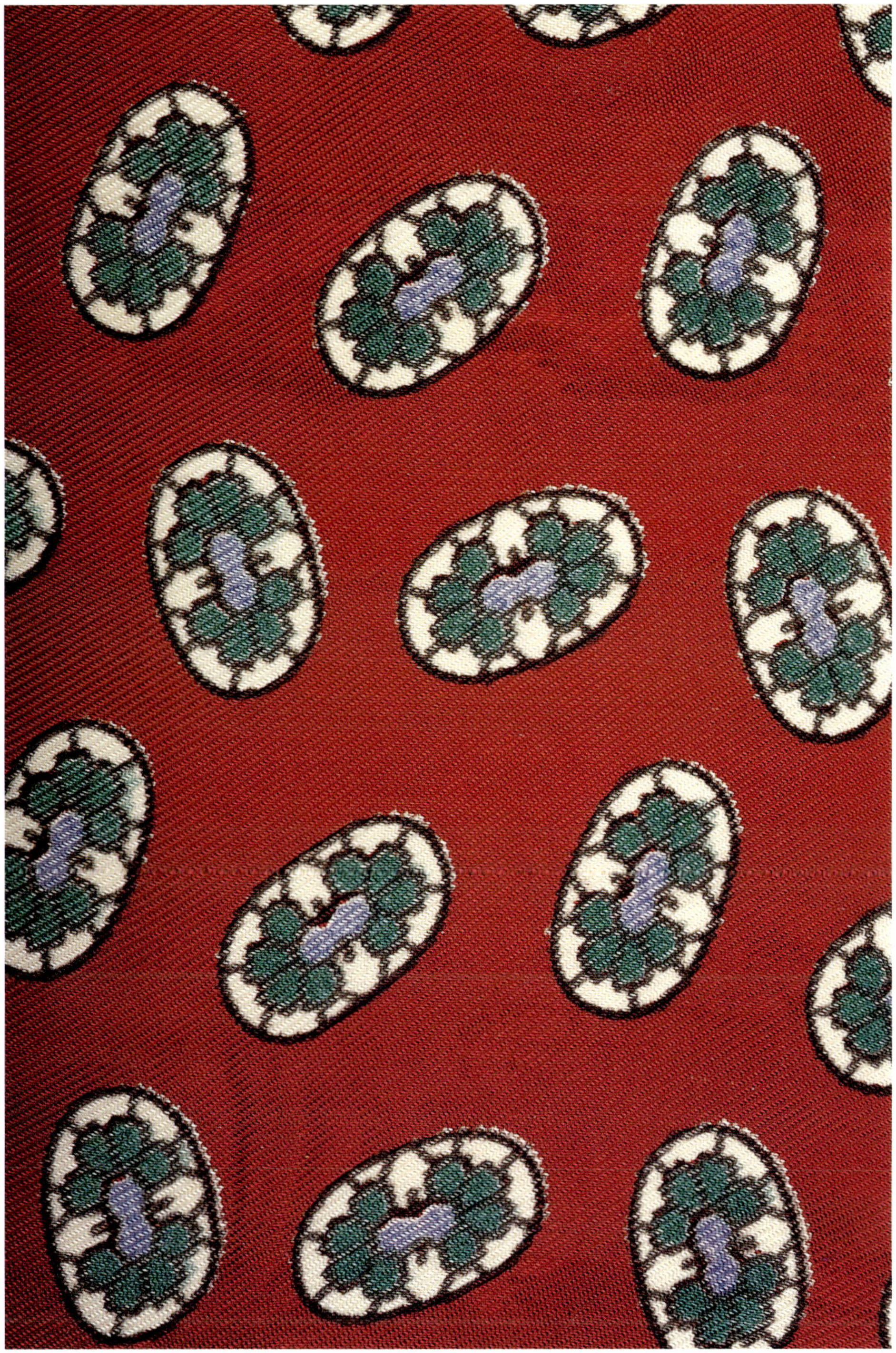

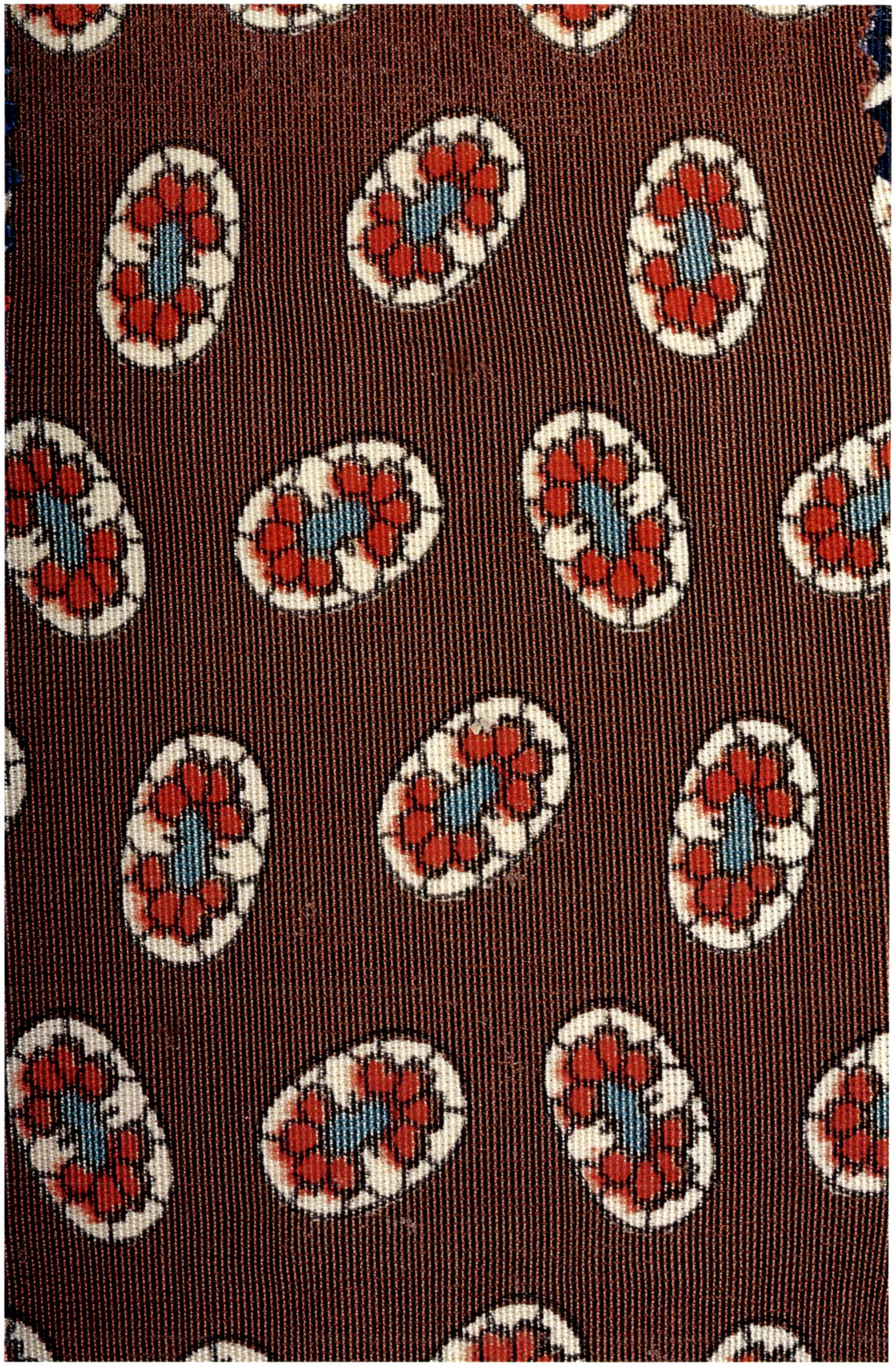

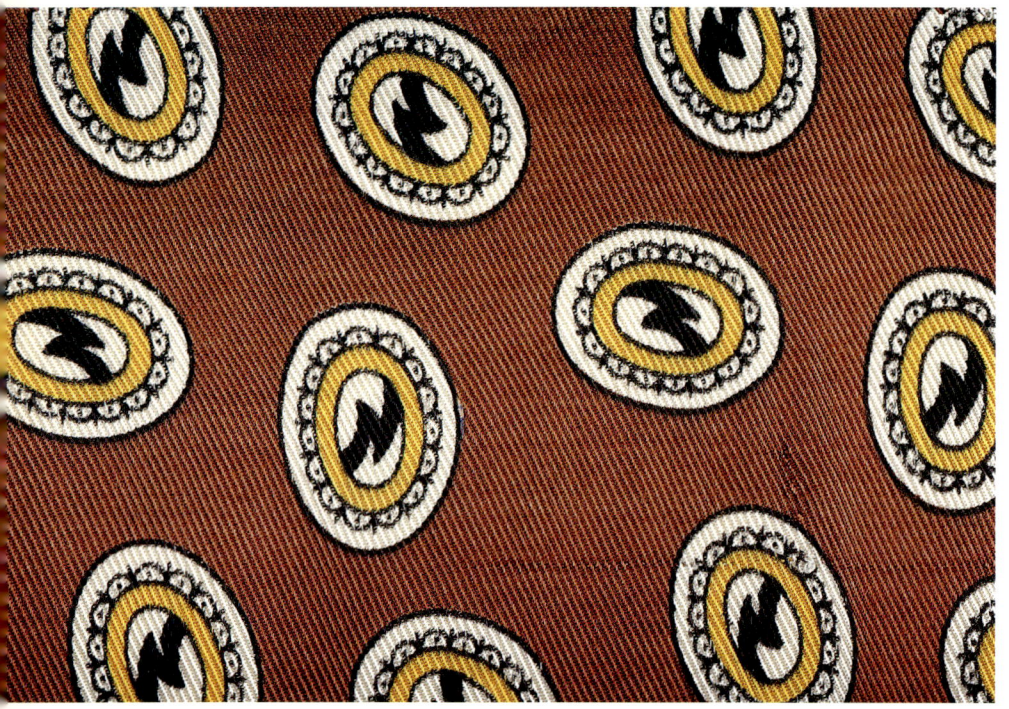
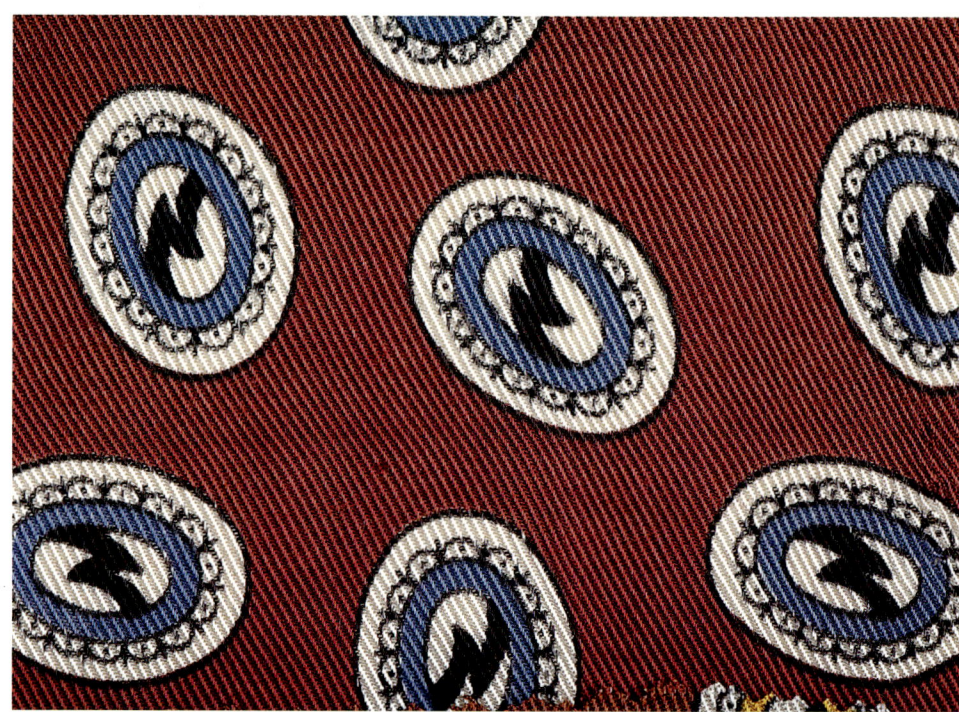
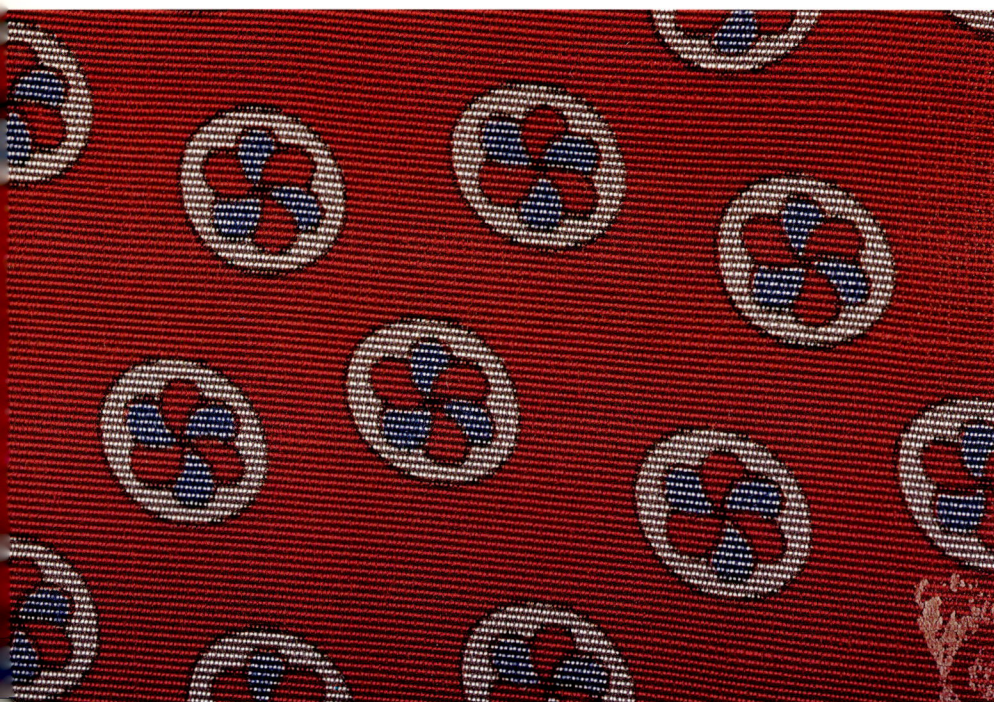
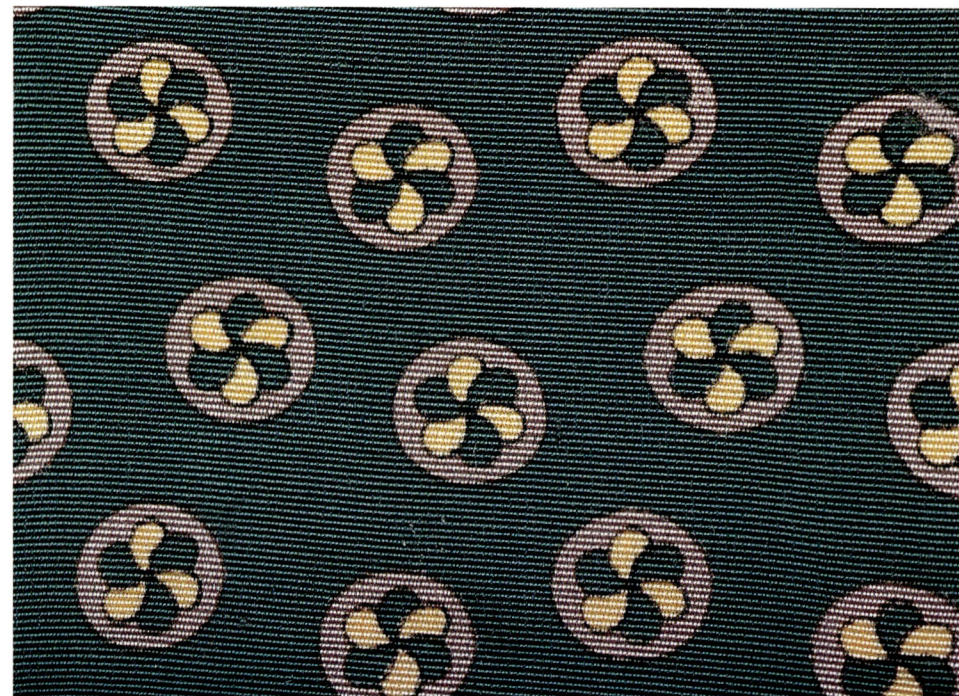

74

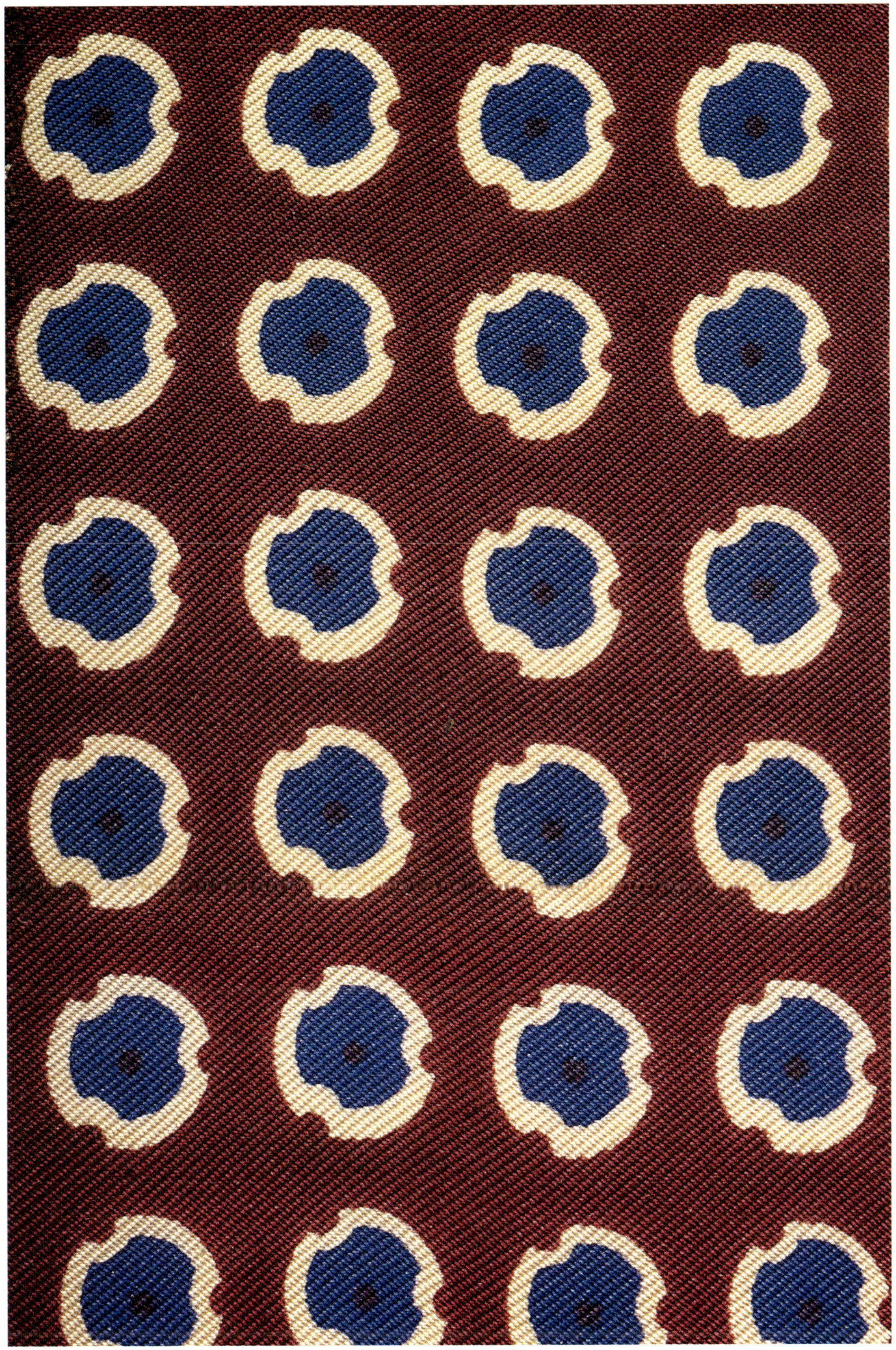

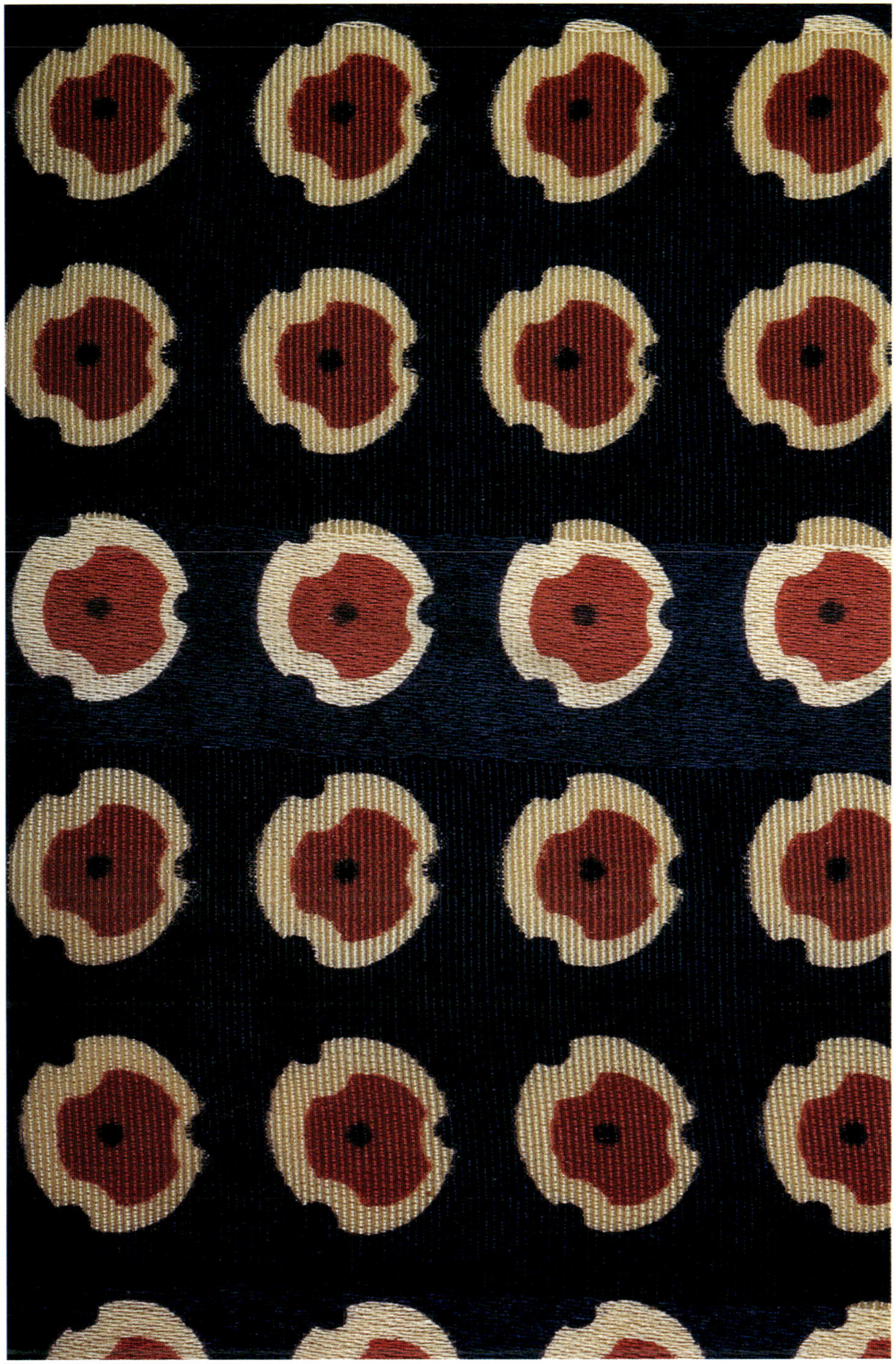

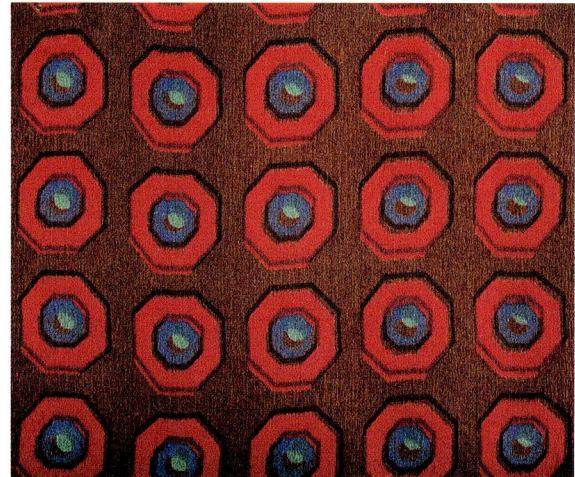

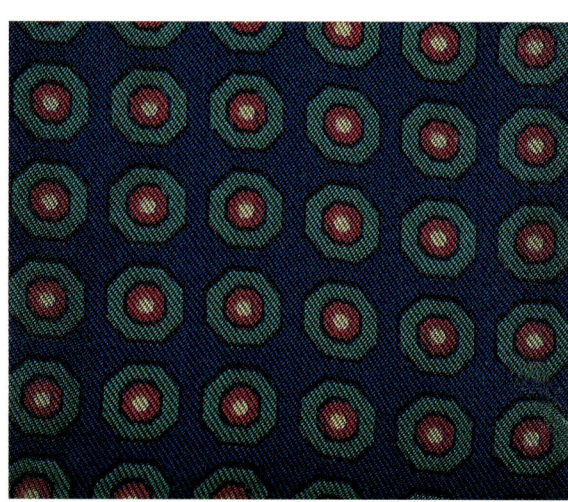

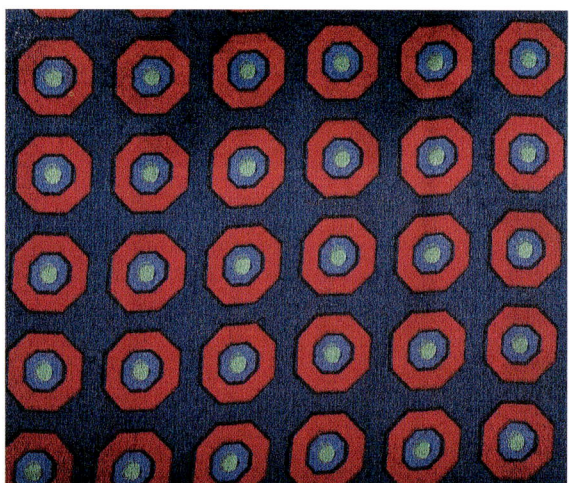

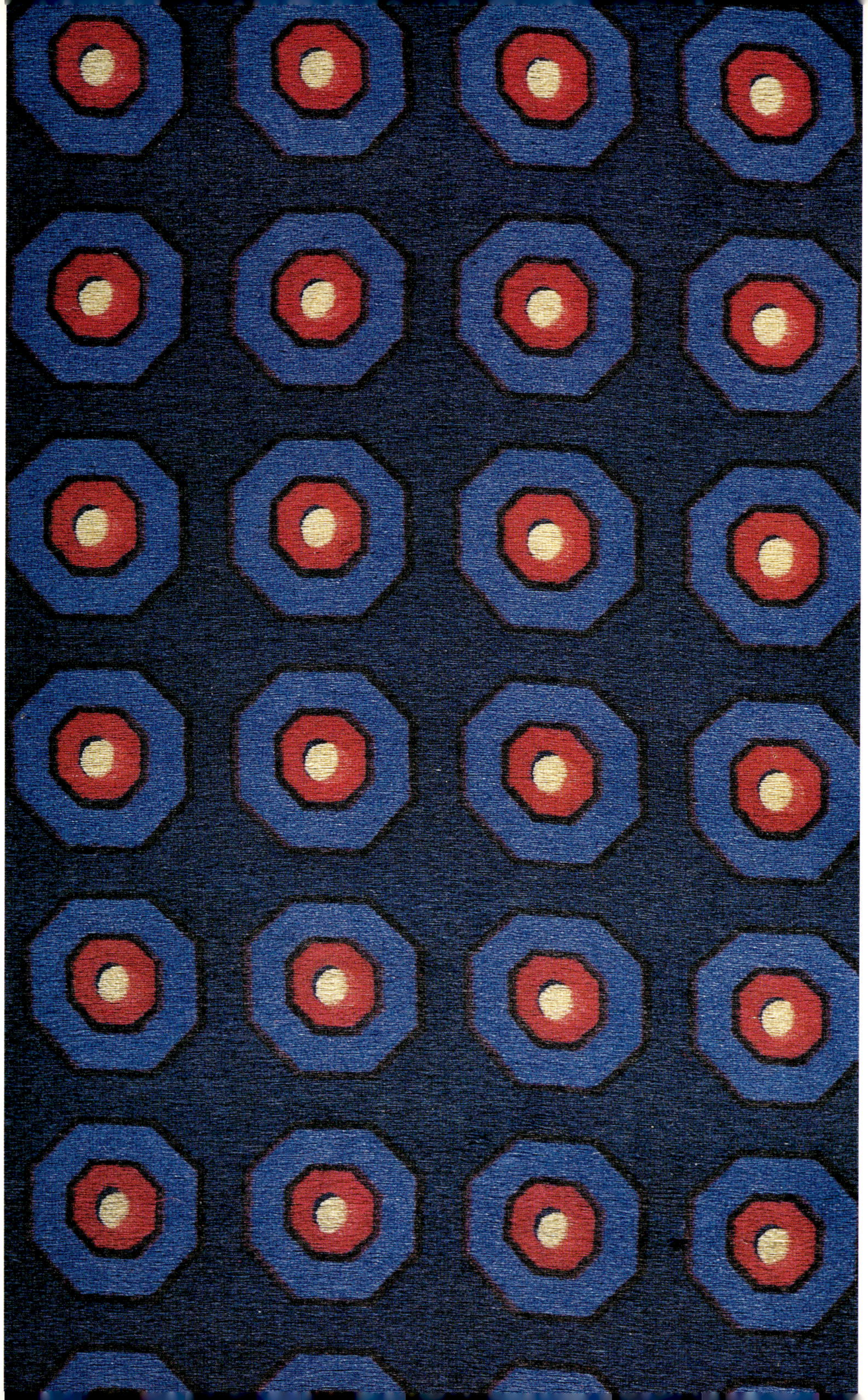

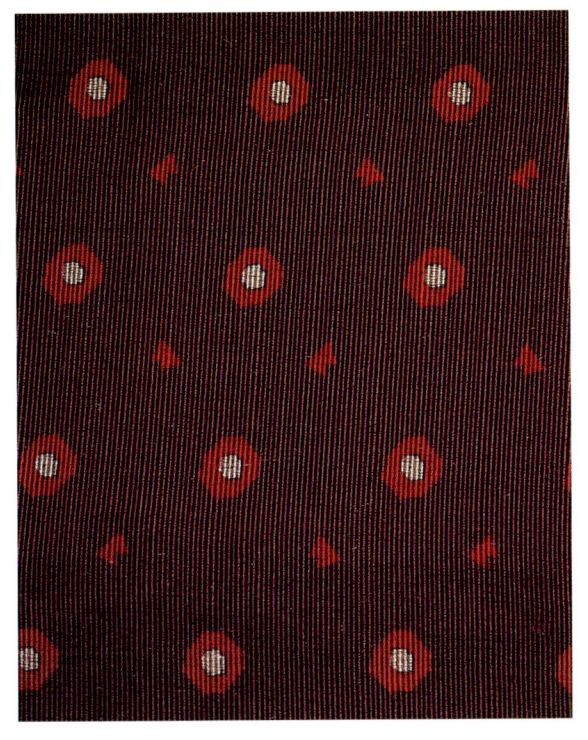

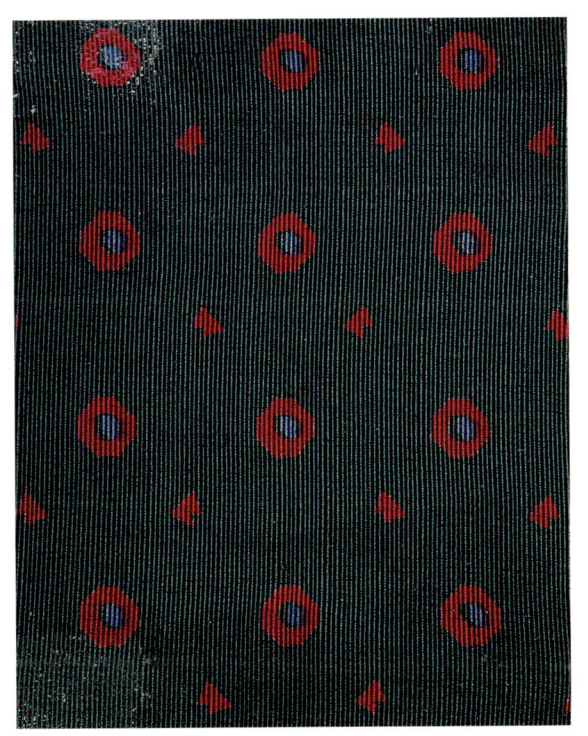
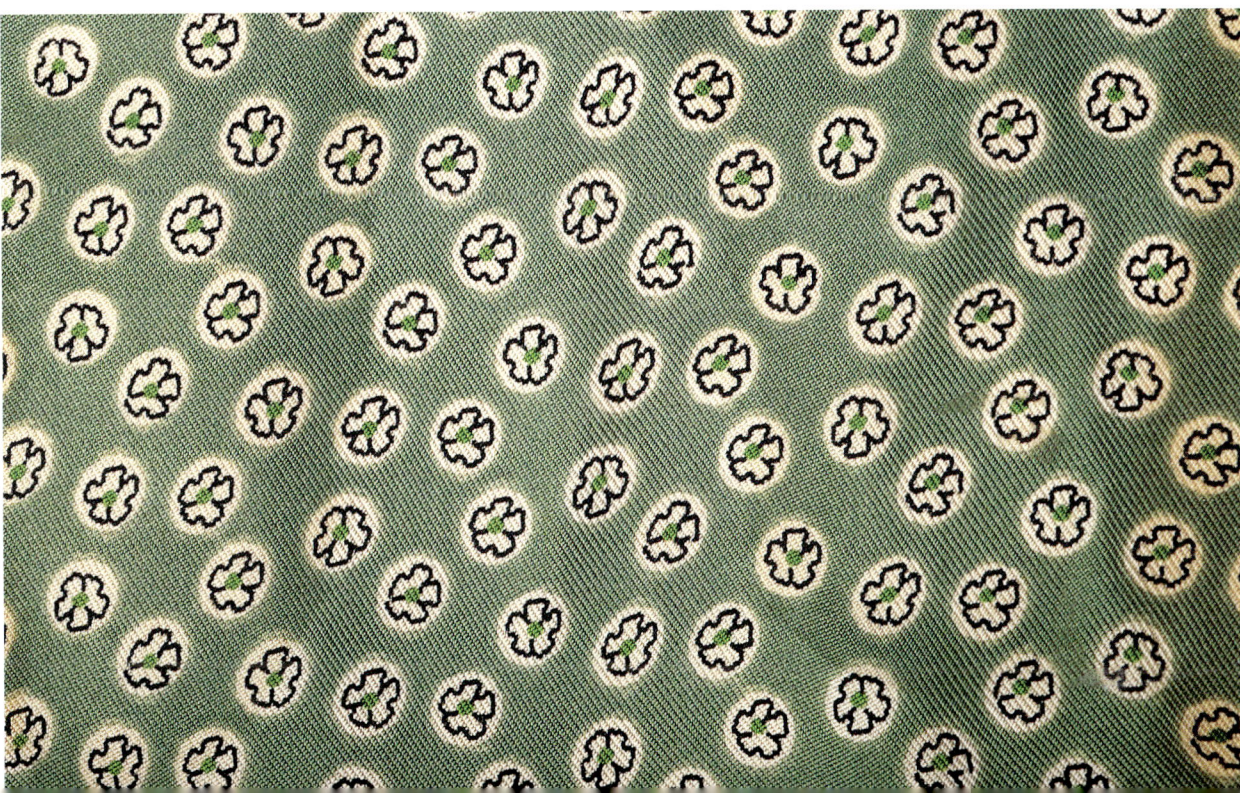

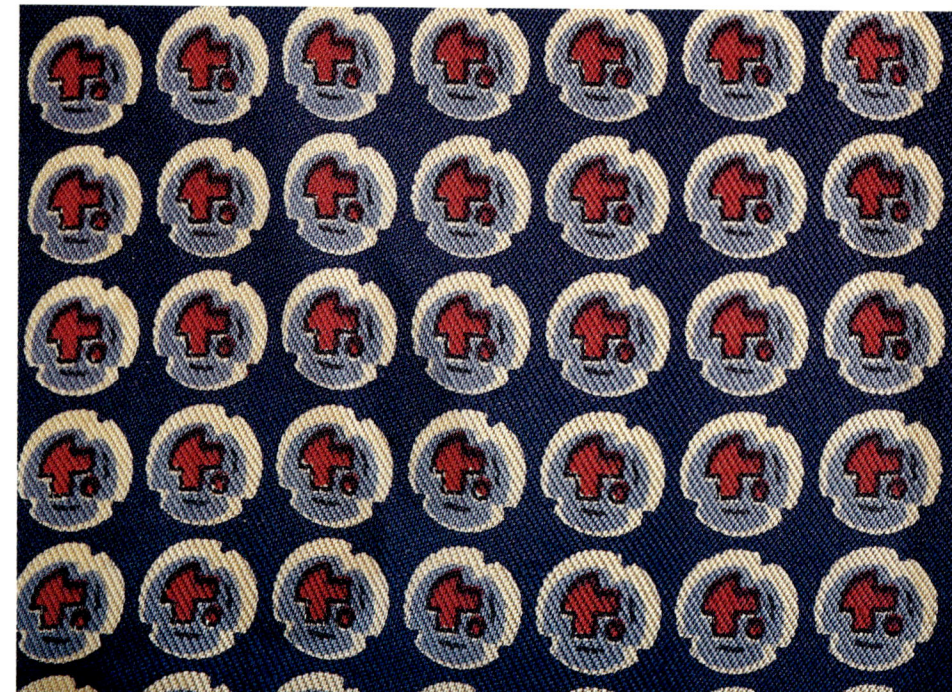

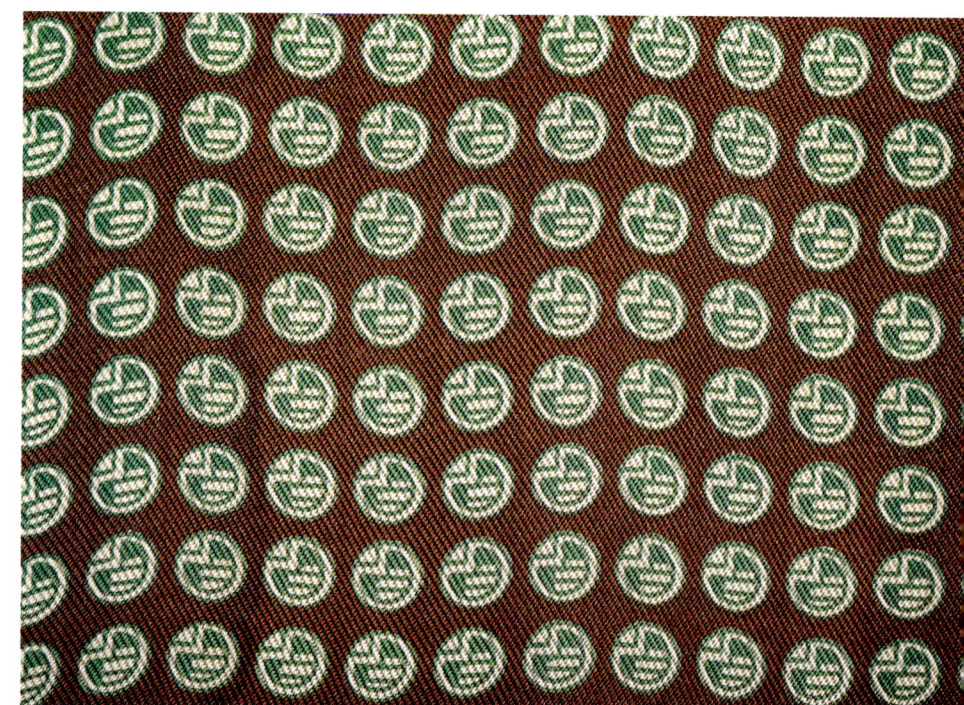

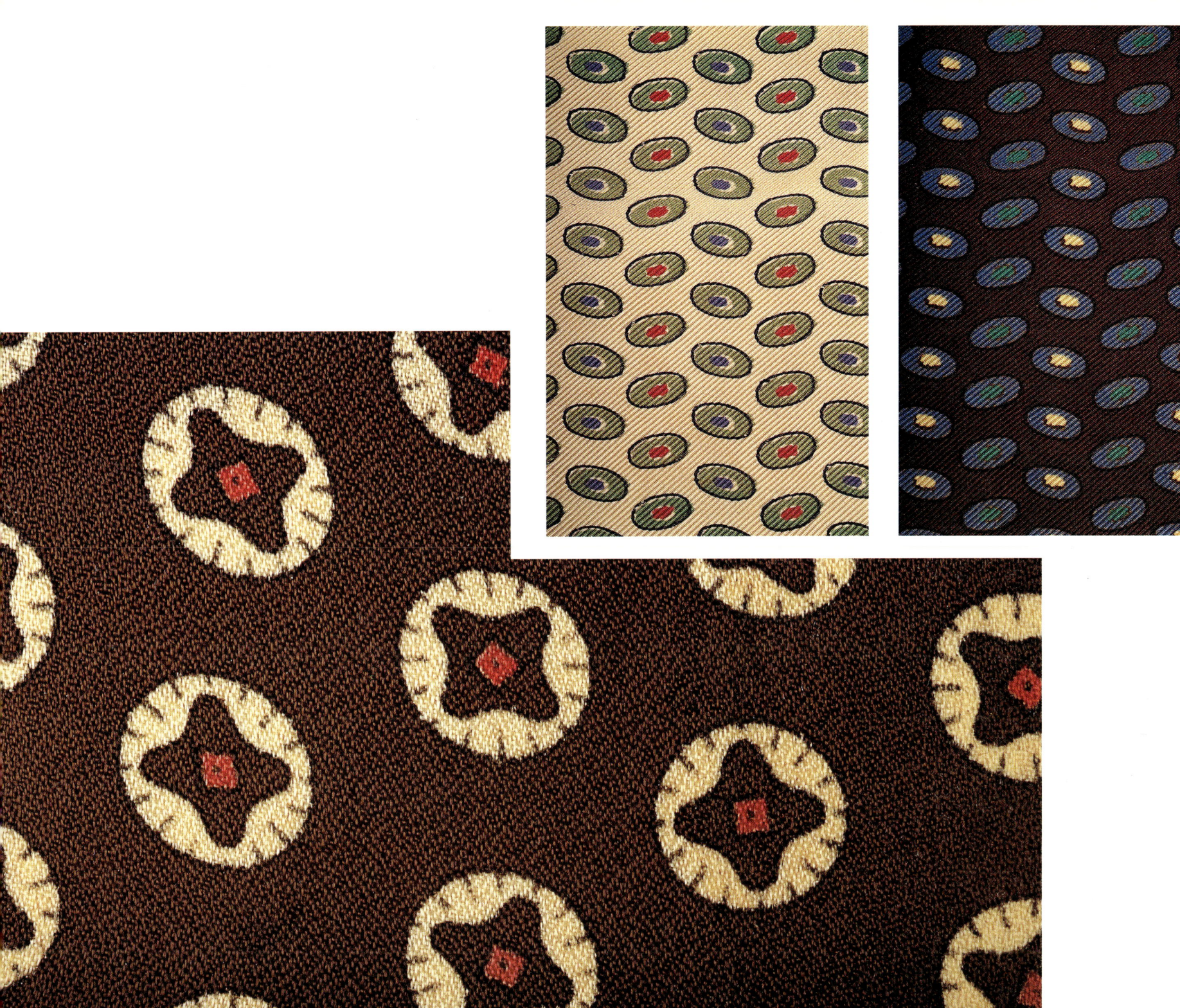

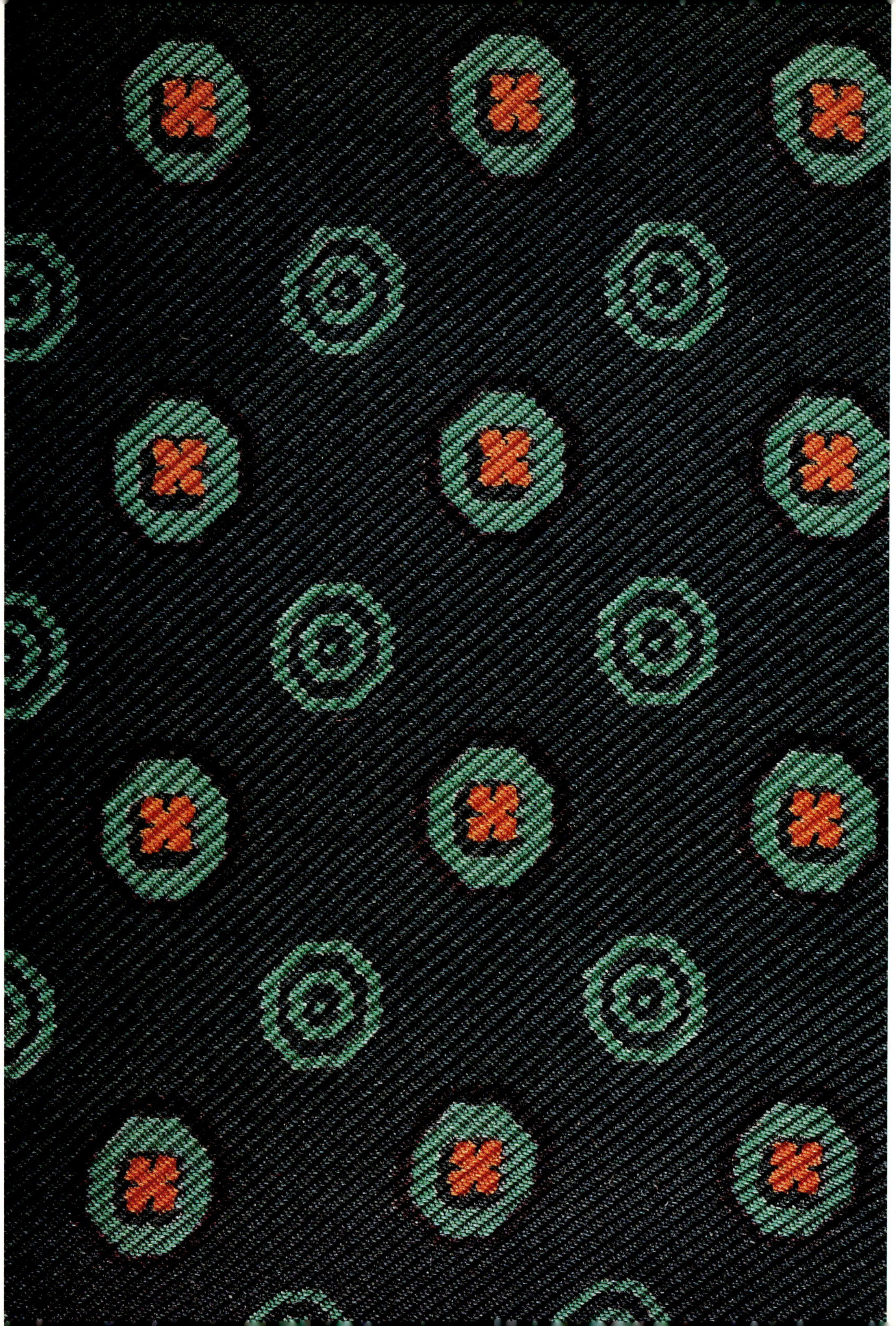

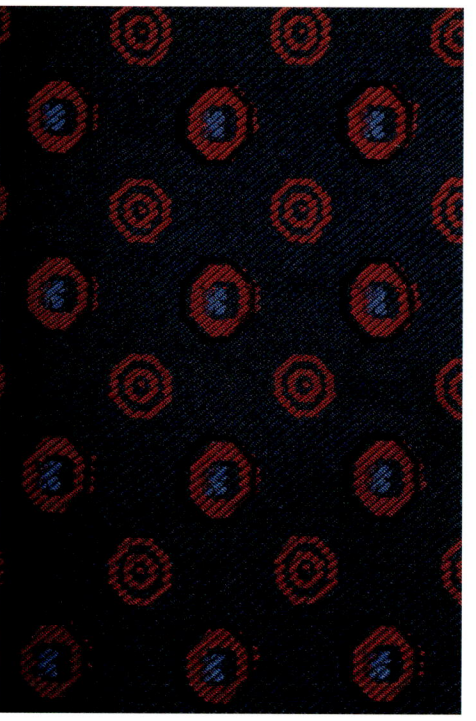

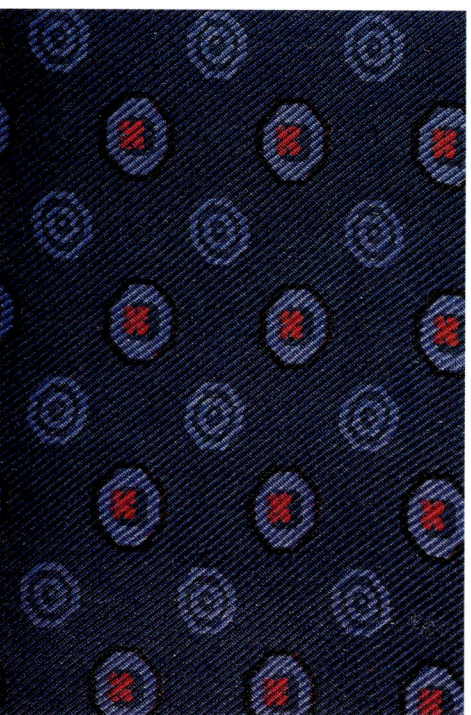

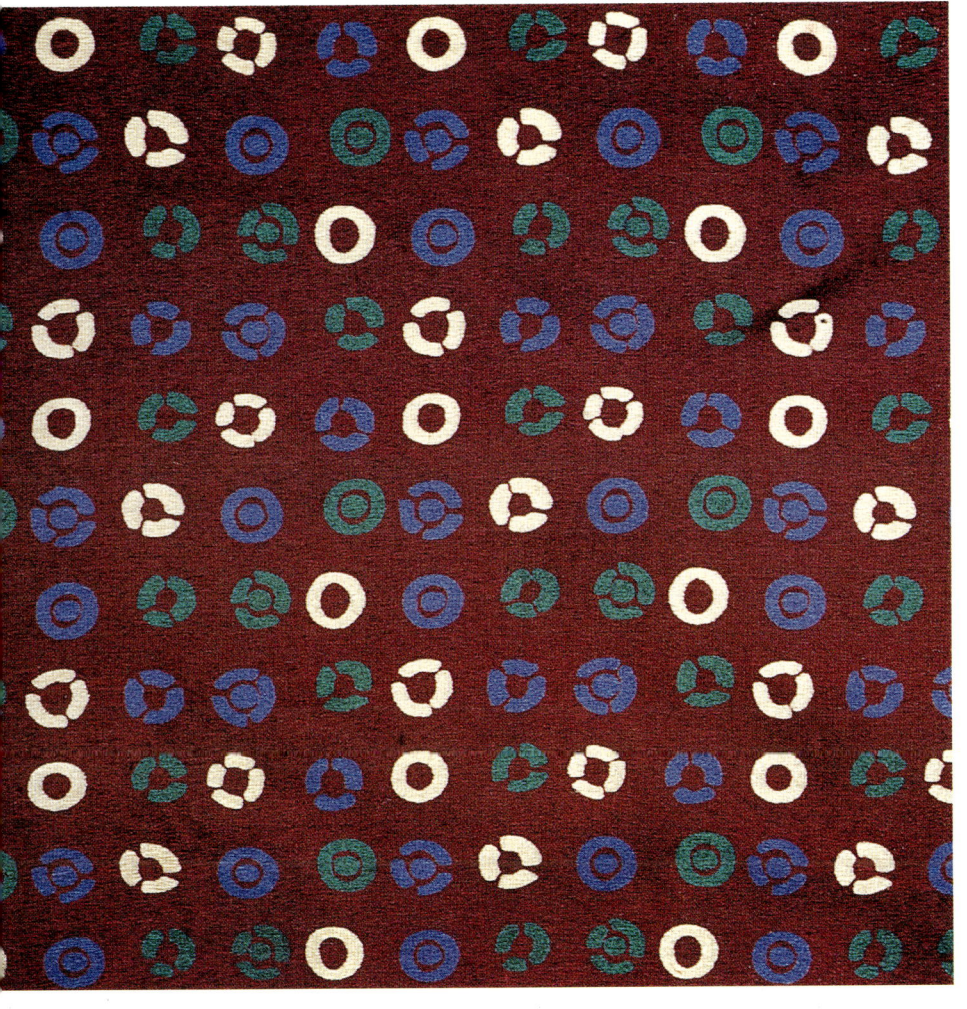
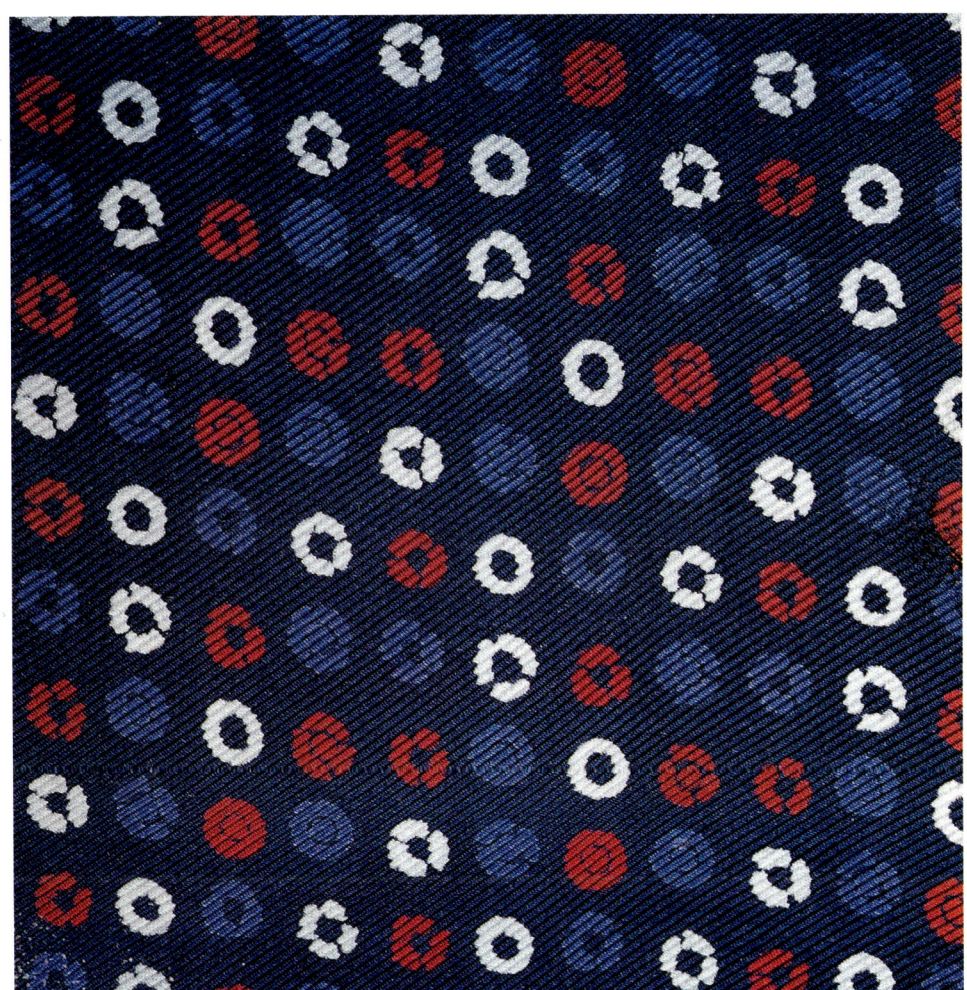

Exaggerated Size

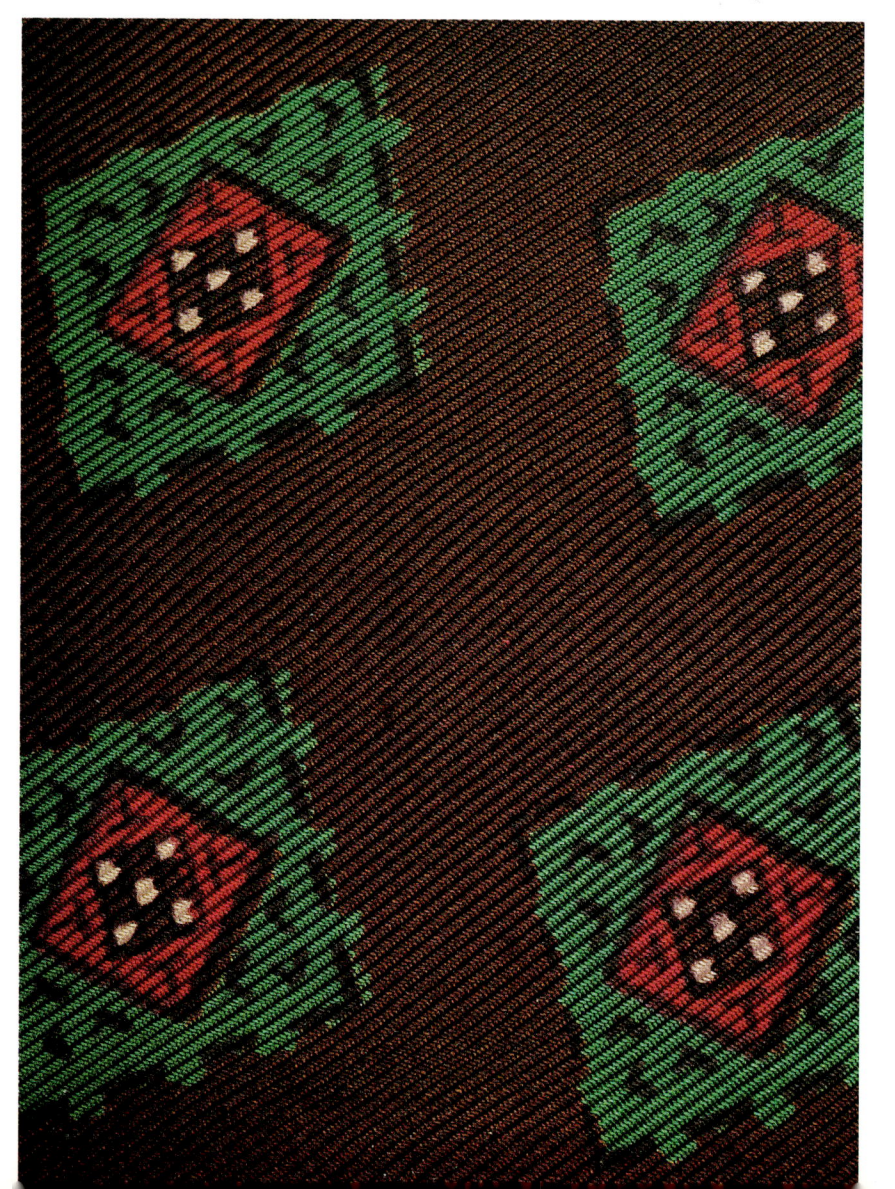

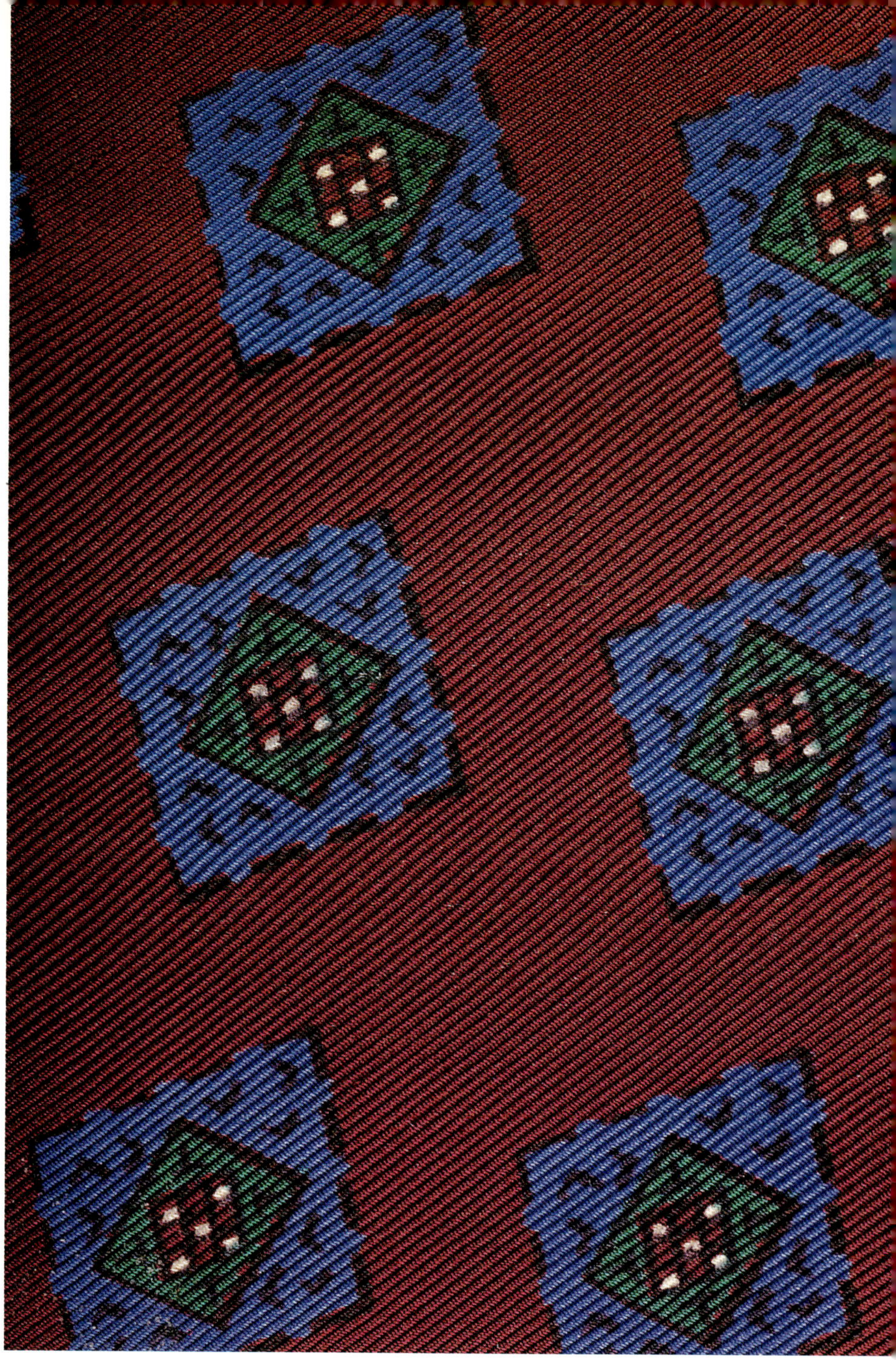

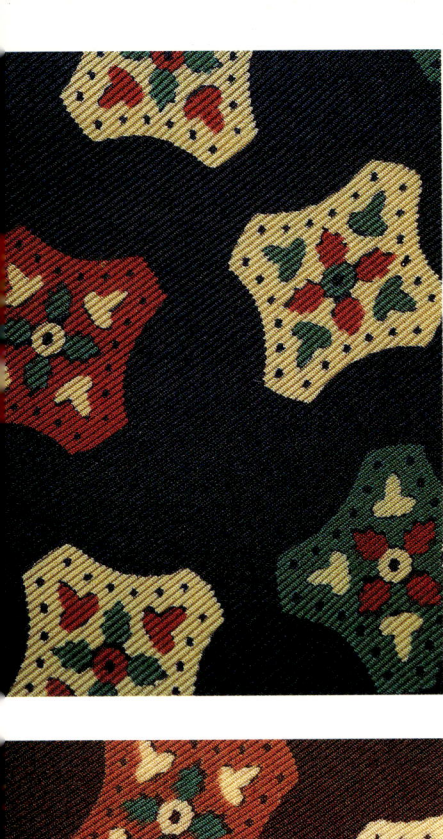
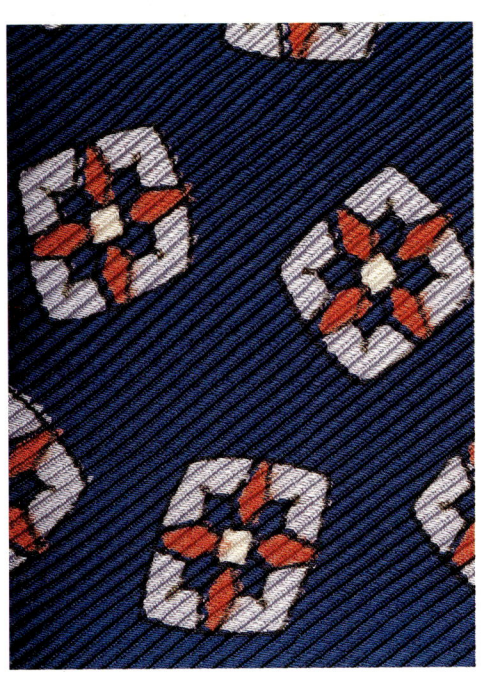
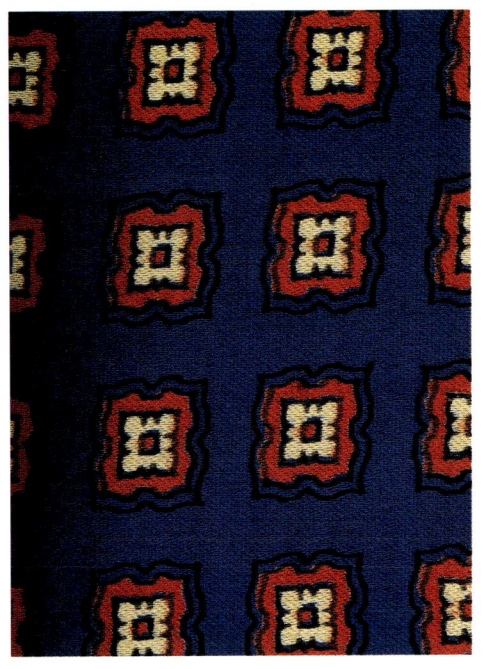
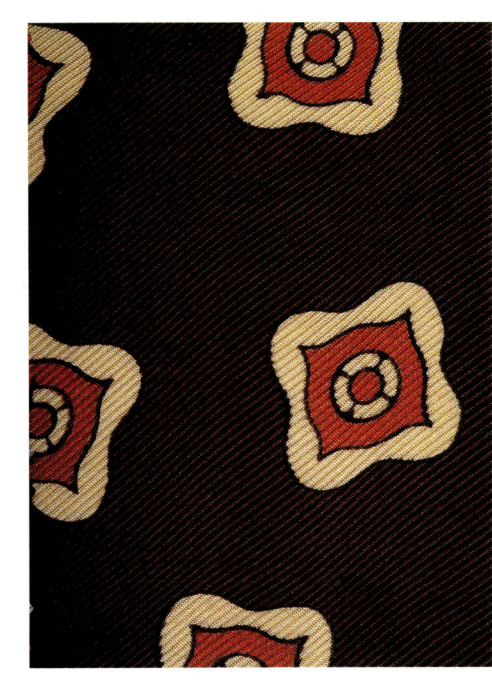
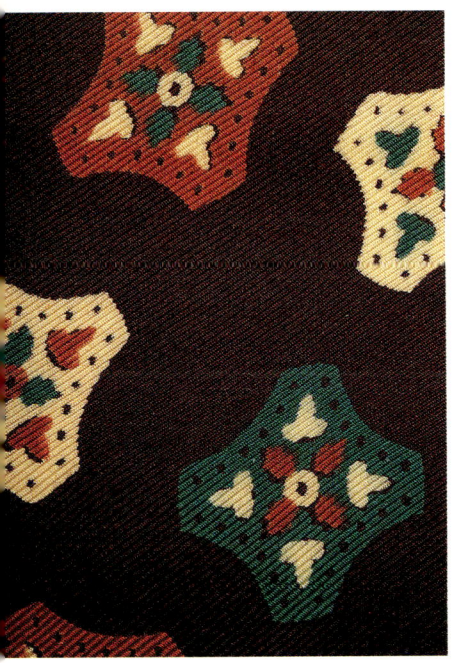
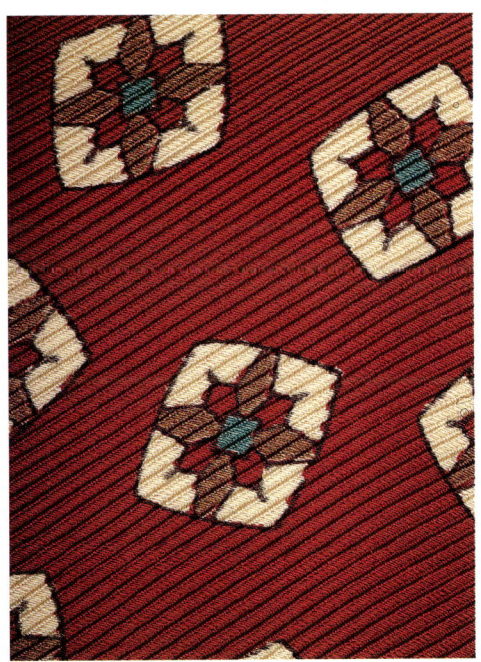
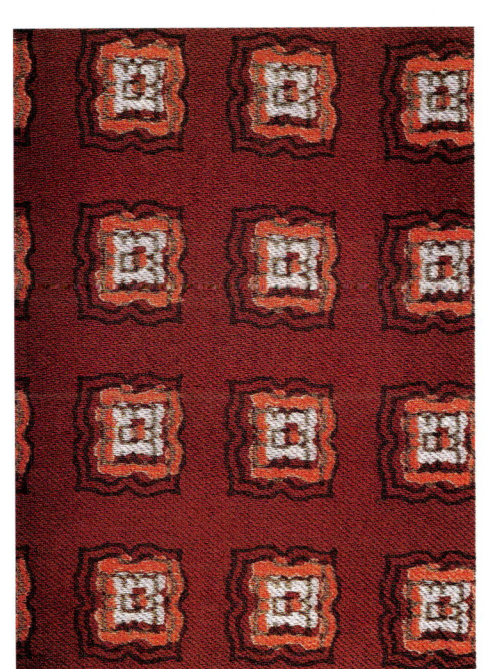

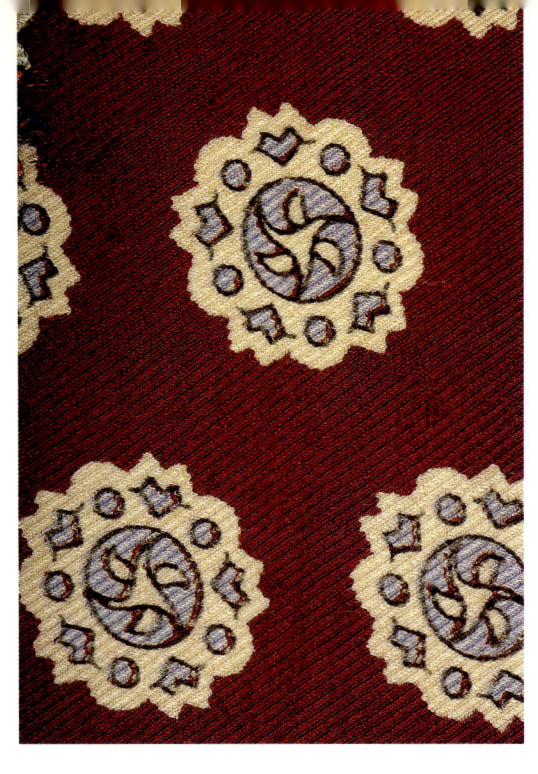
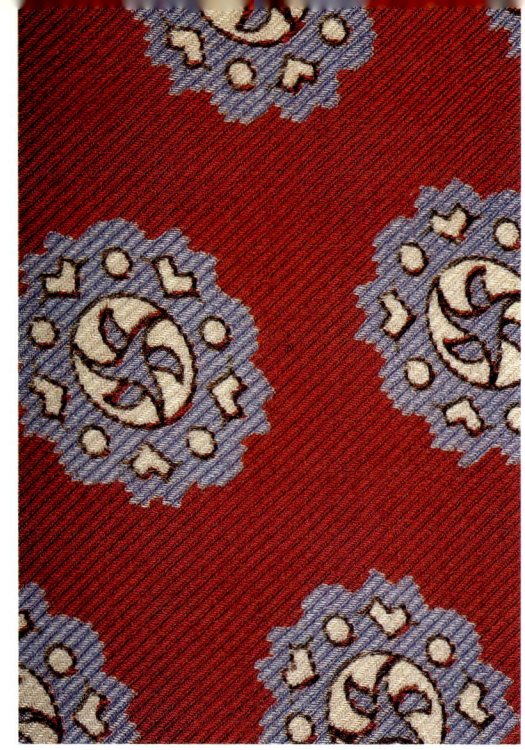
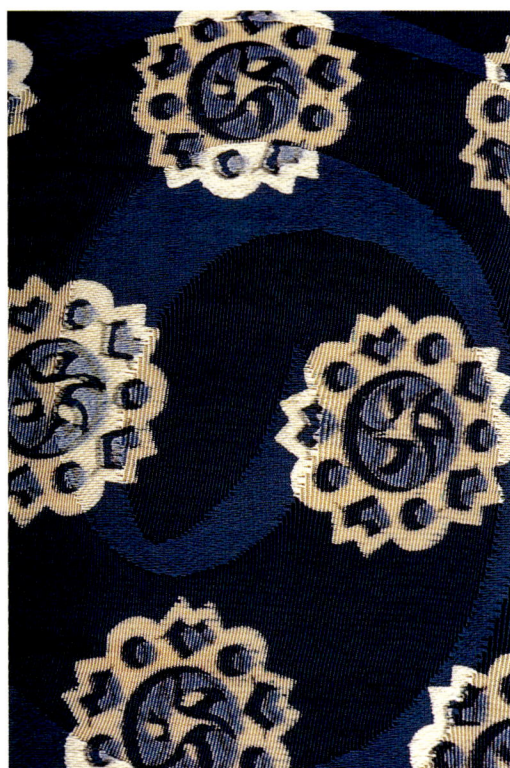
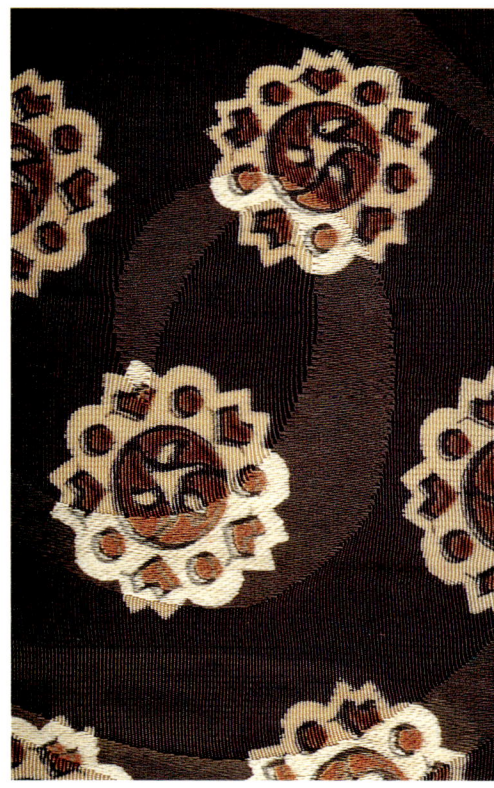

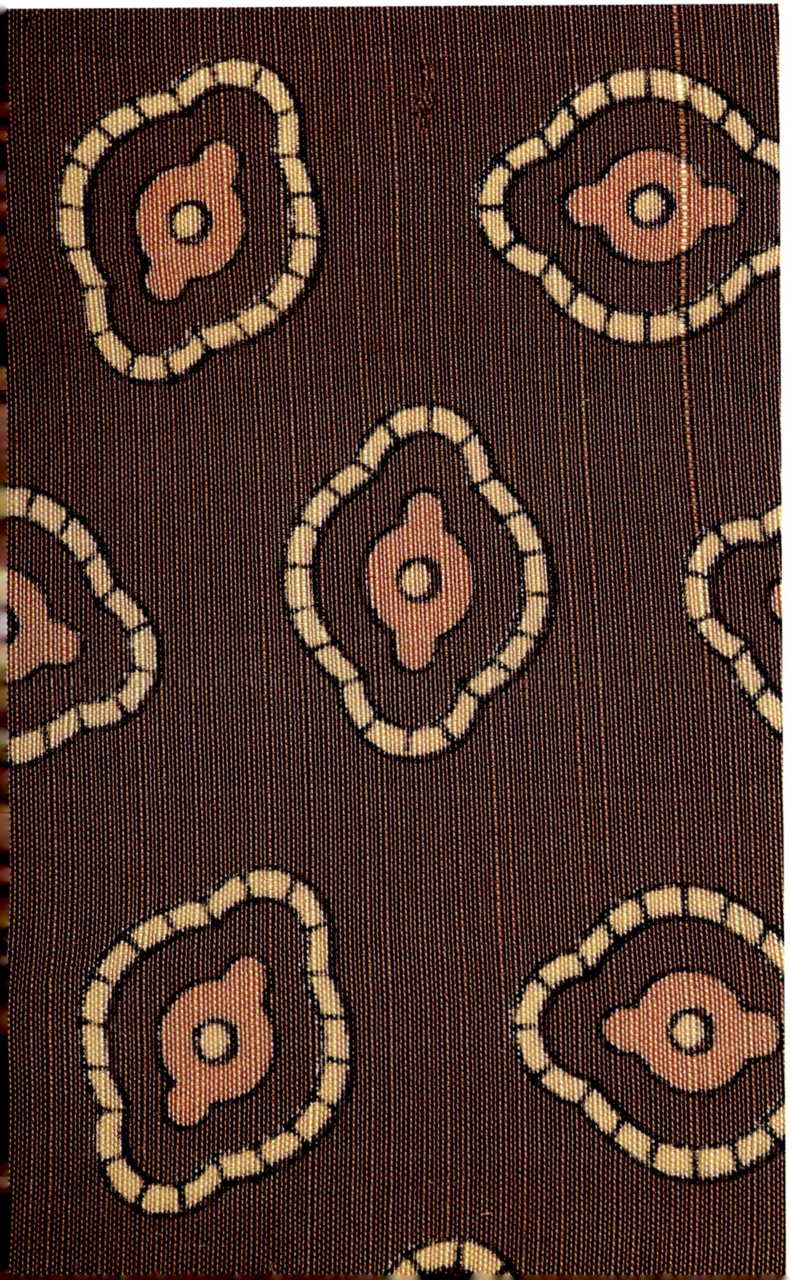

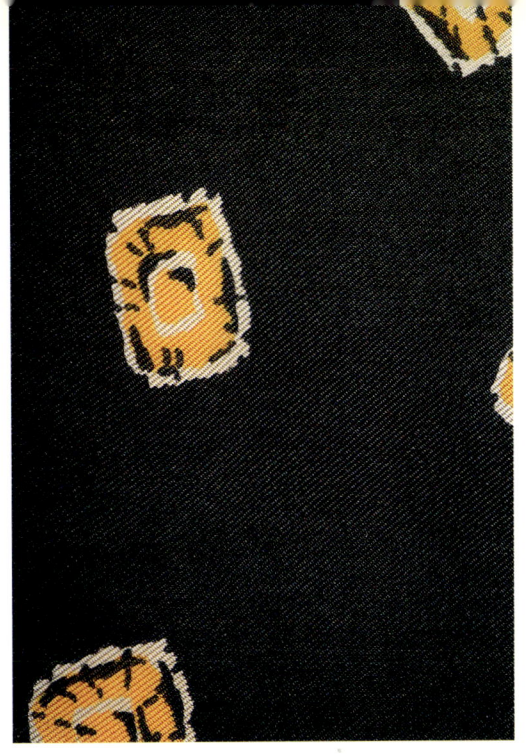
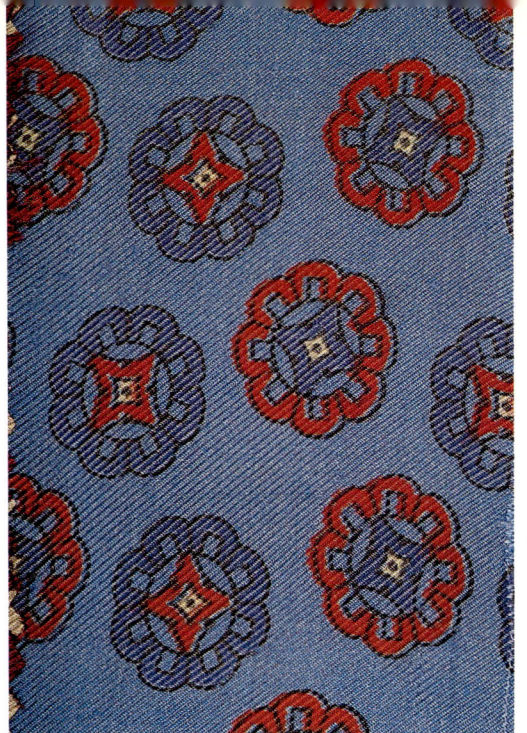
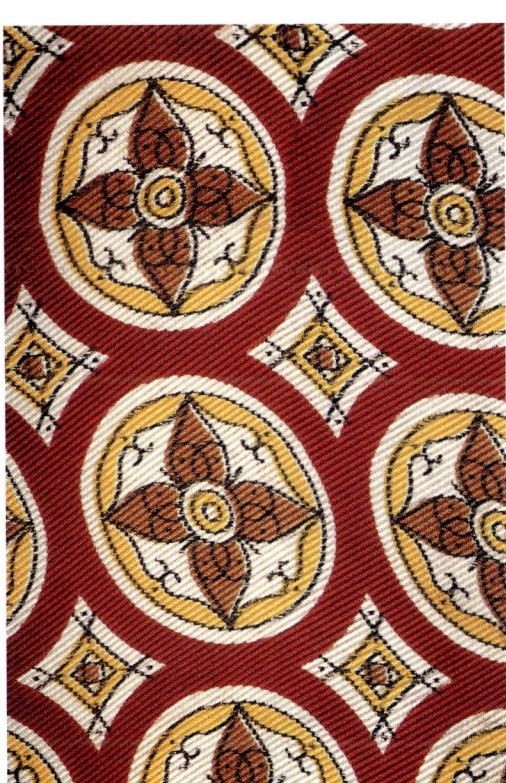
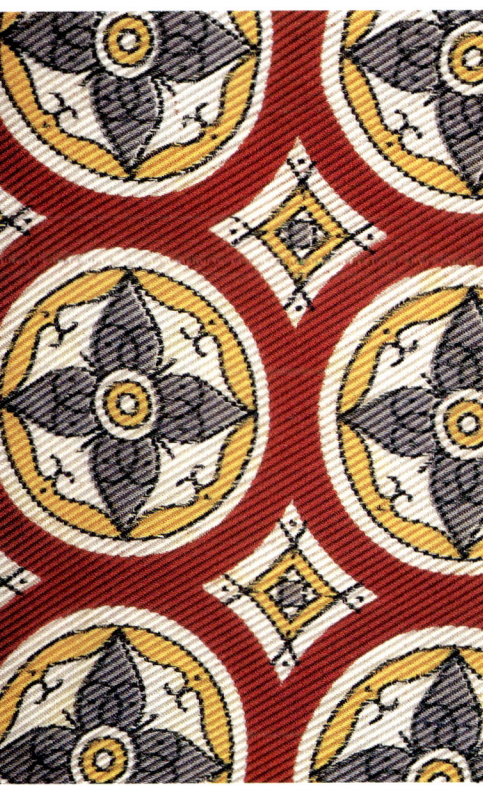

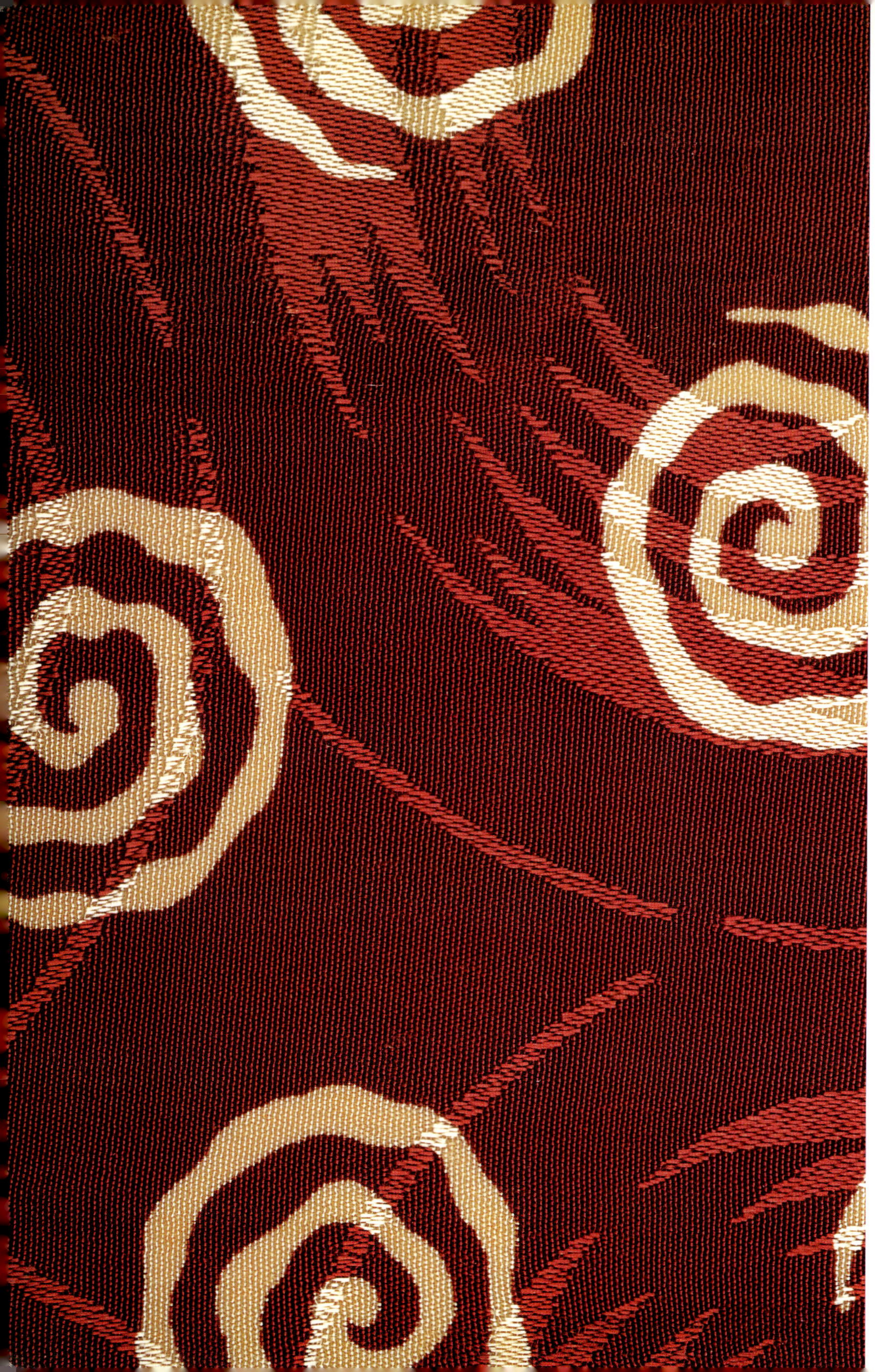

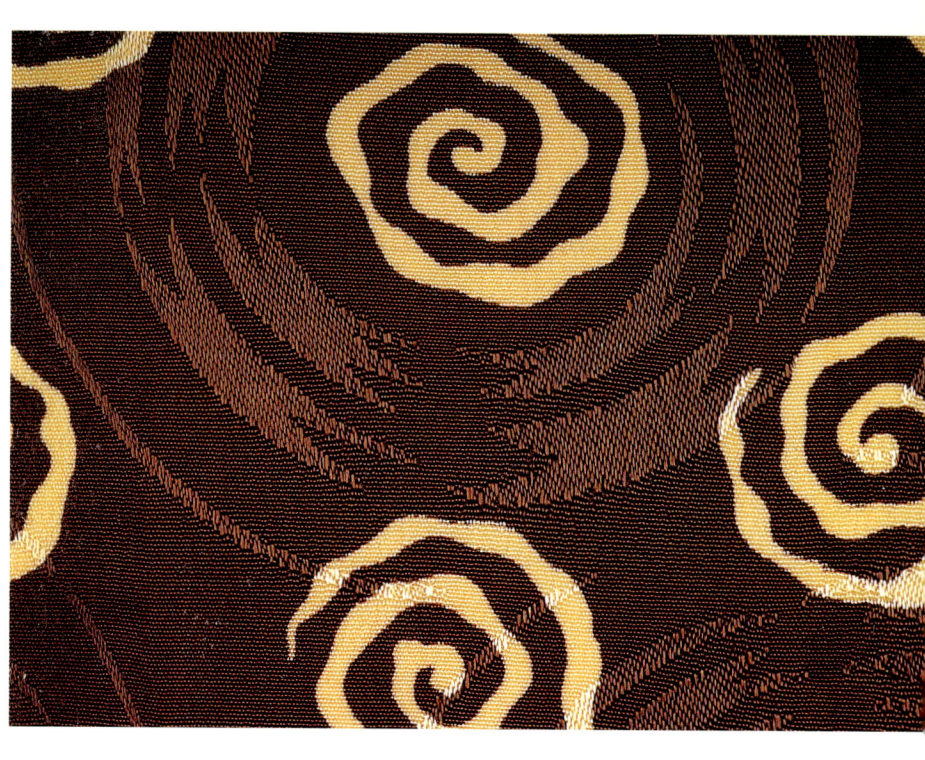
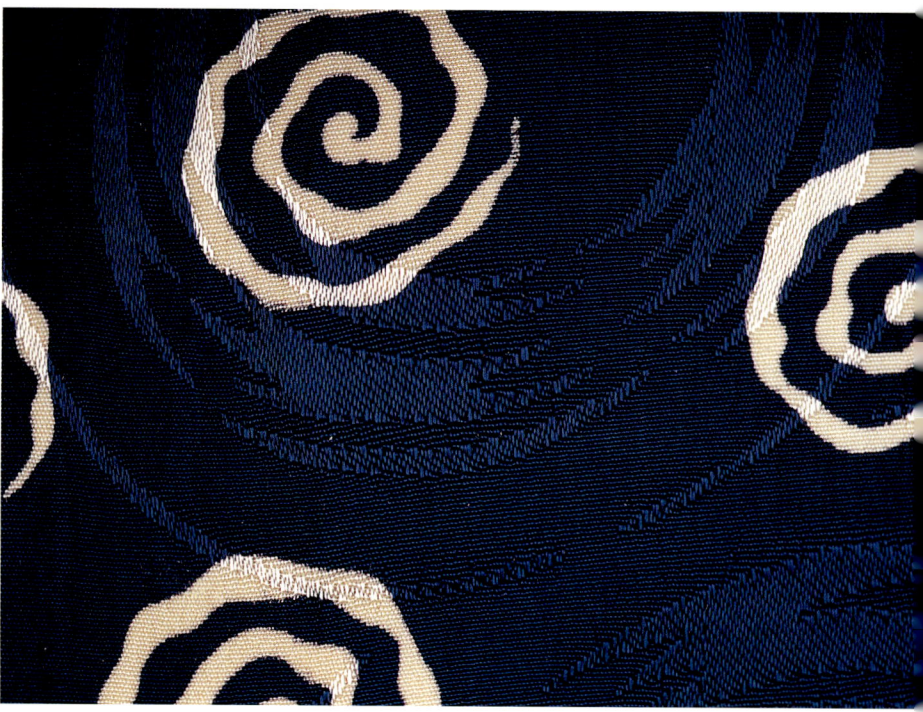

Simplified

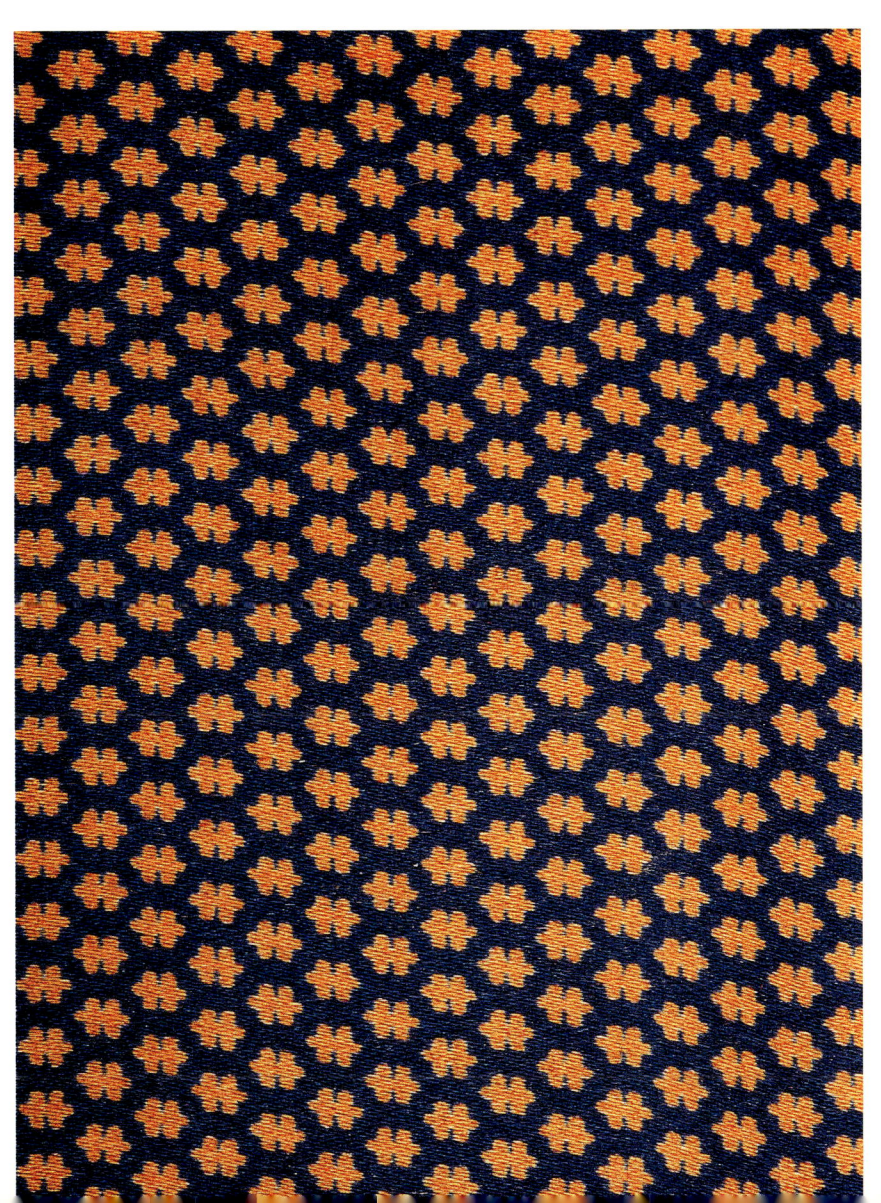

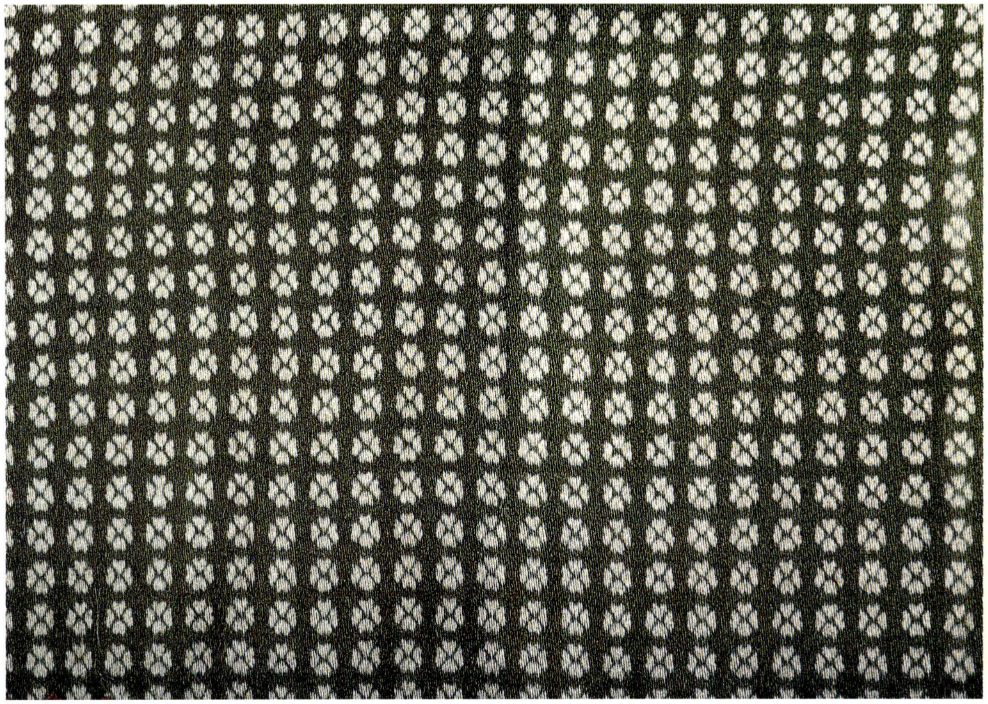

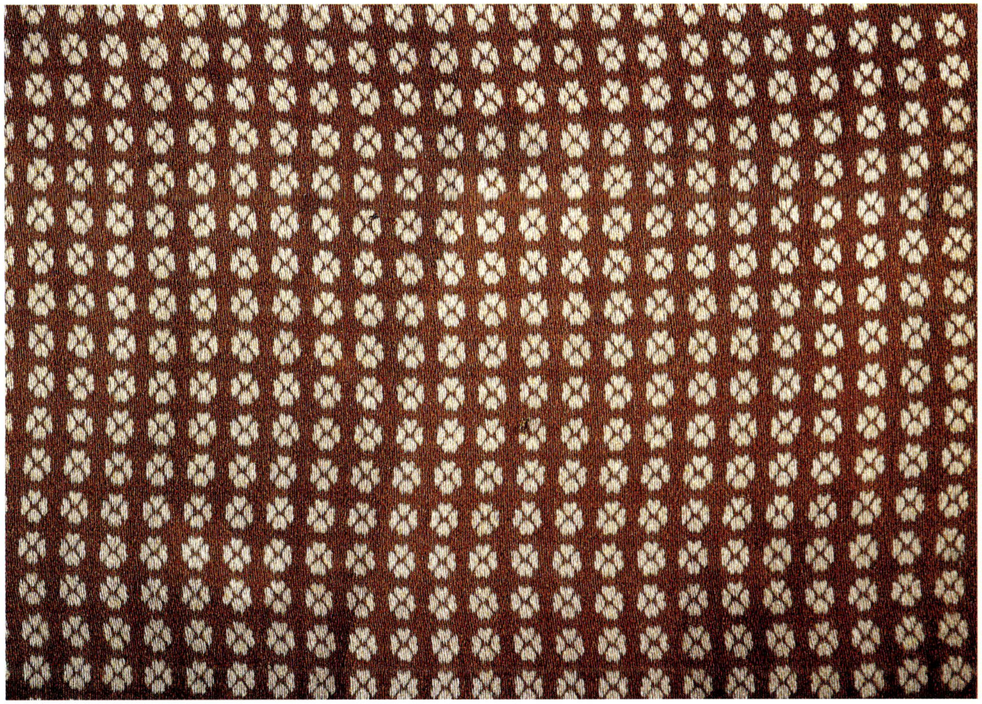

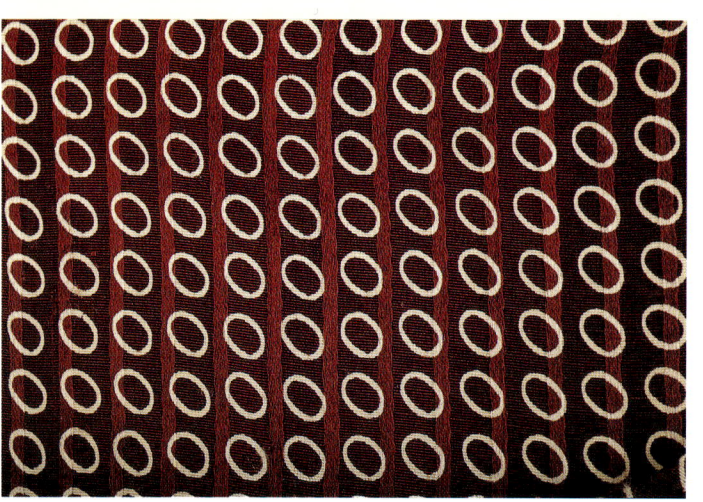

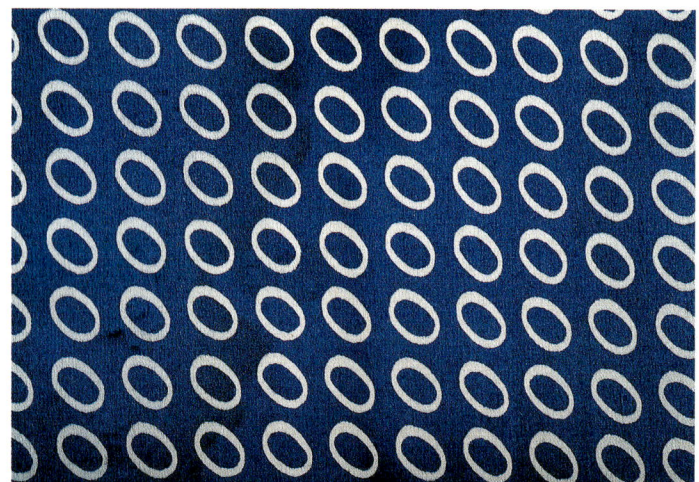

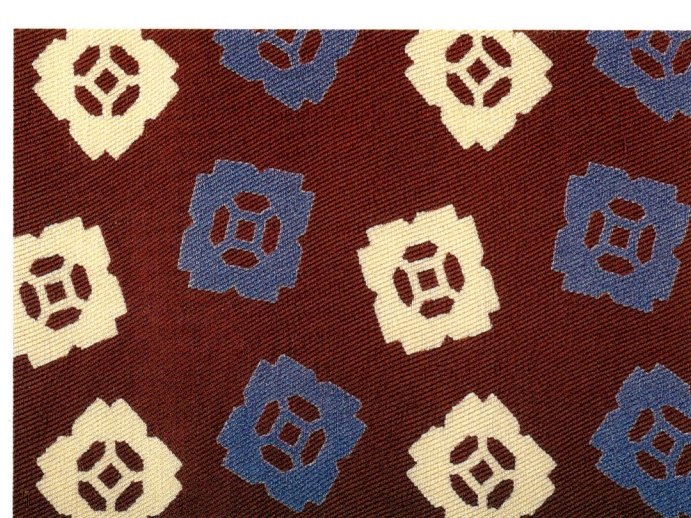

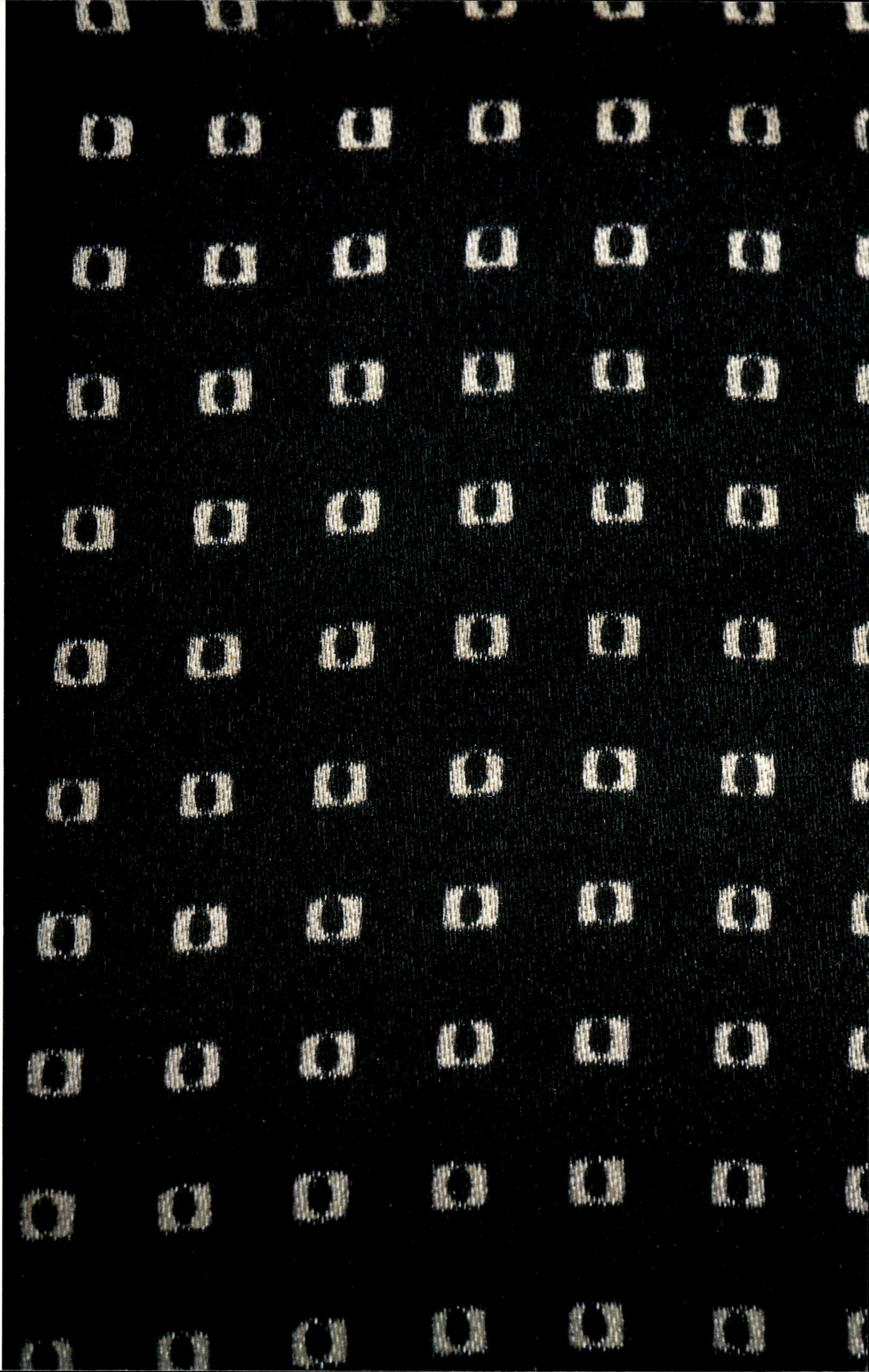

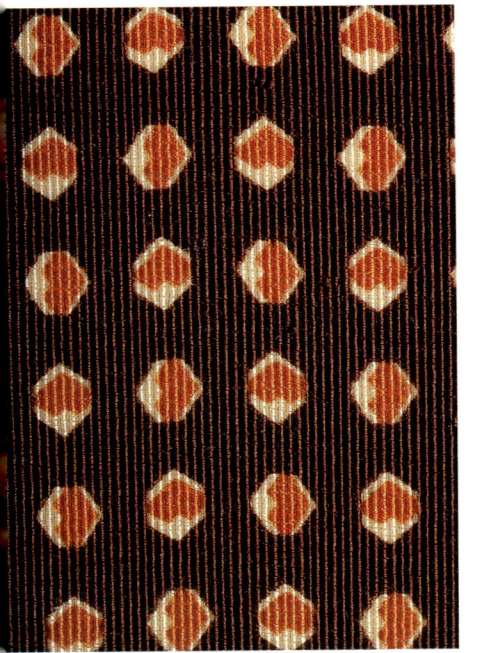
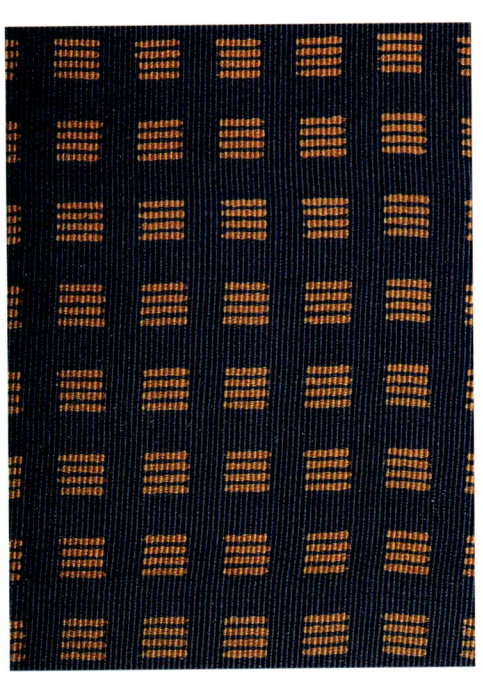
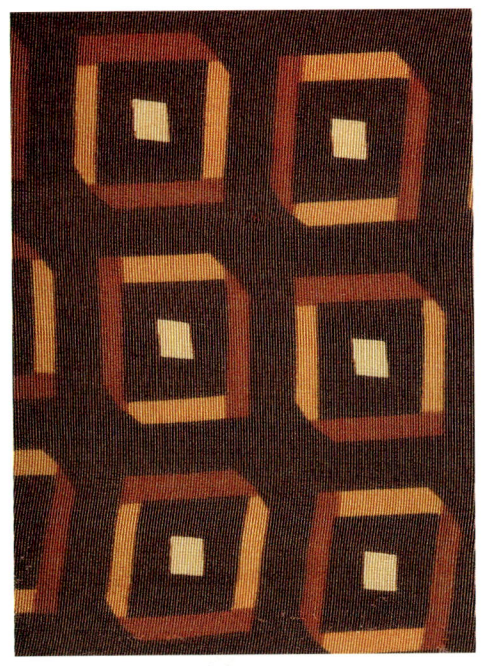
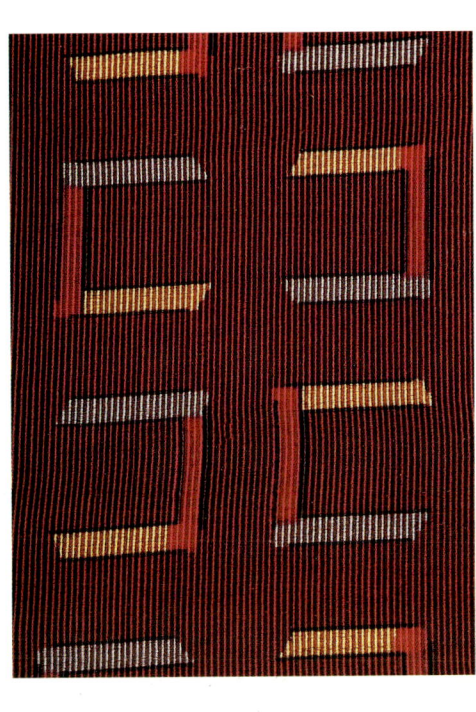
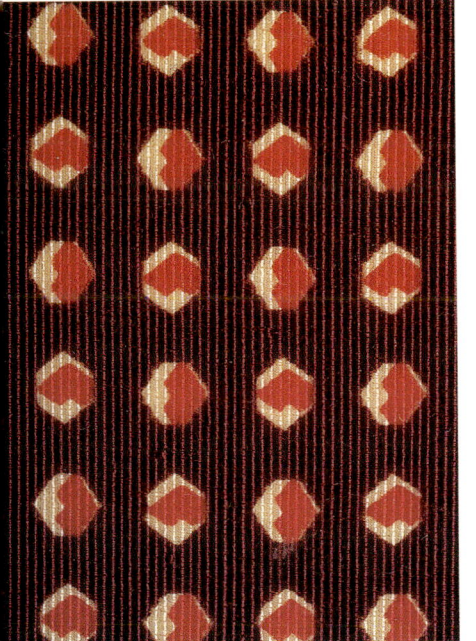
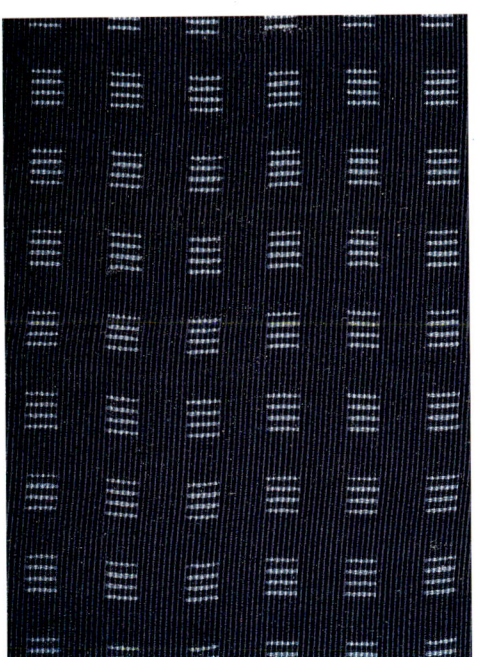
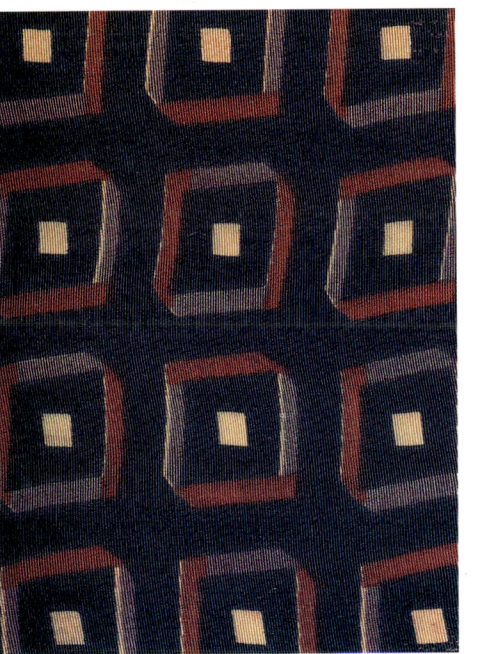
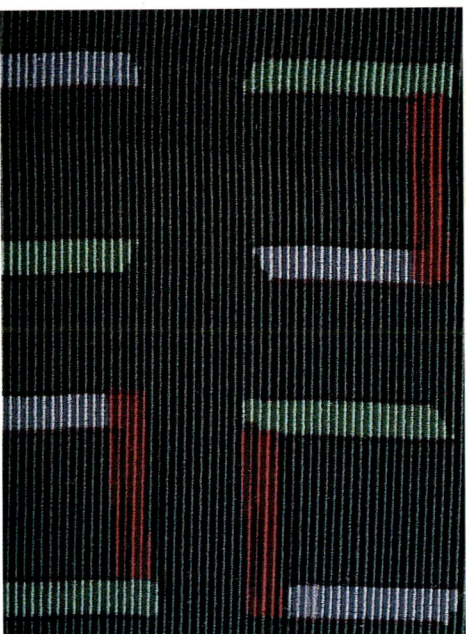

Space Age
Far Out

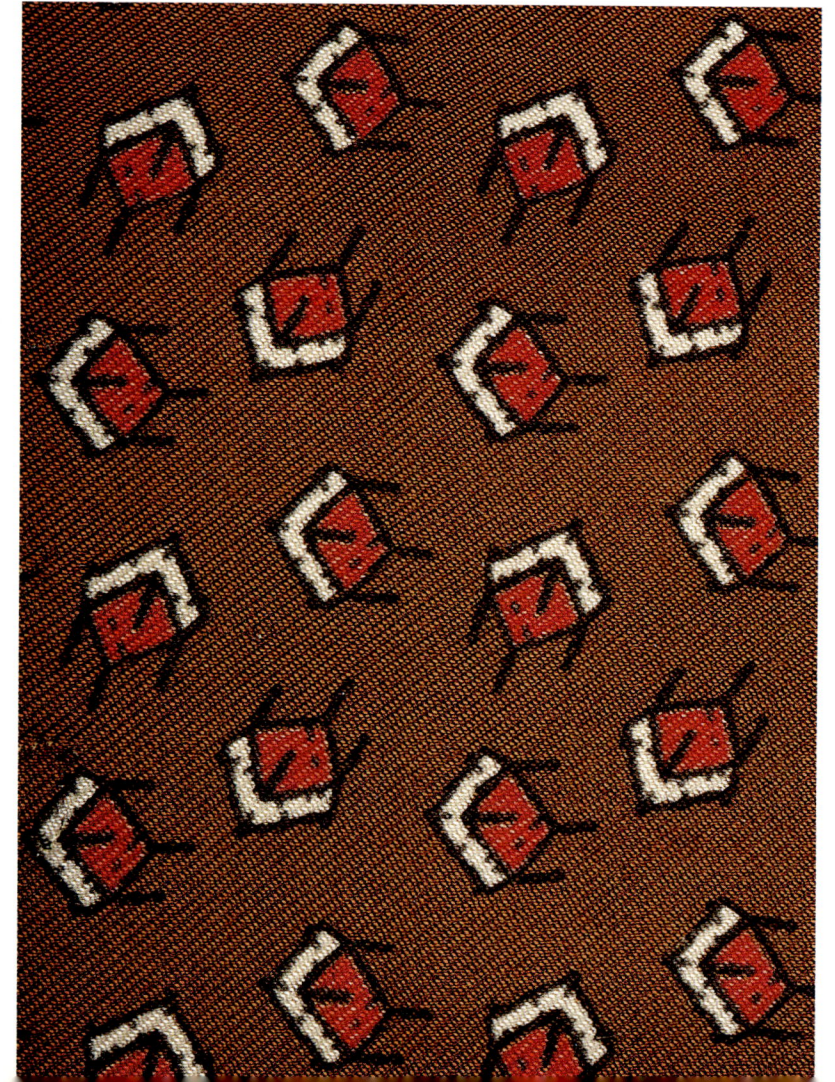

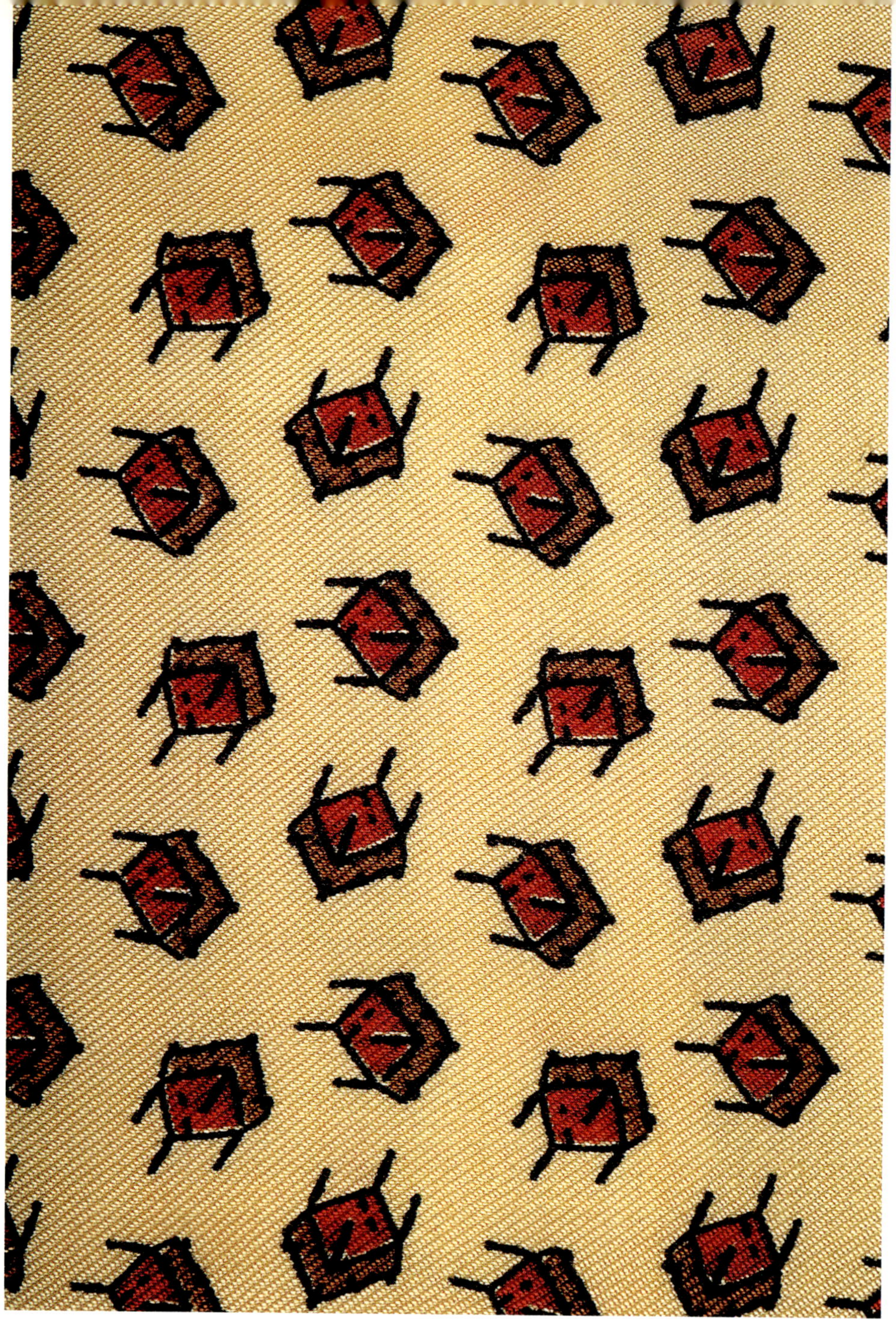

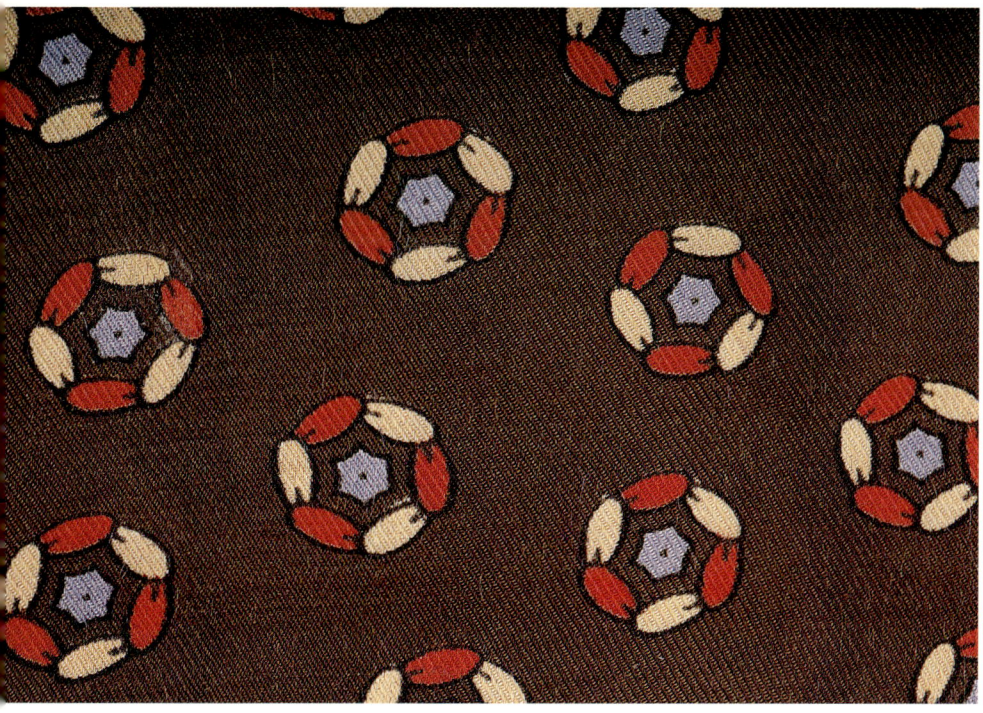
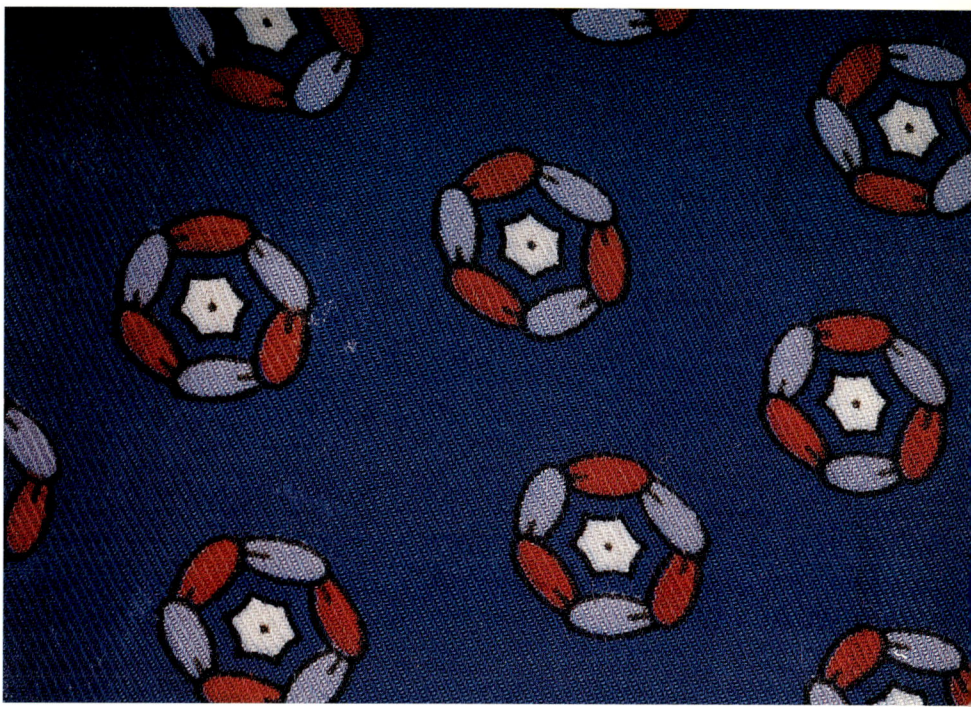
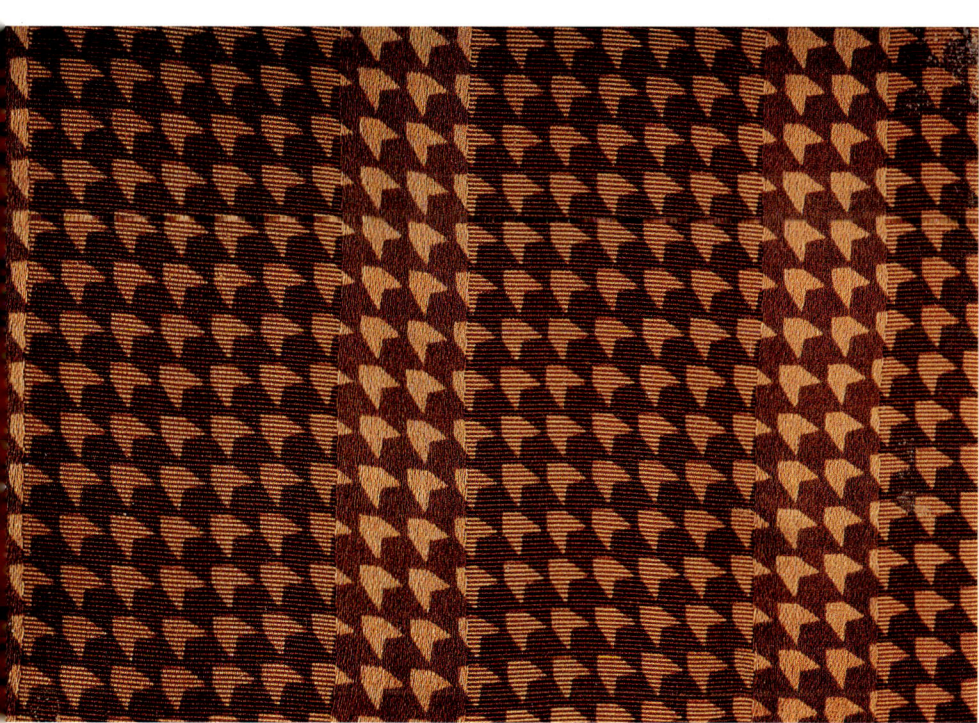
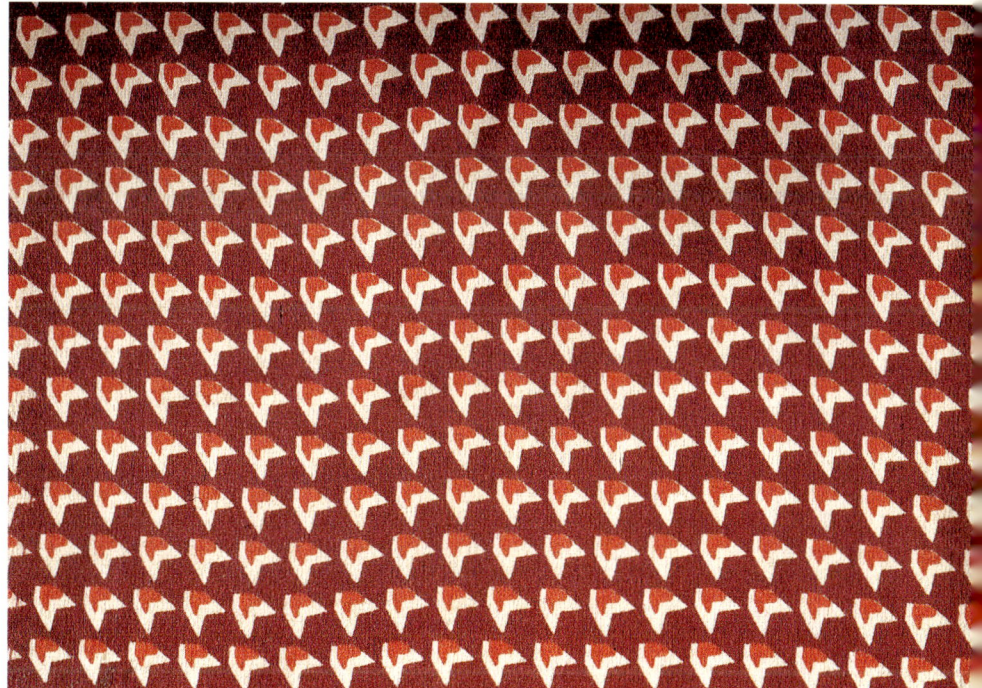

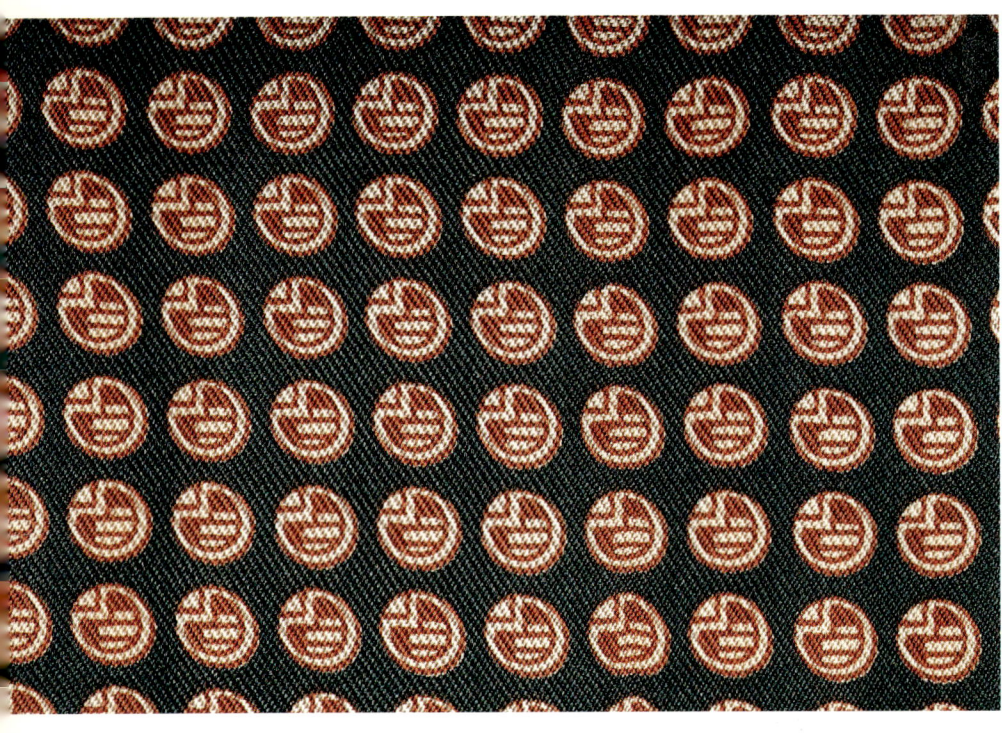
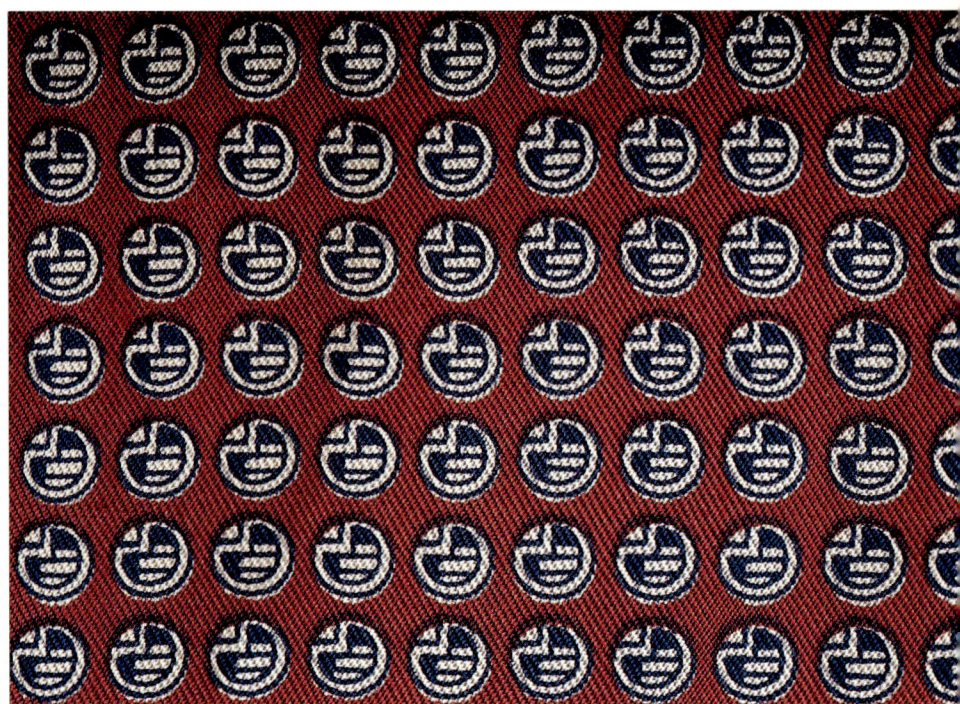
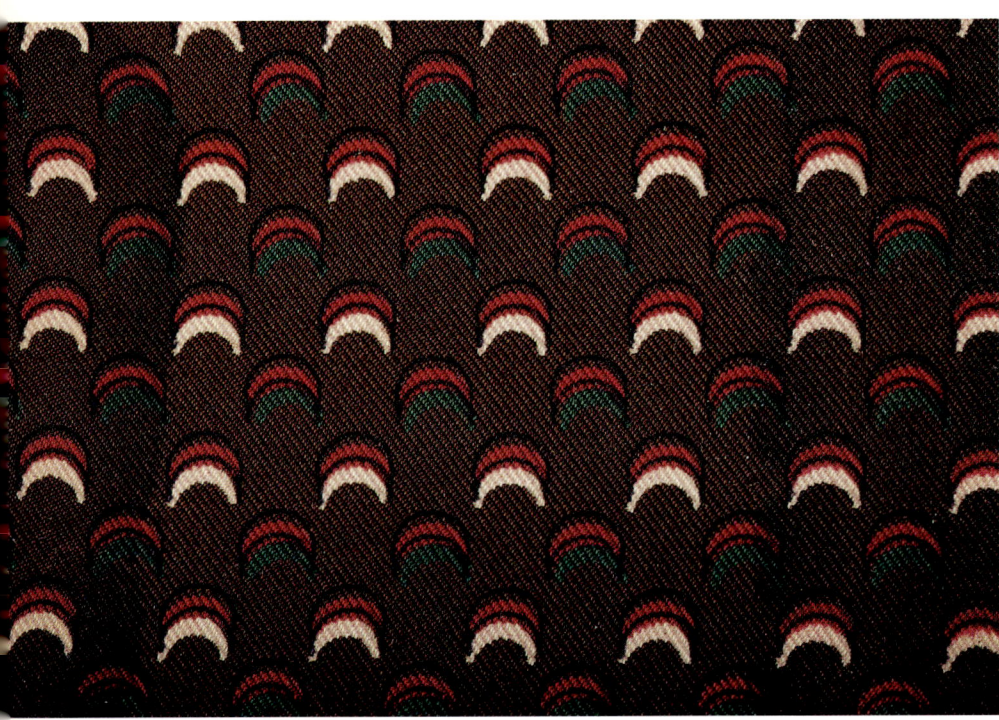
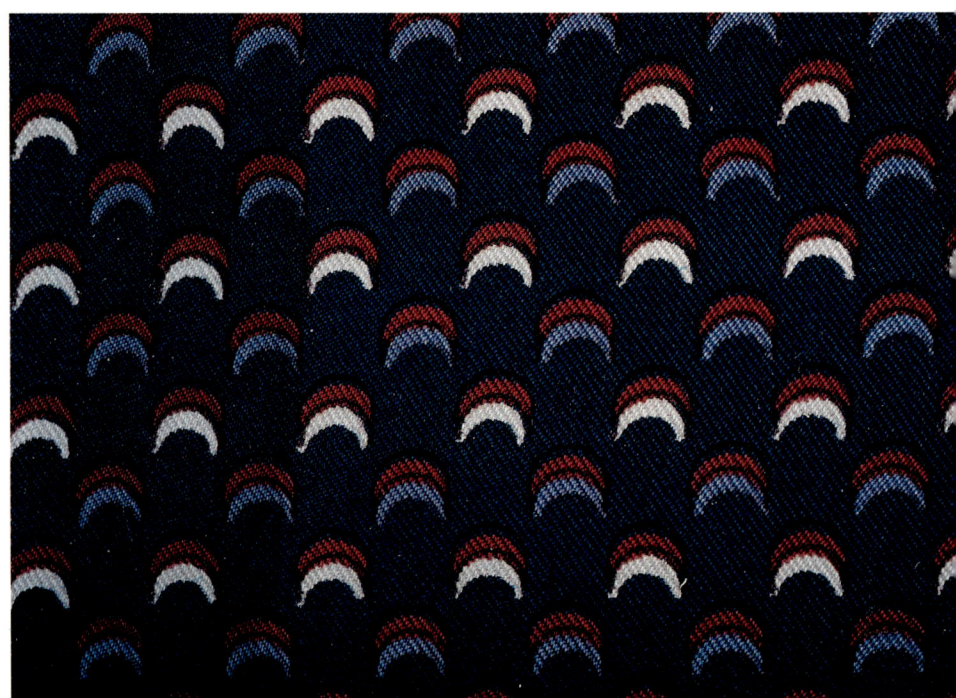

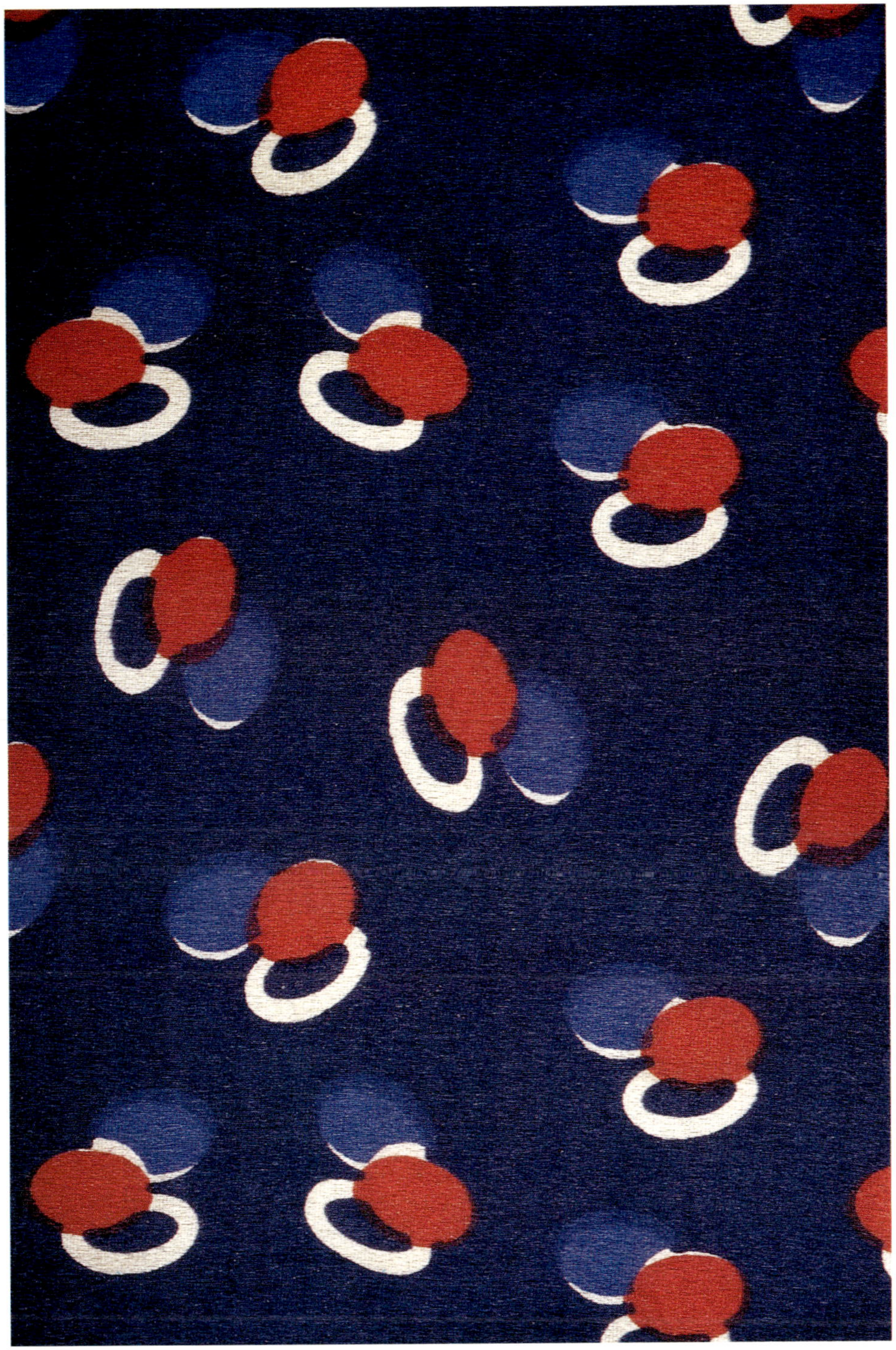

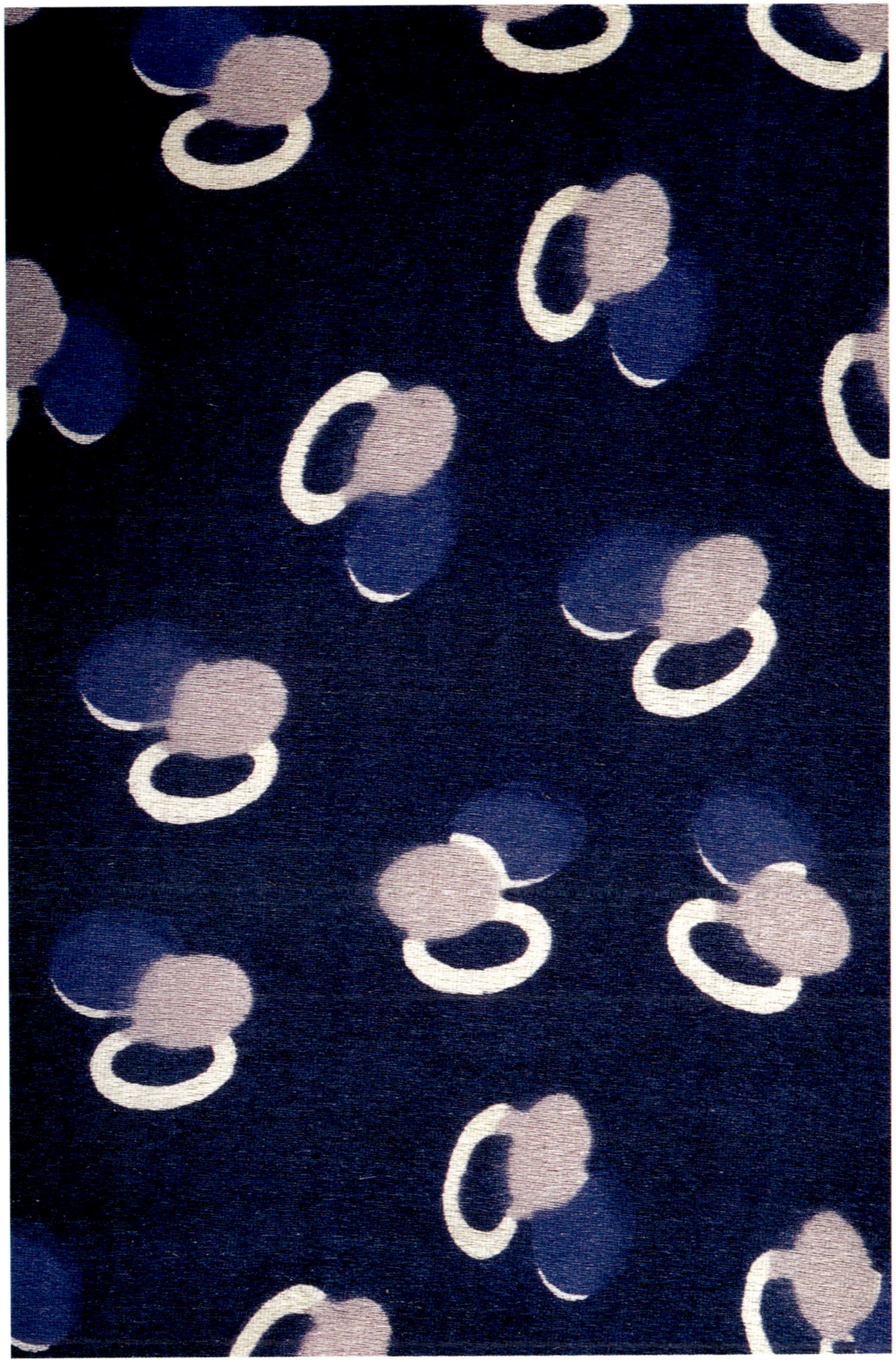

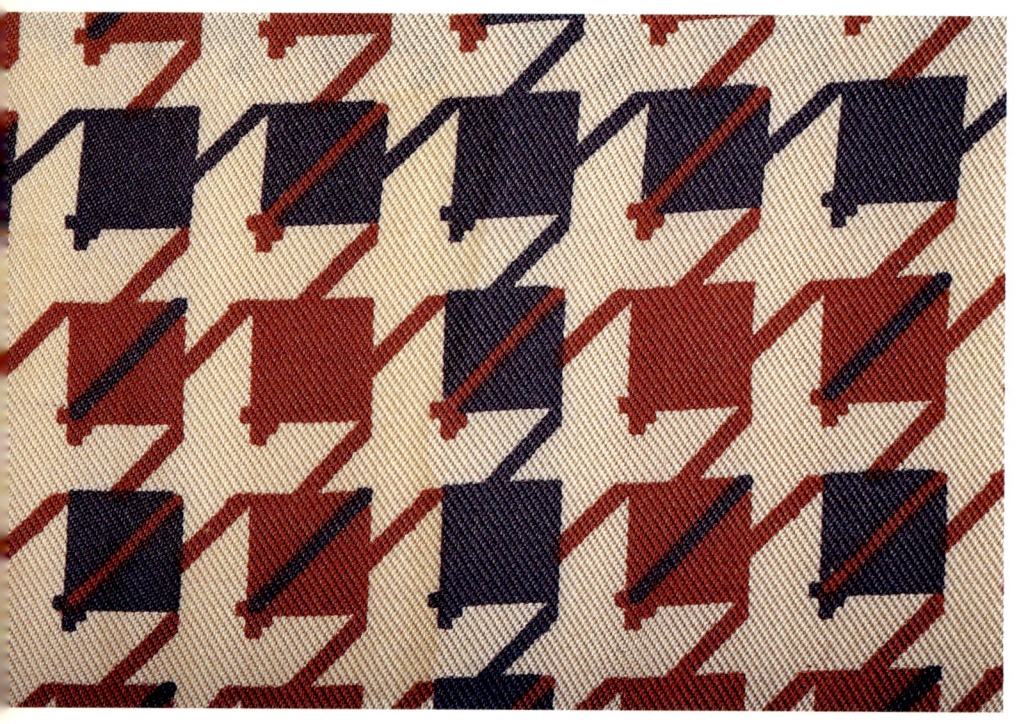
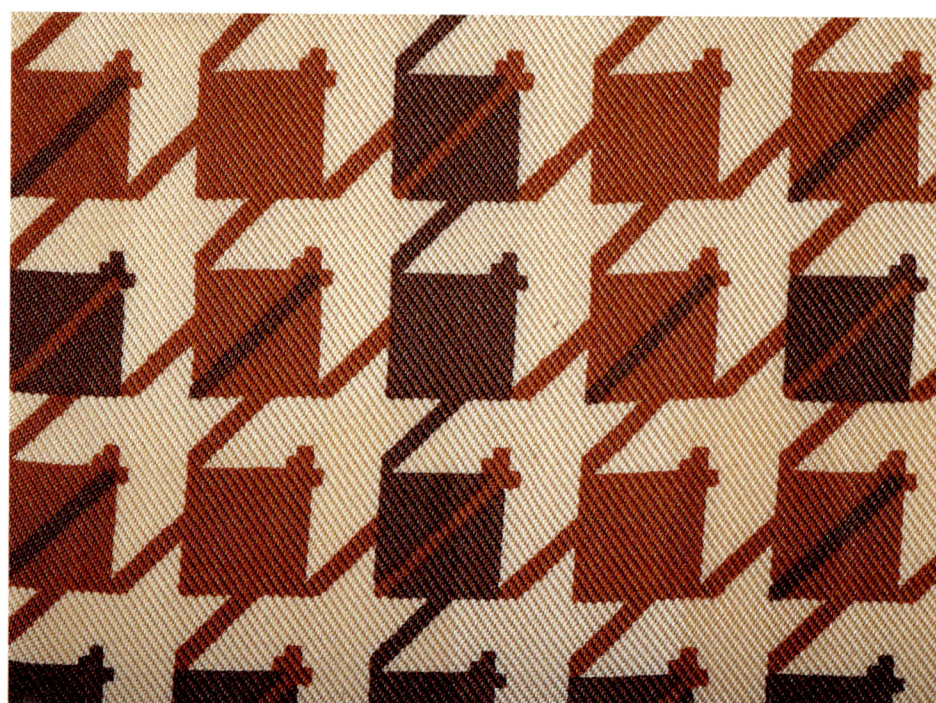
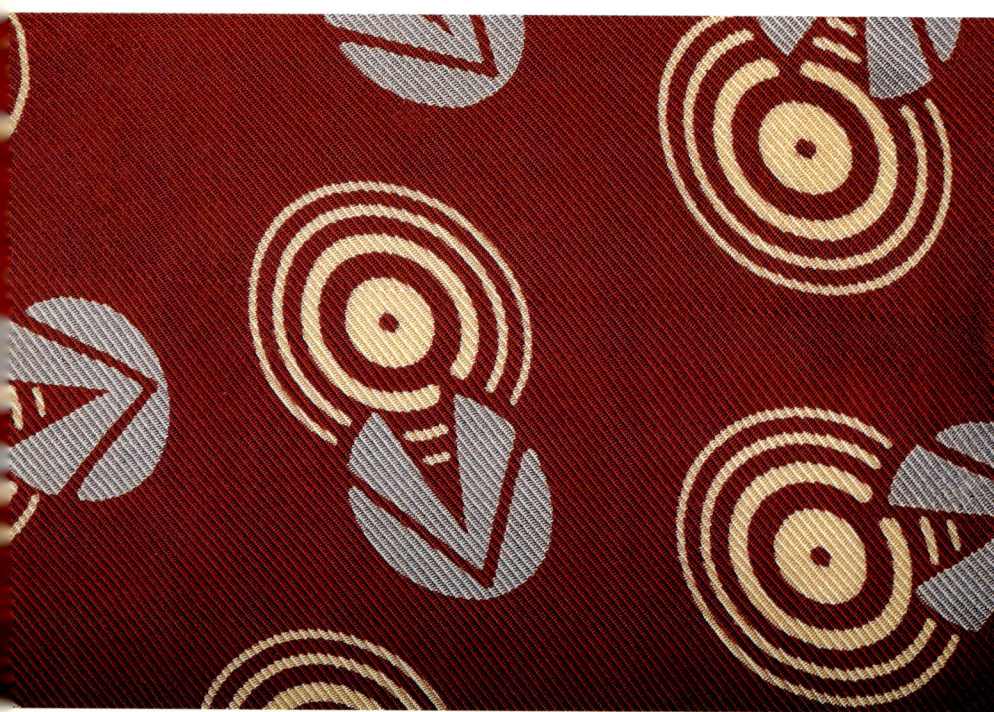
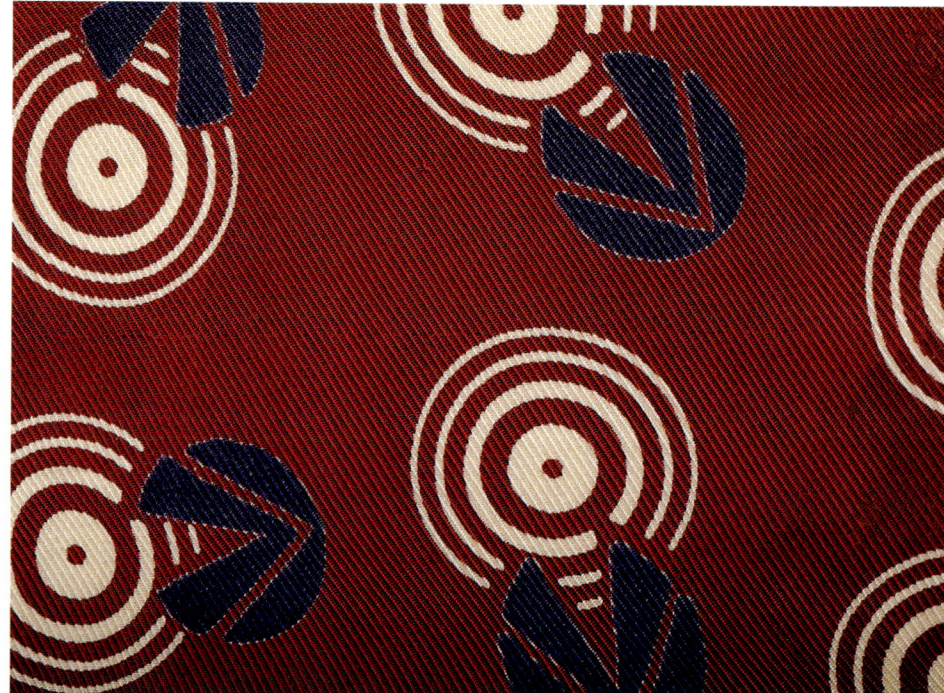

94

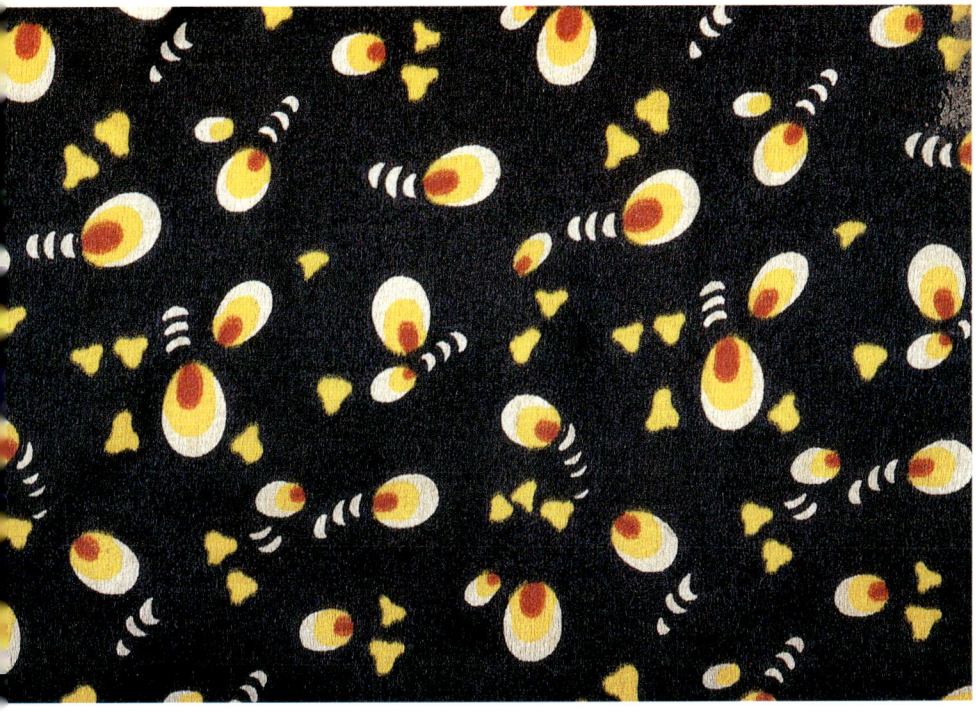
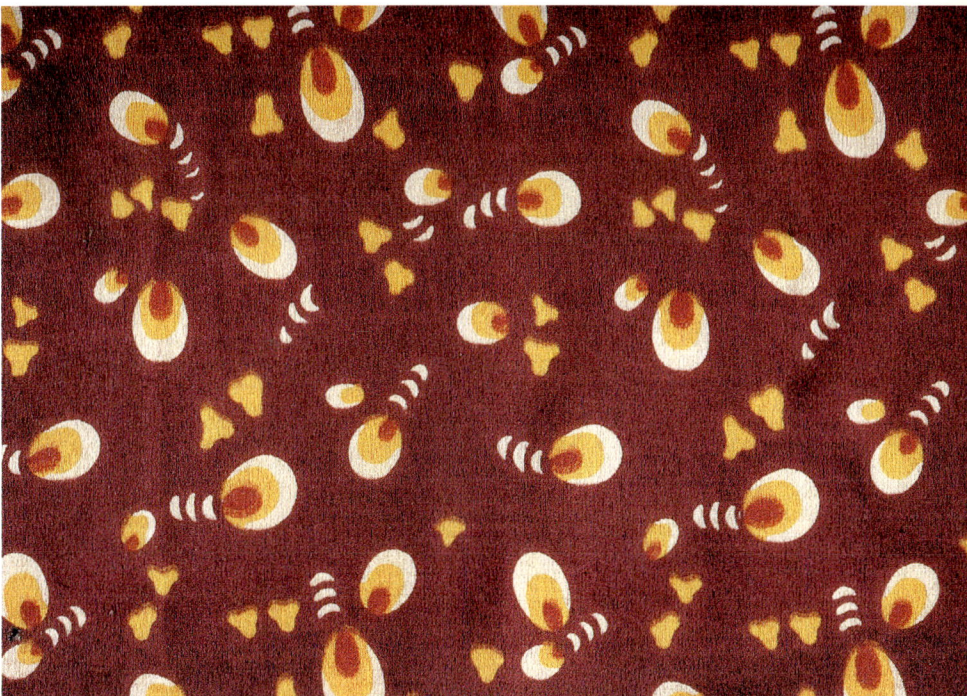
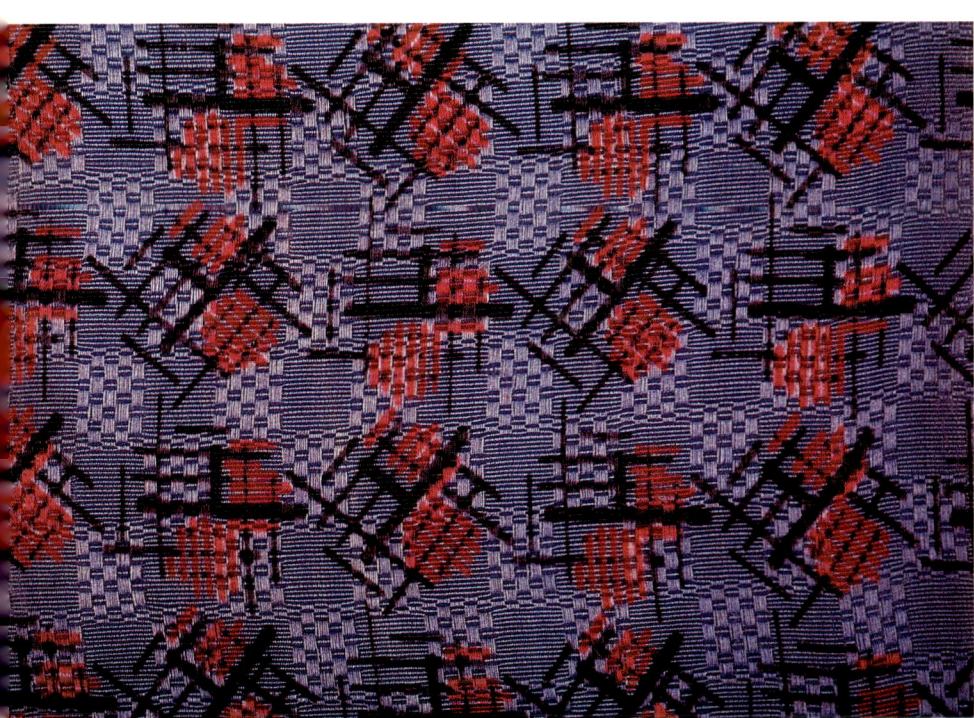
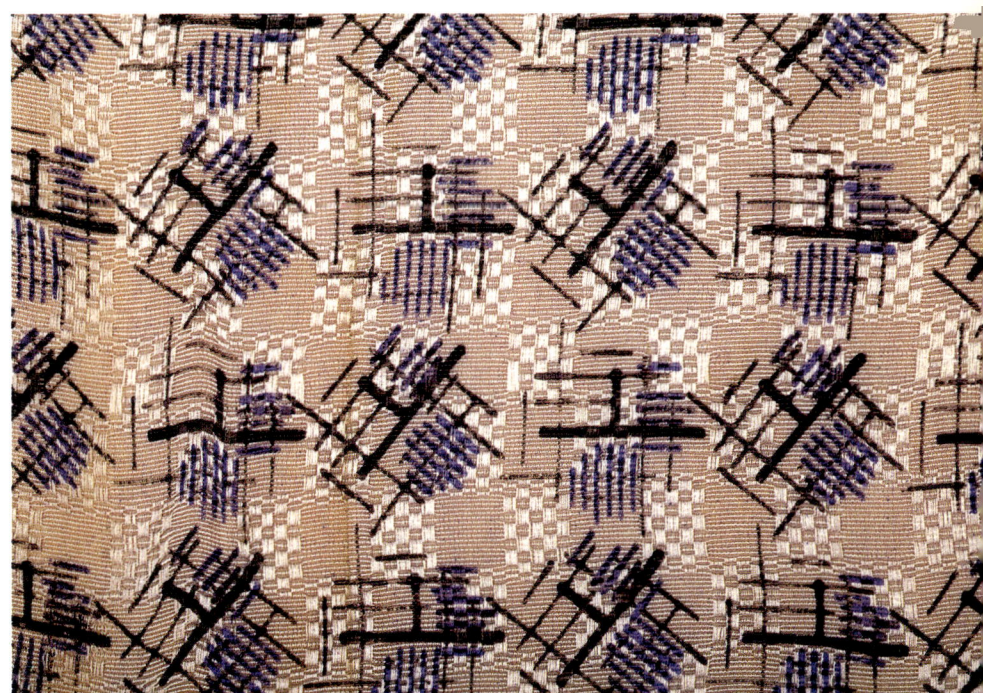

Conversational

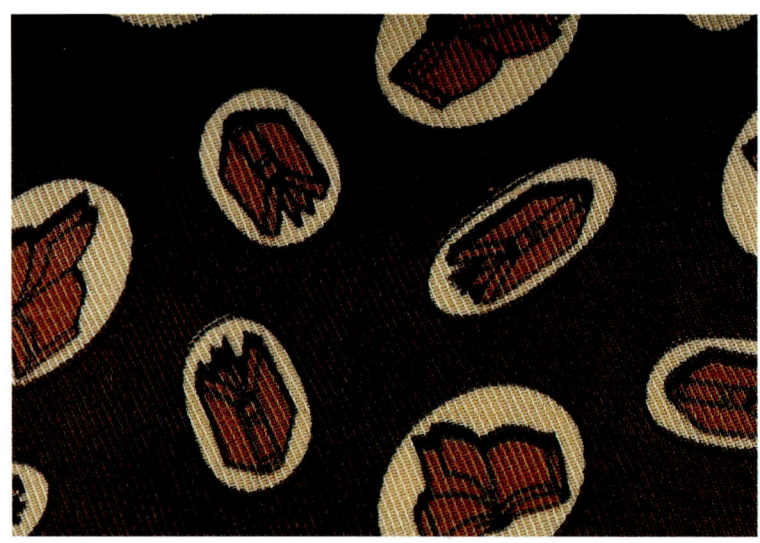

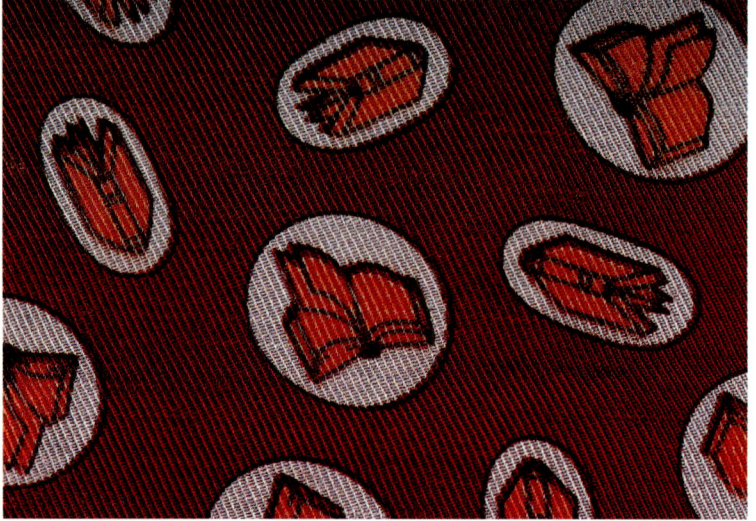

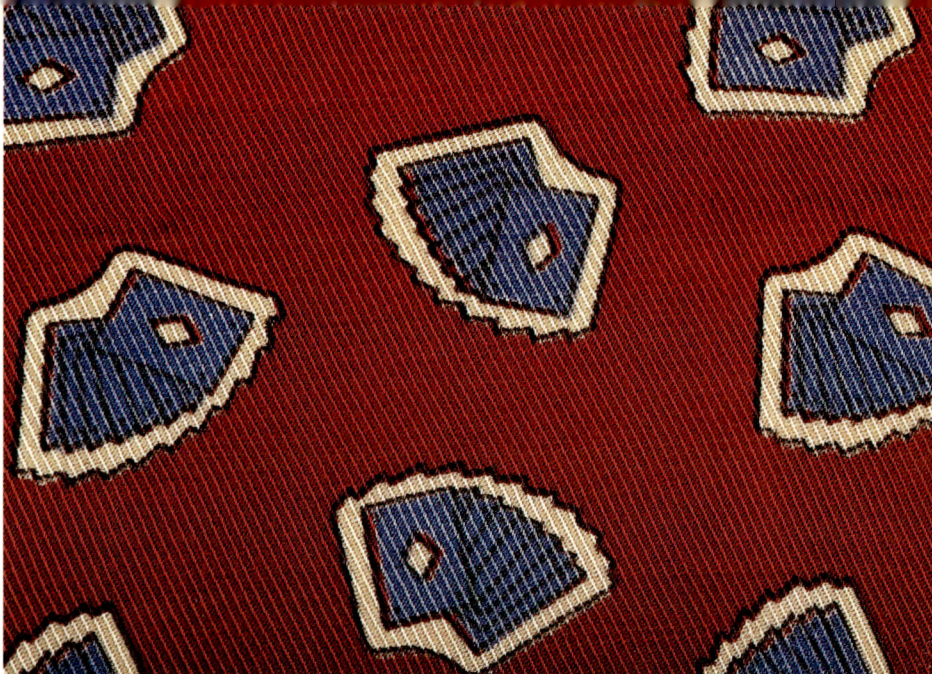

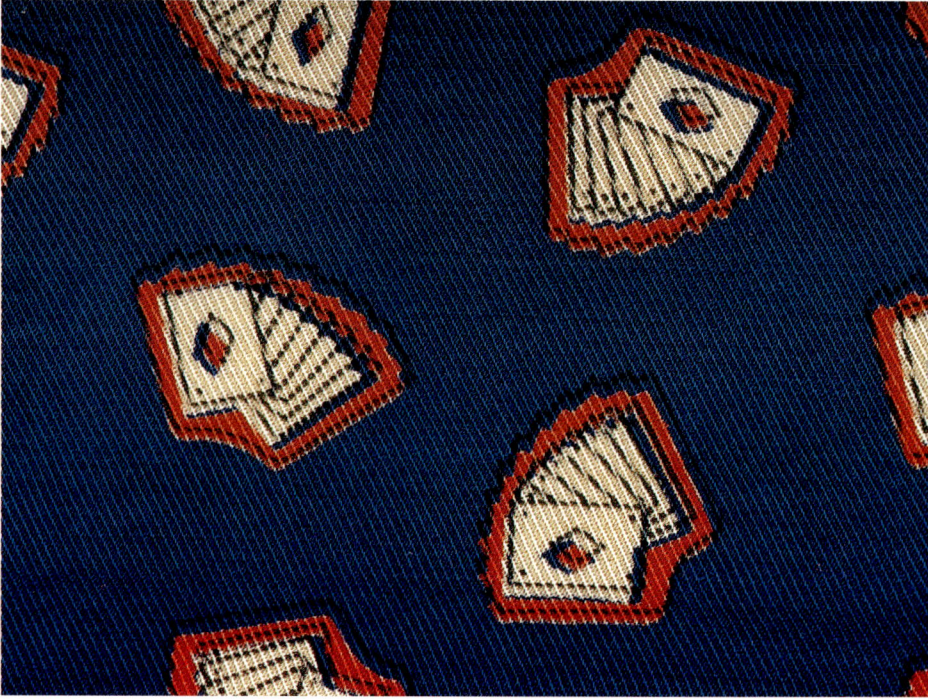

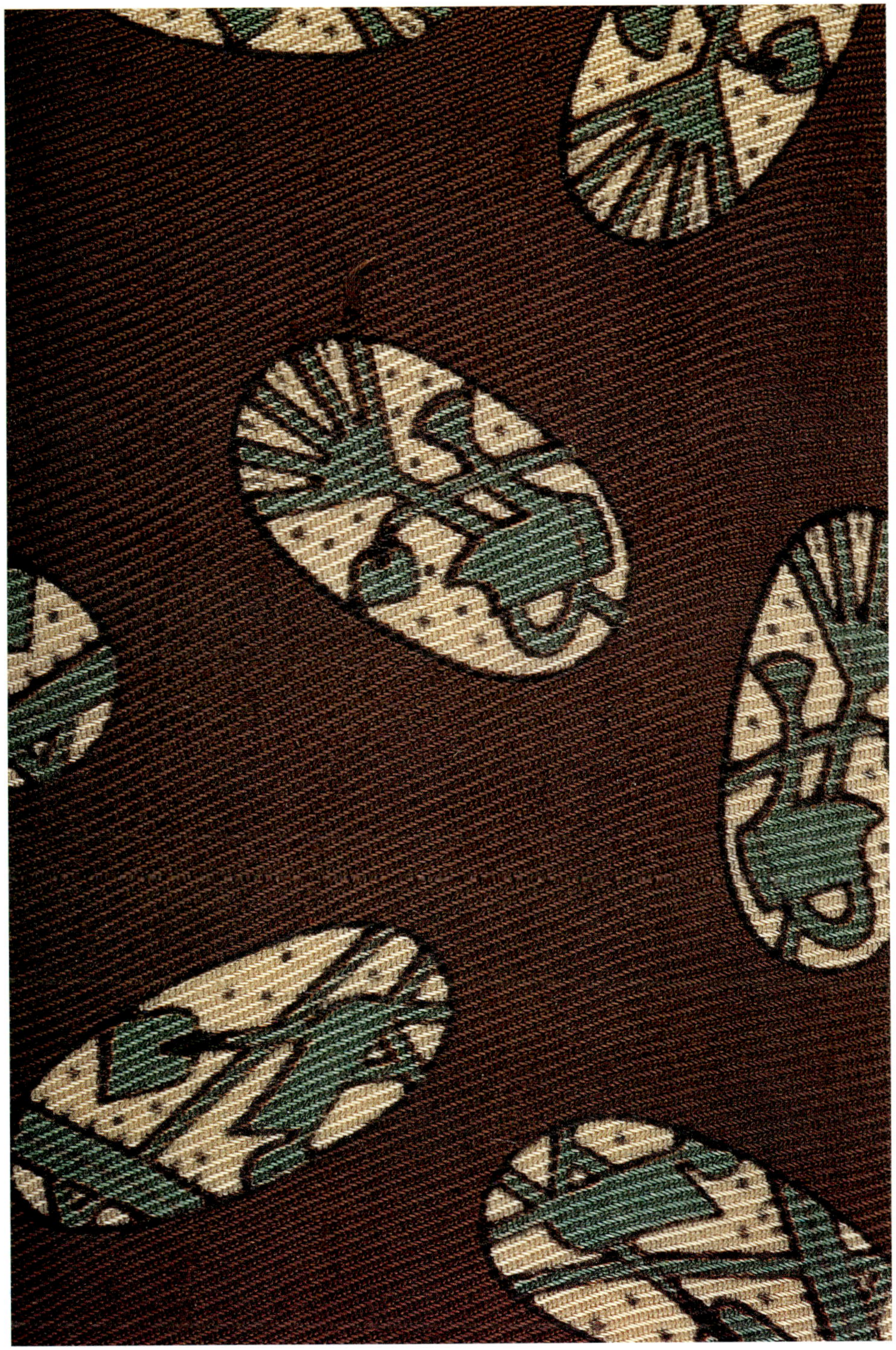

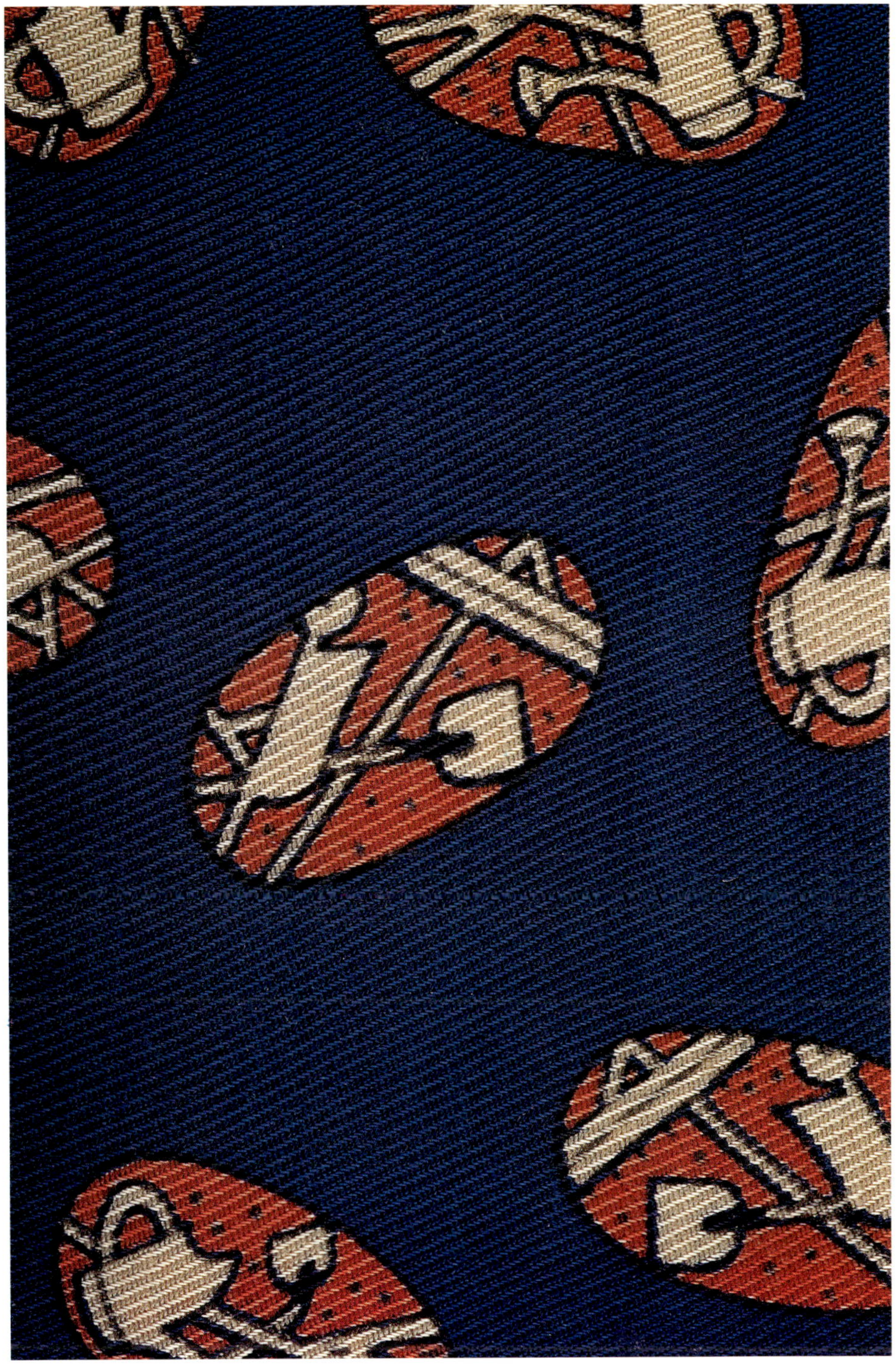

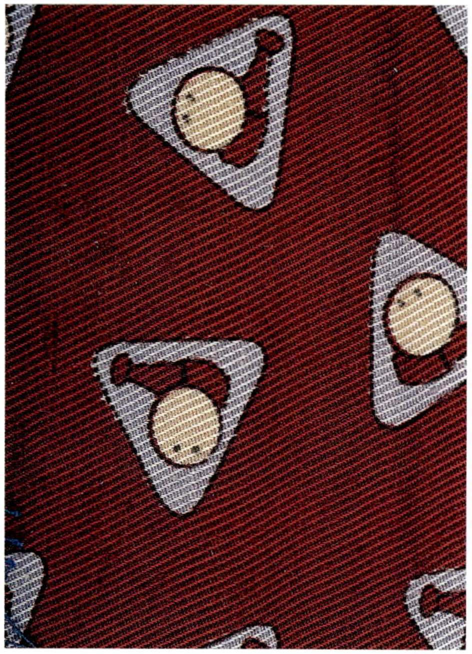
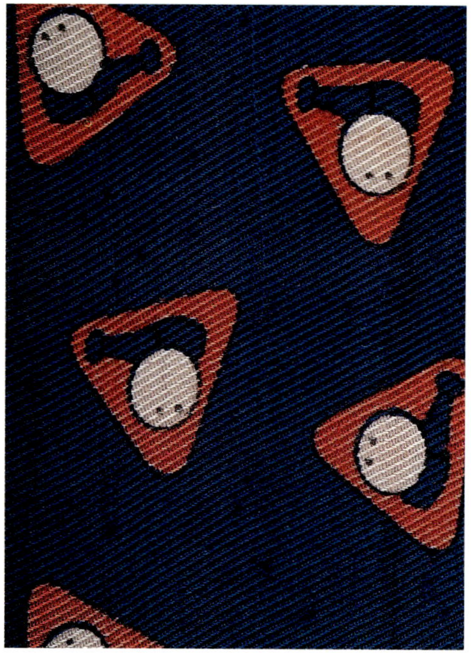
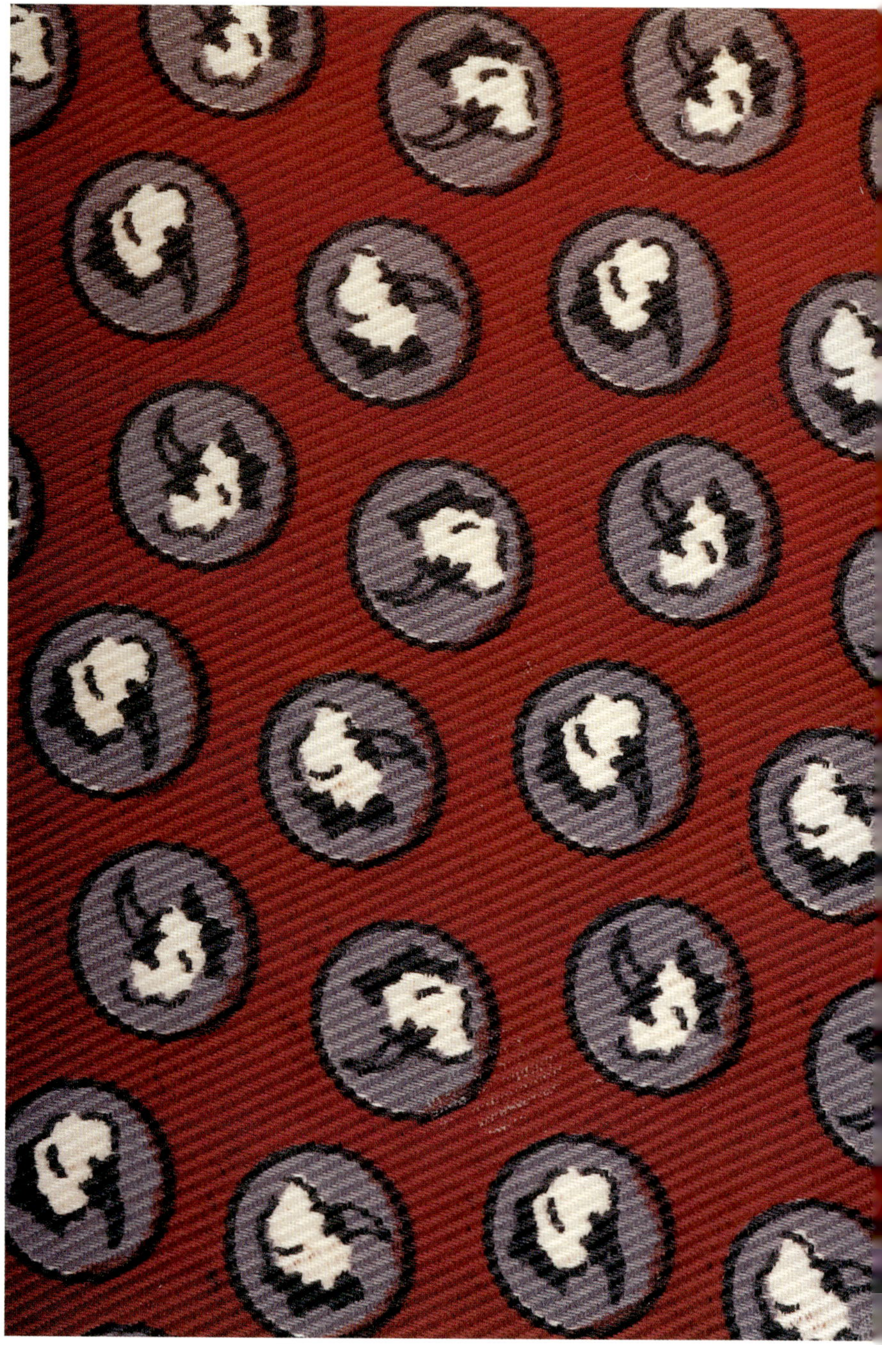

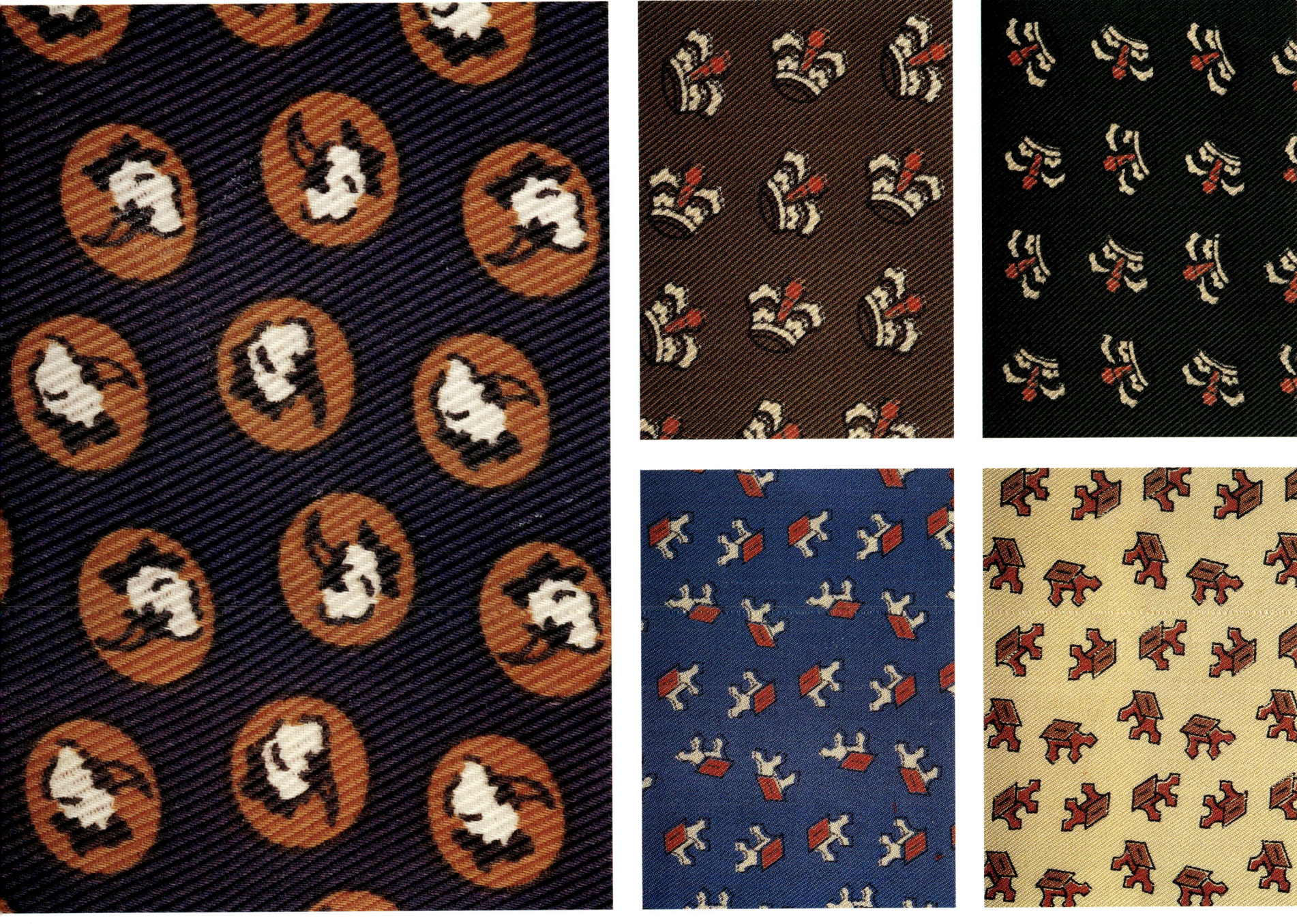

Paisley

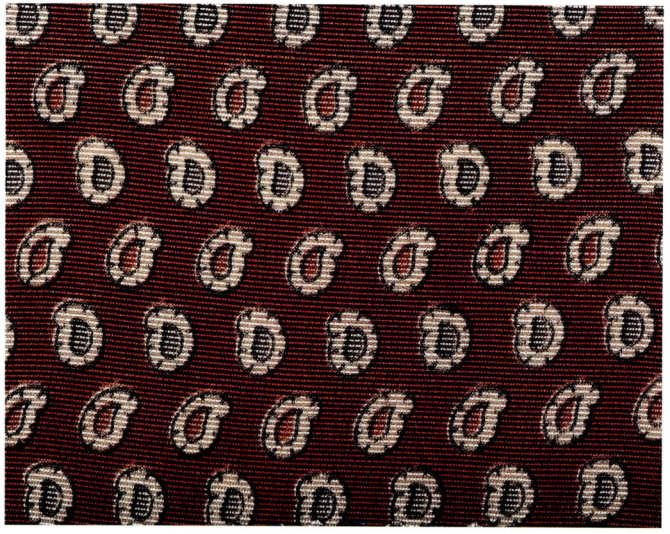

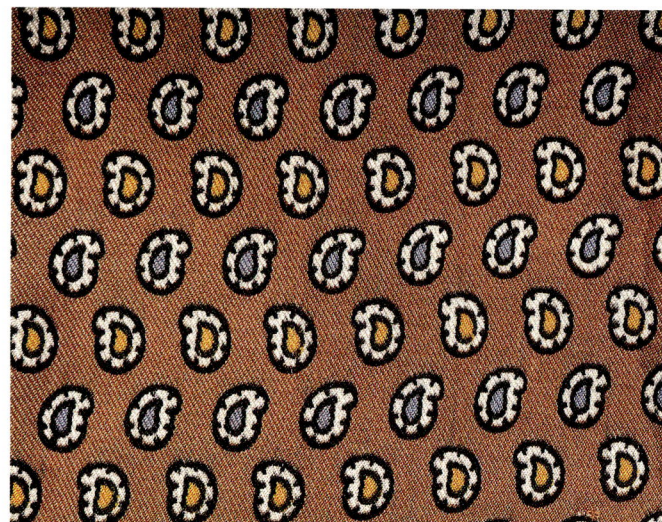

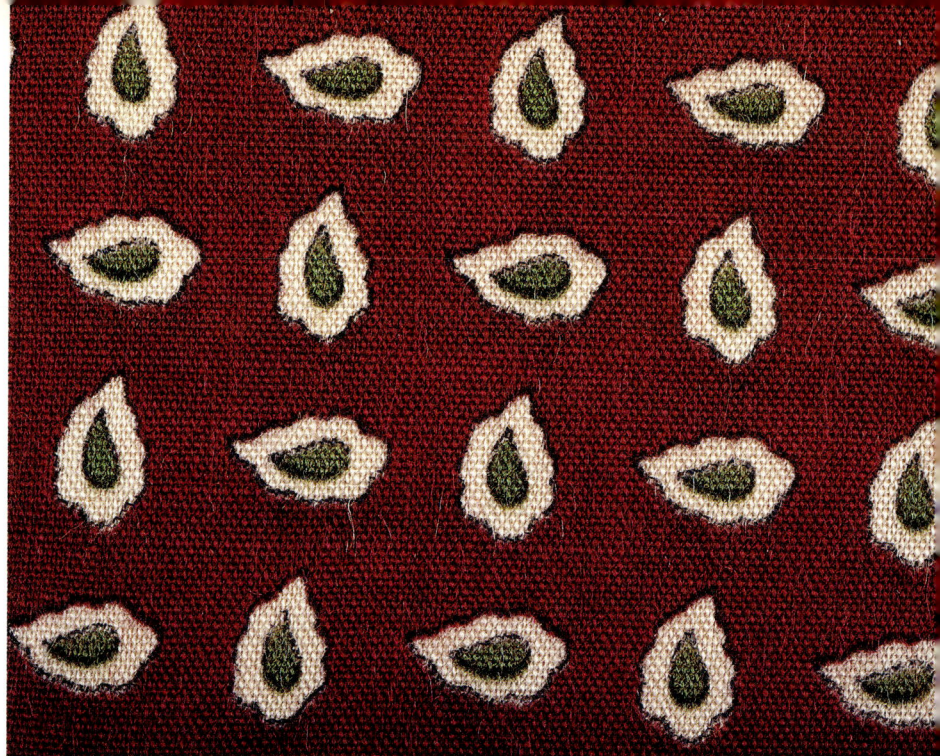

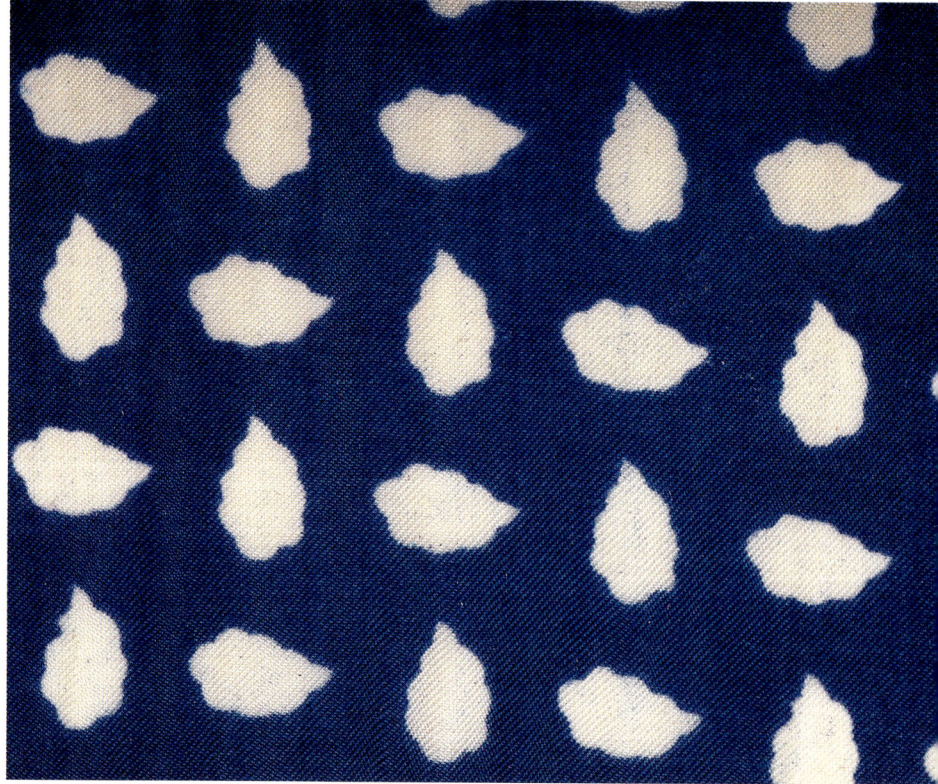

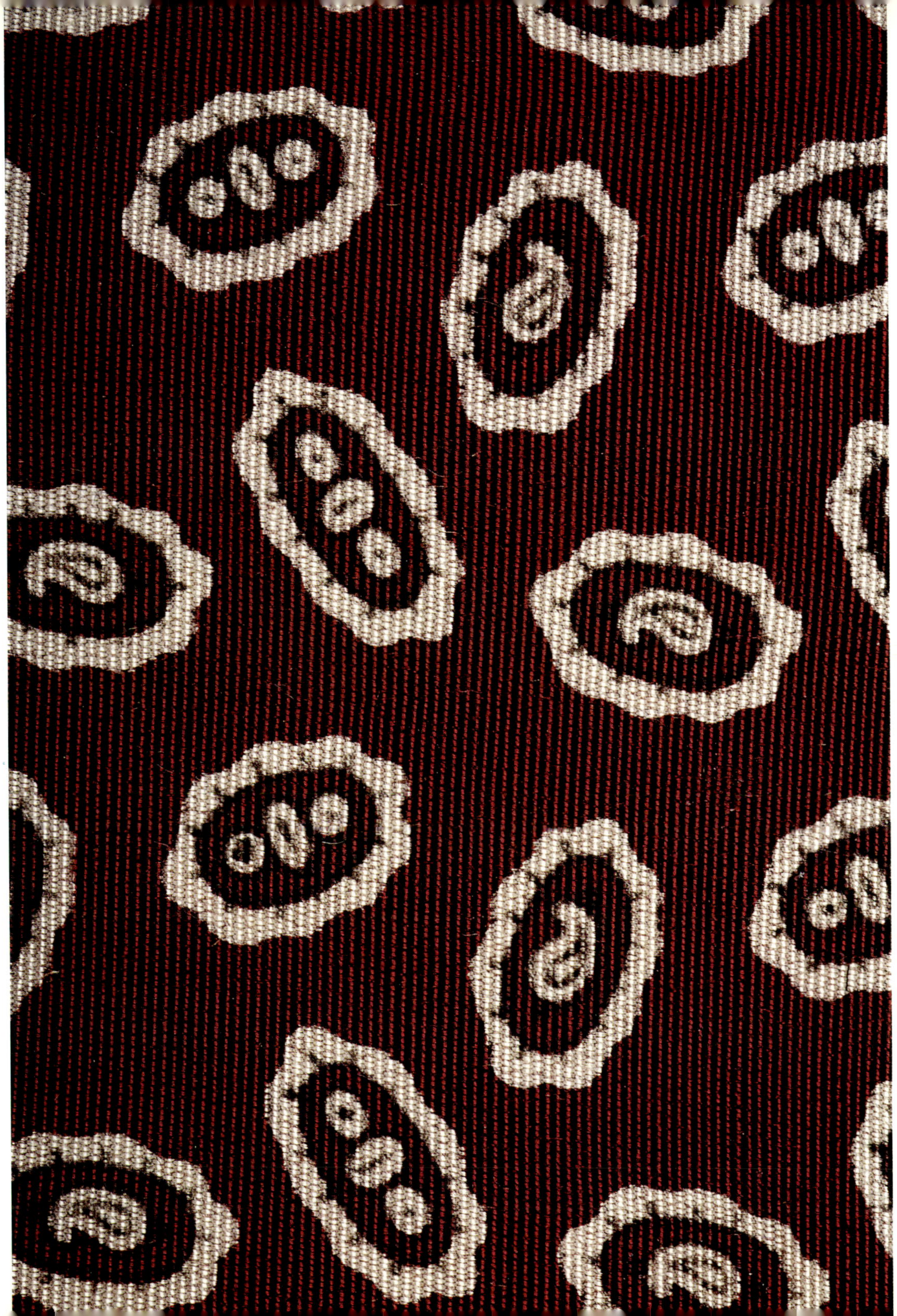

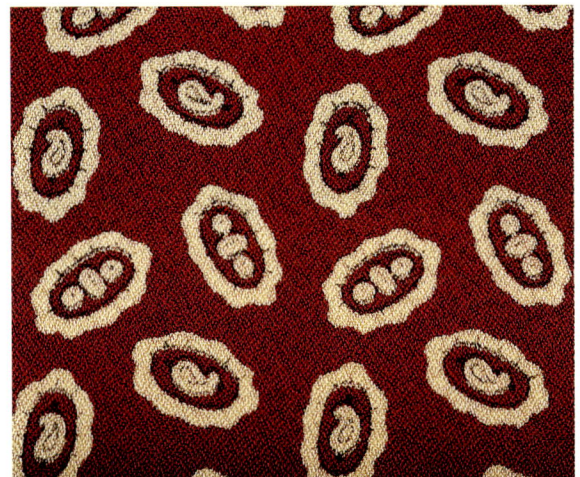

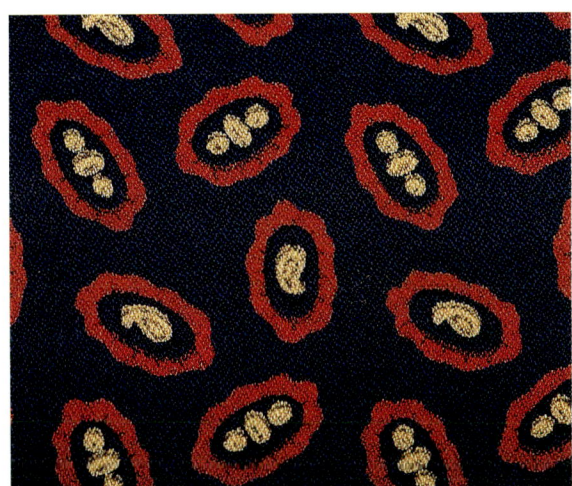

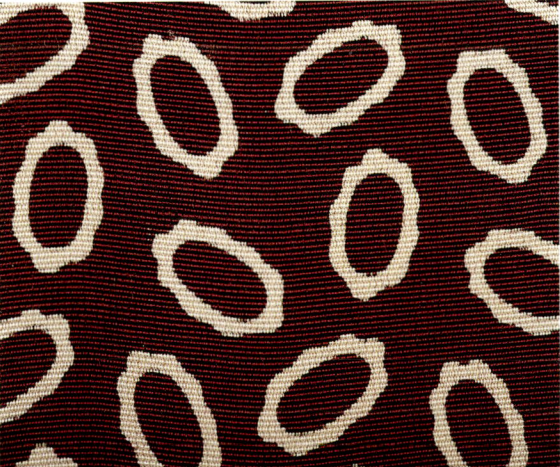

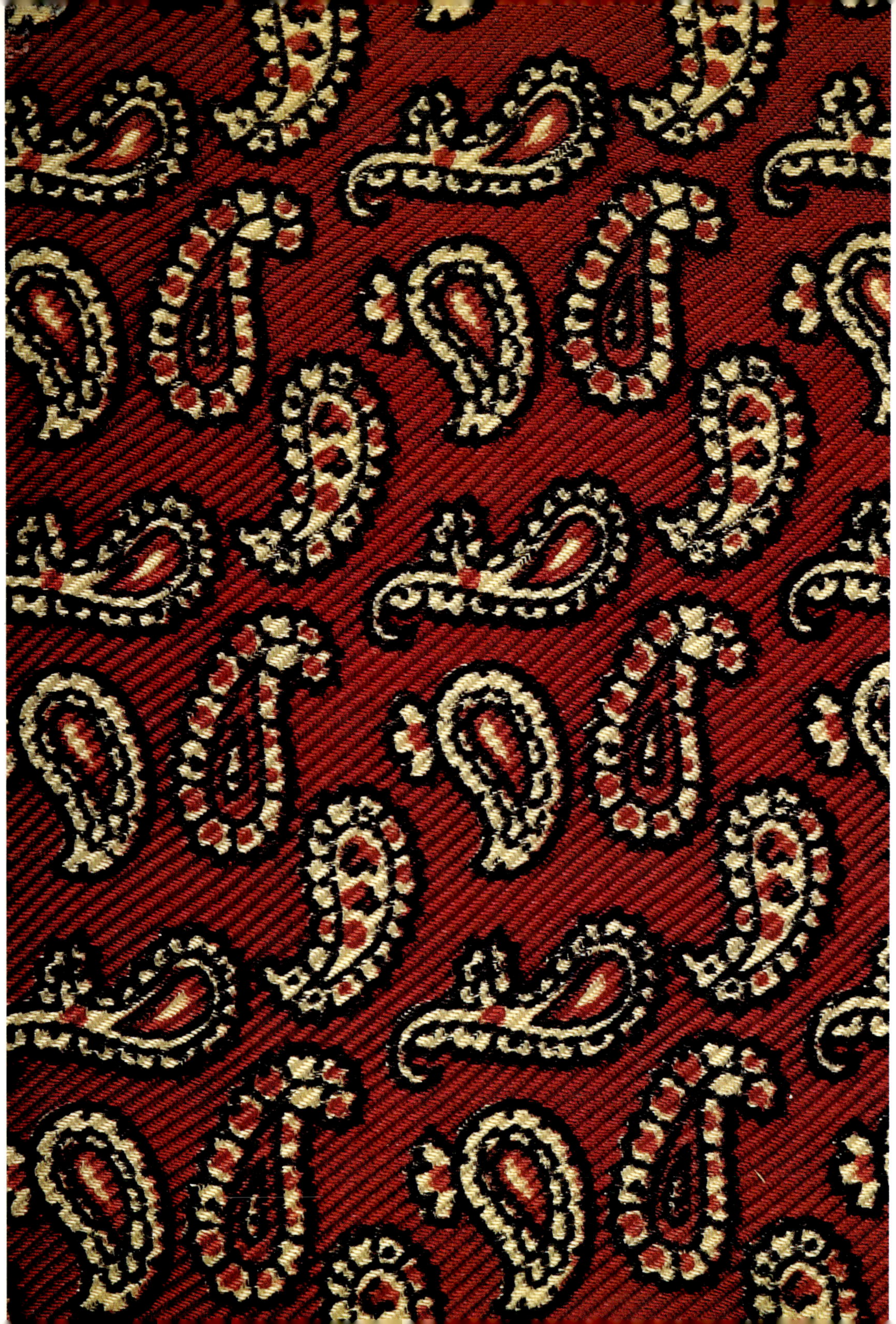

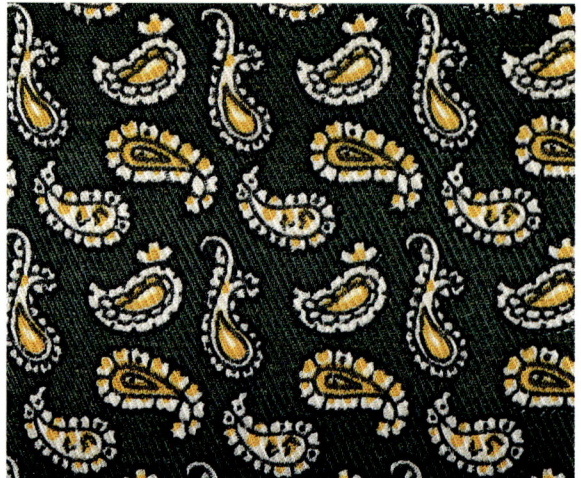

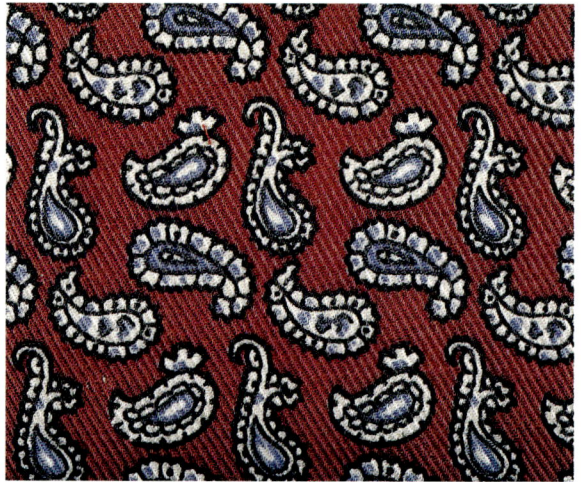

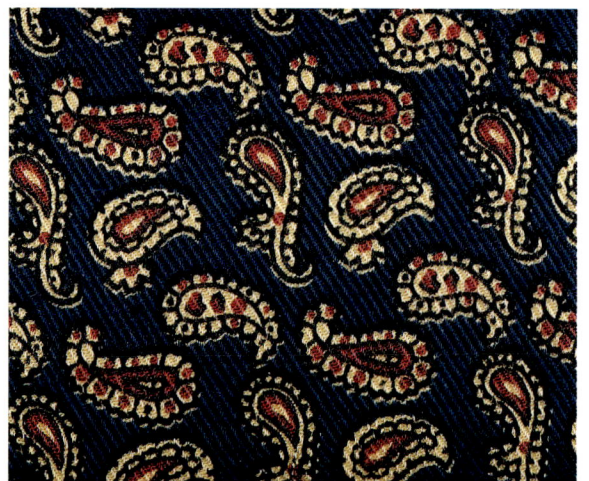

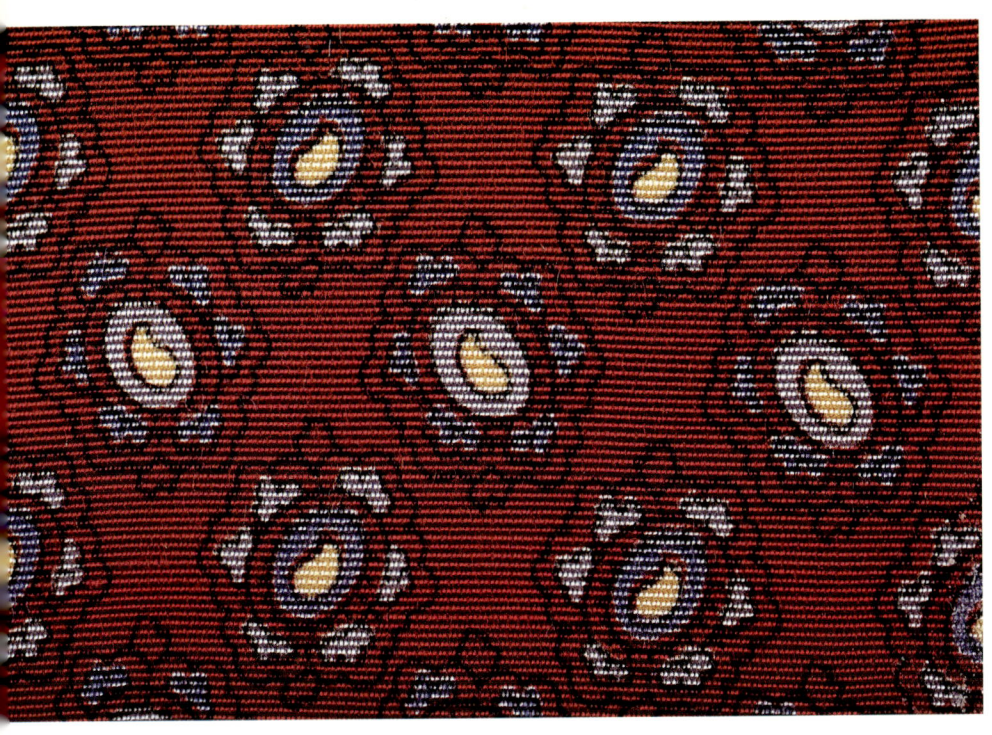
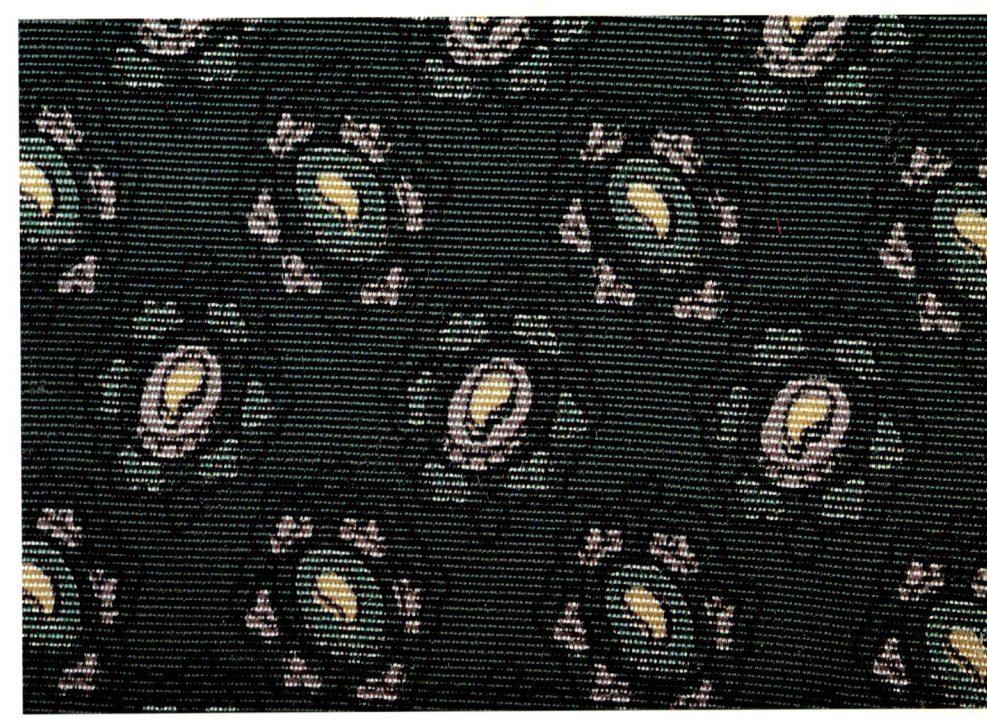
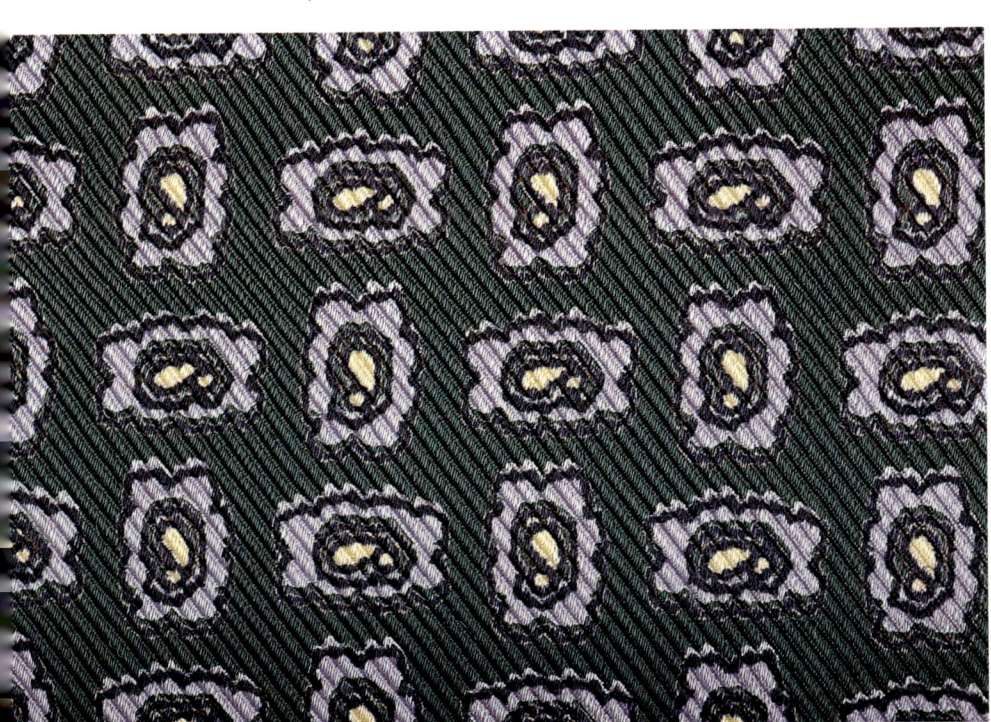
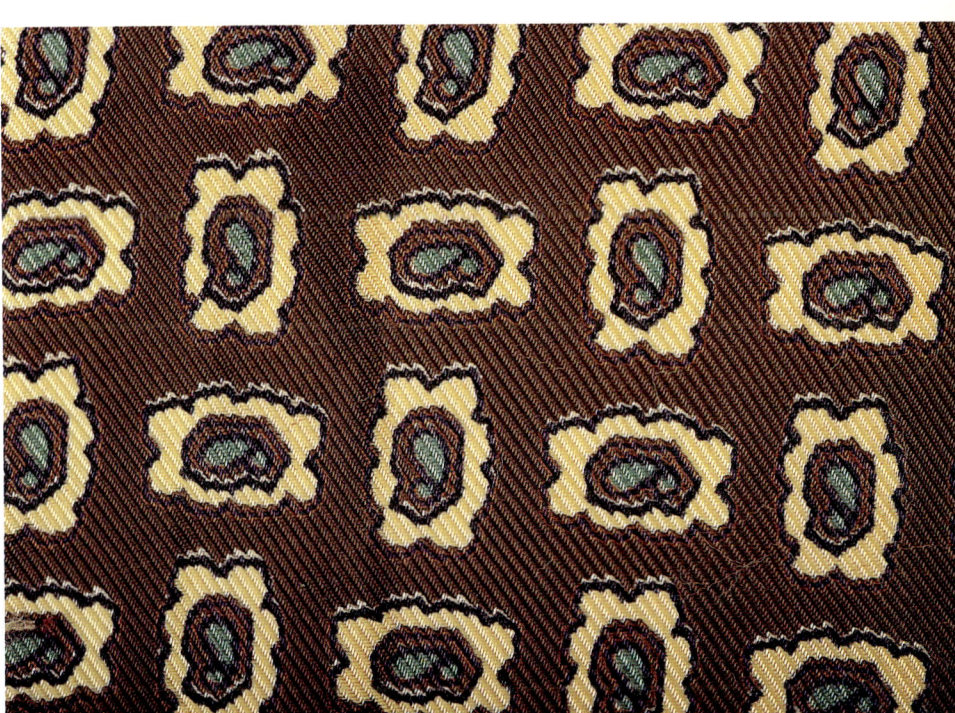

103

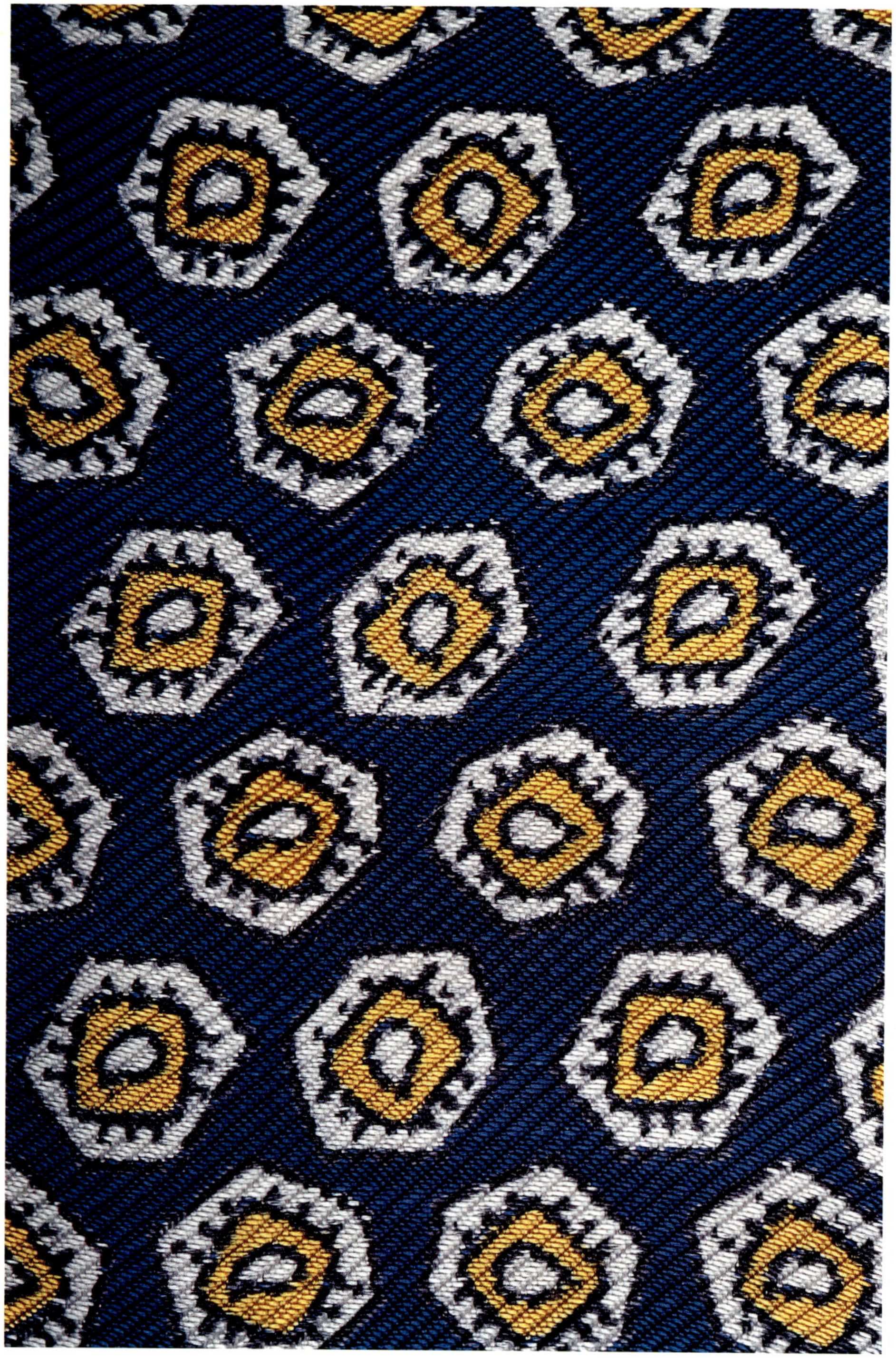

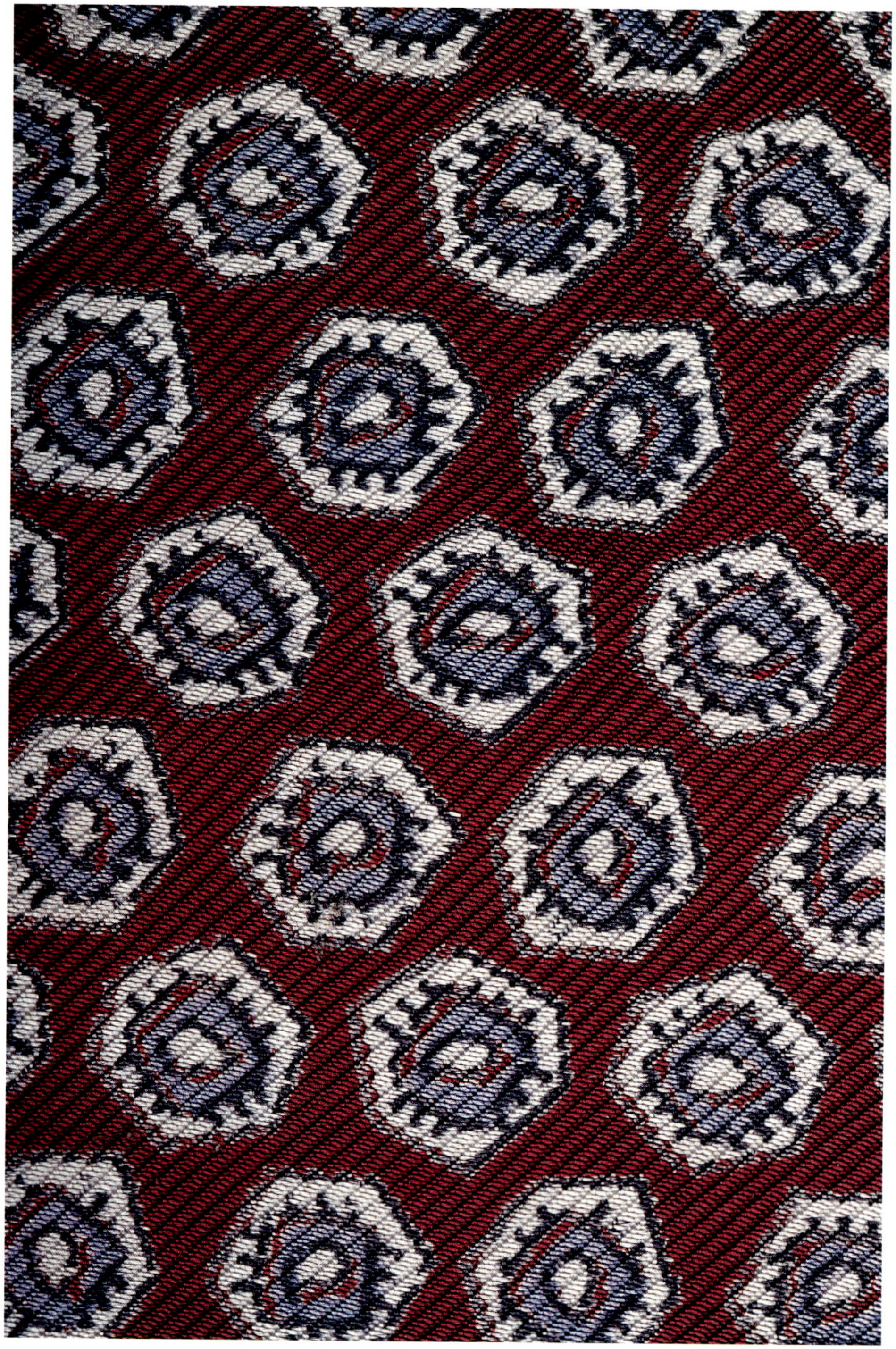

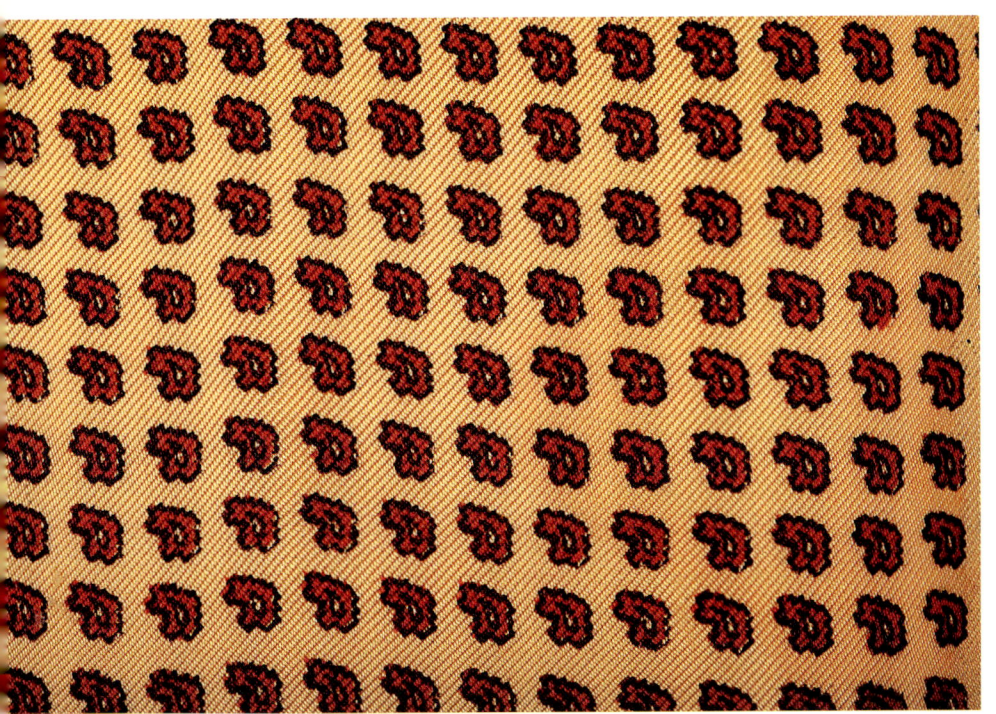
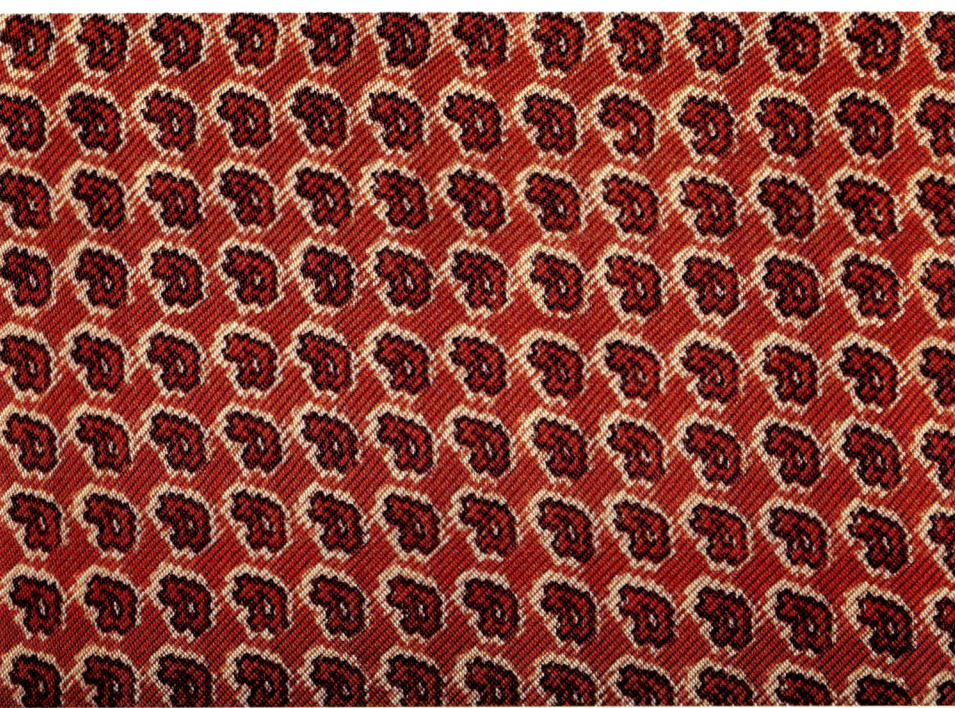
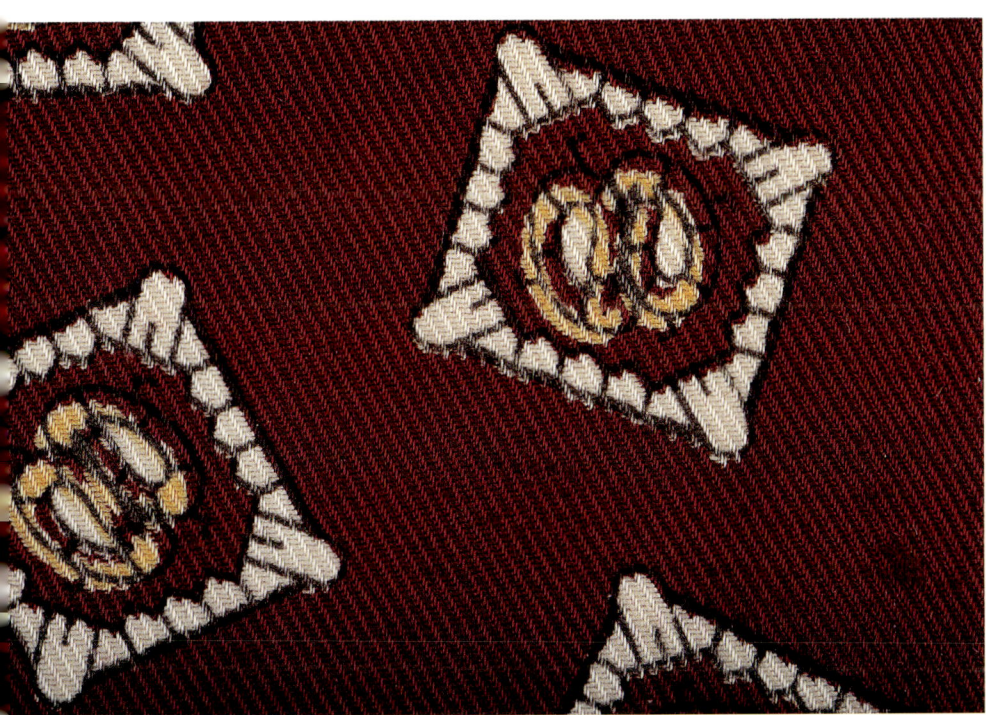
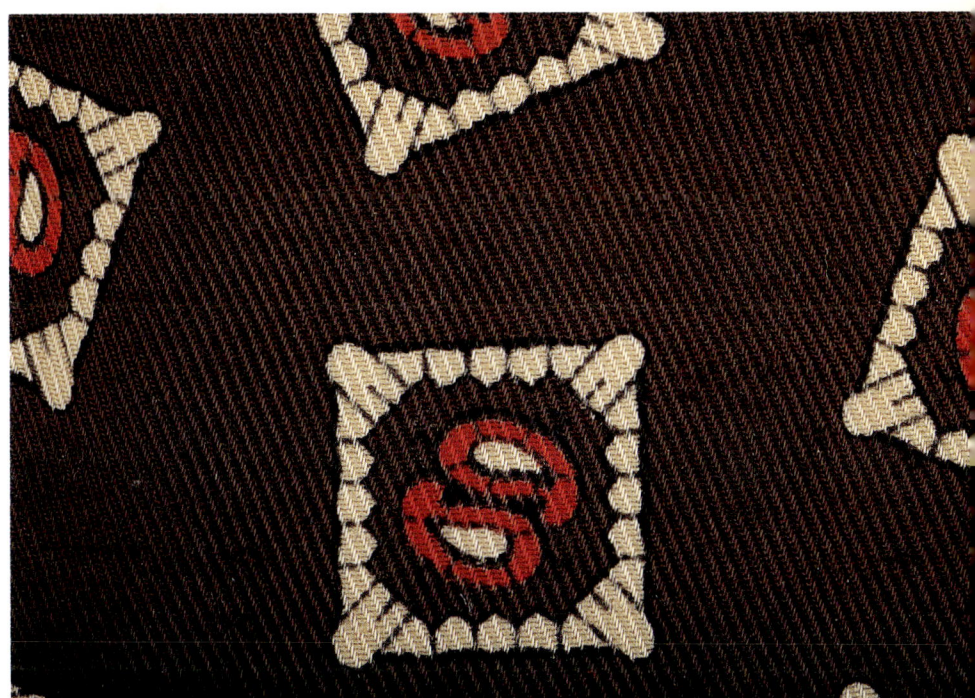

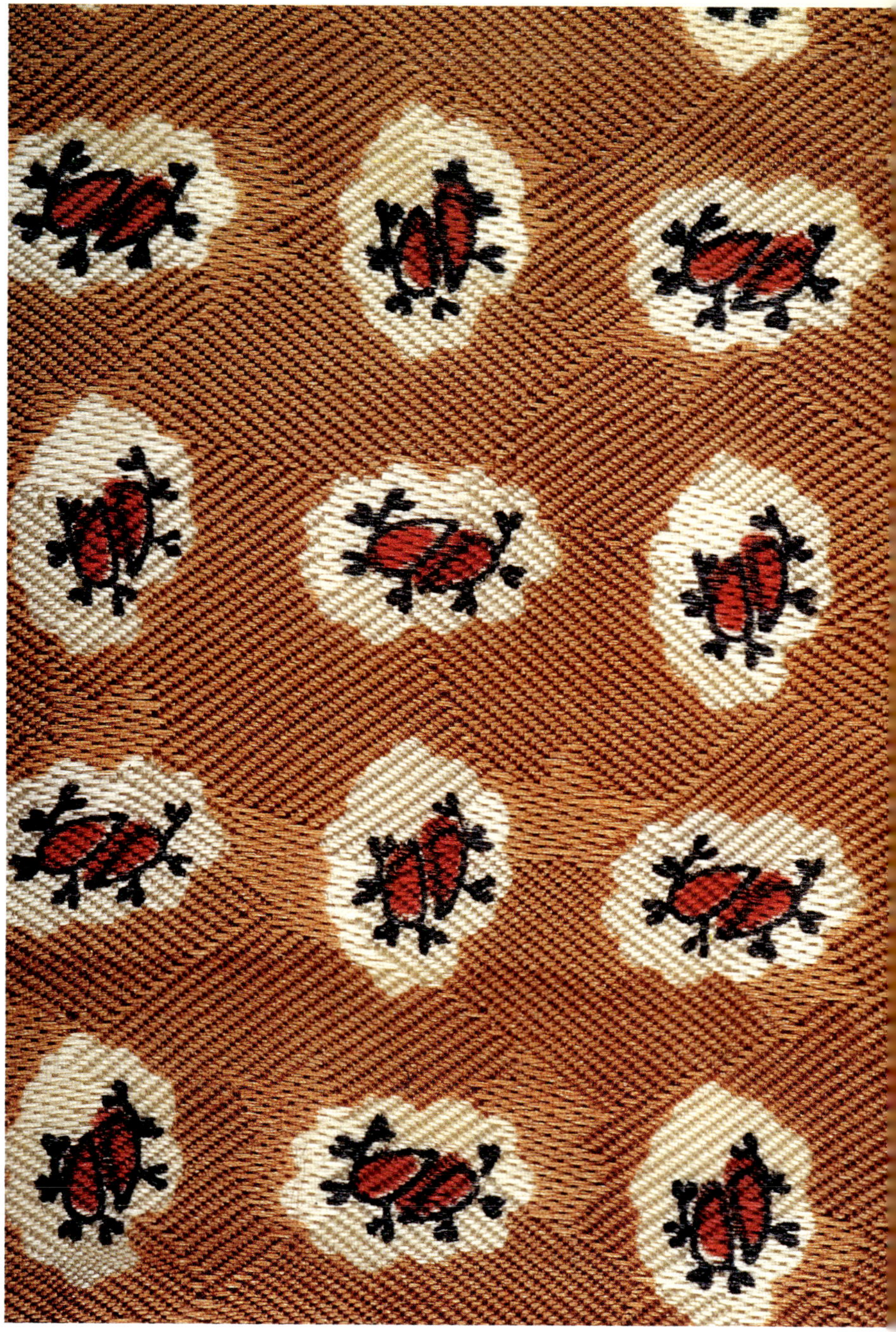

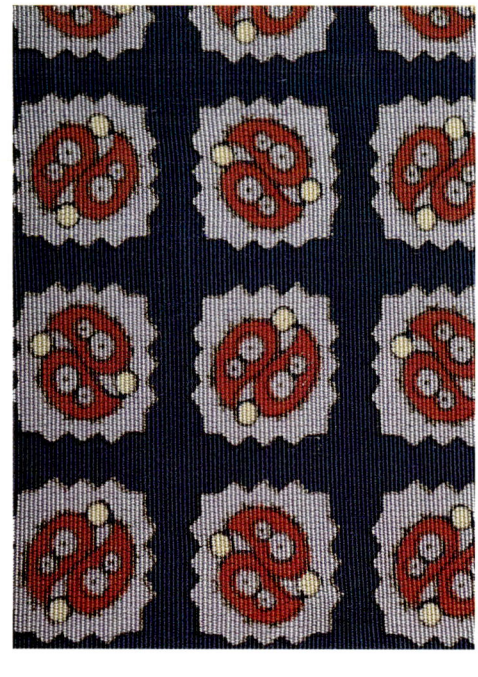
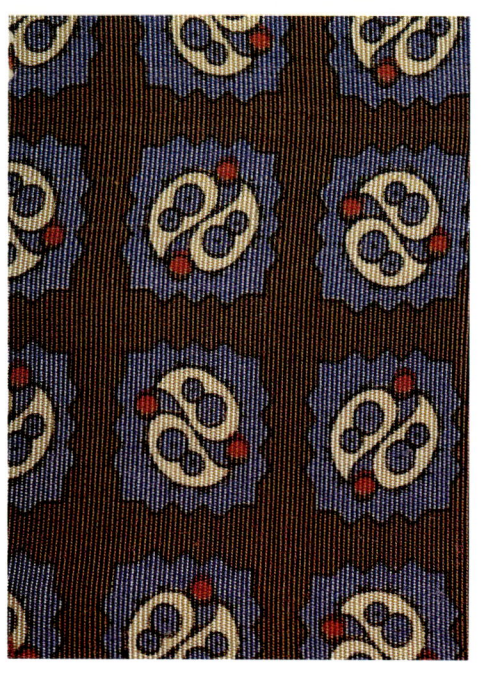
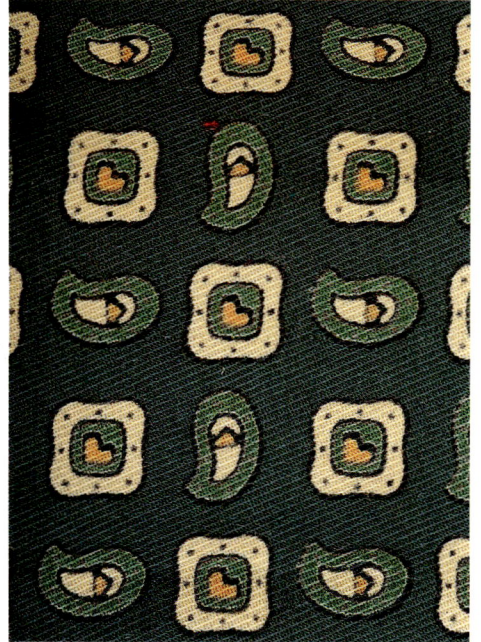
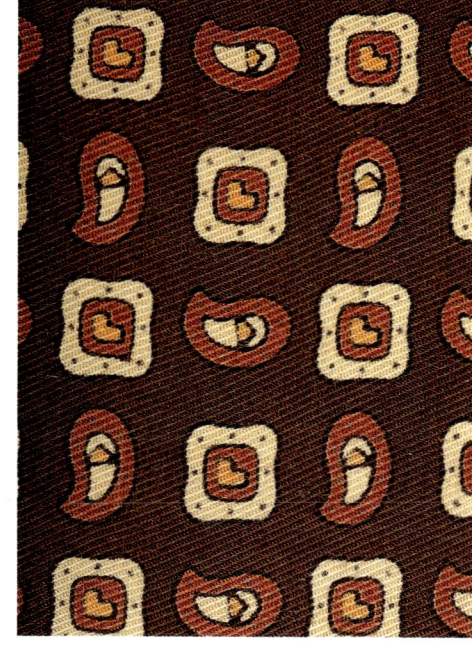

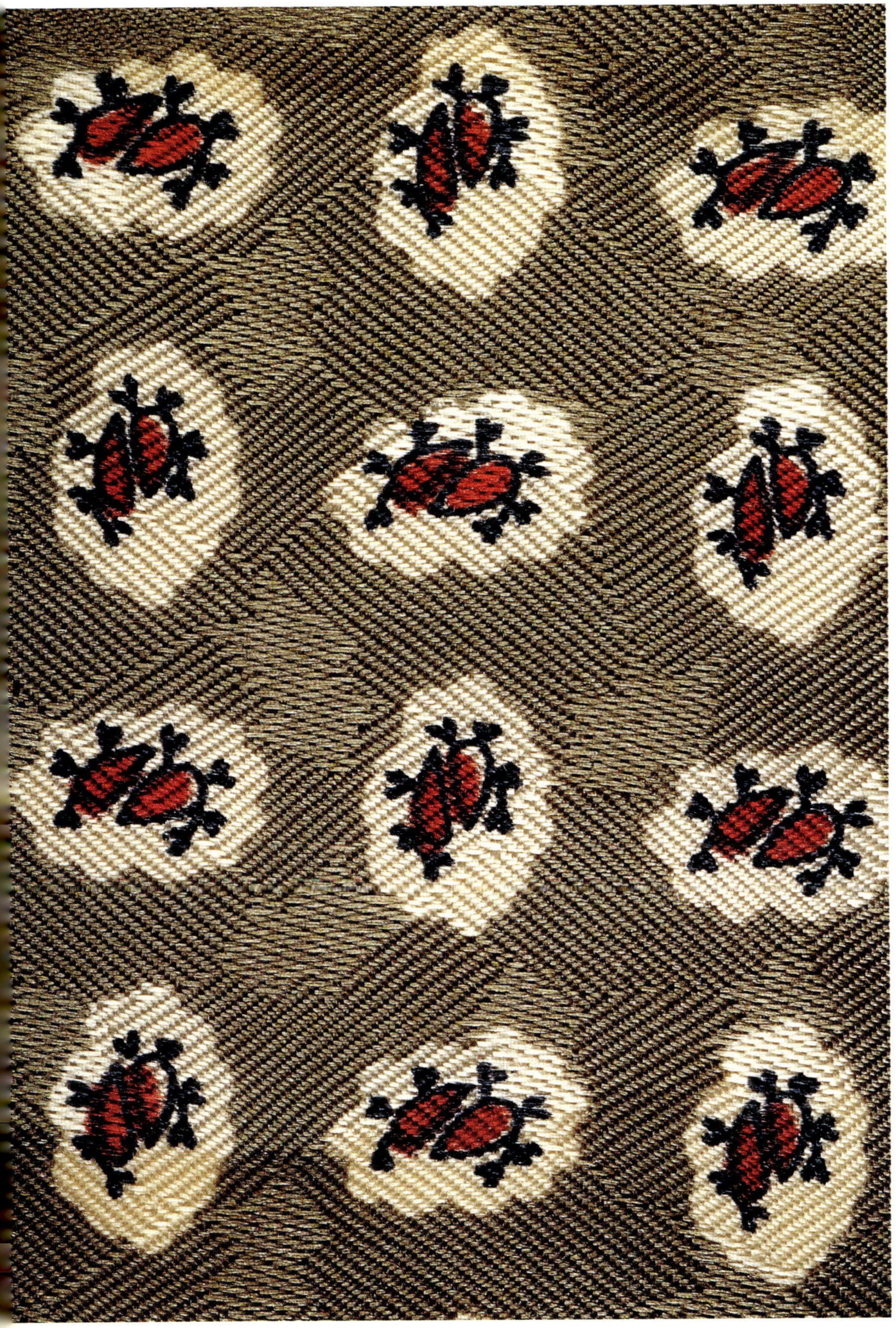

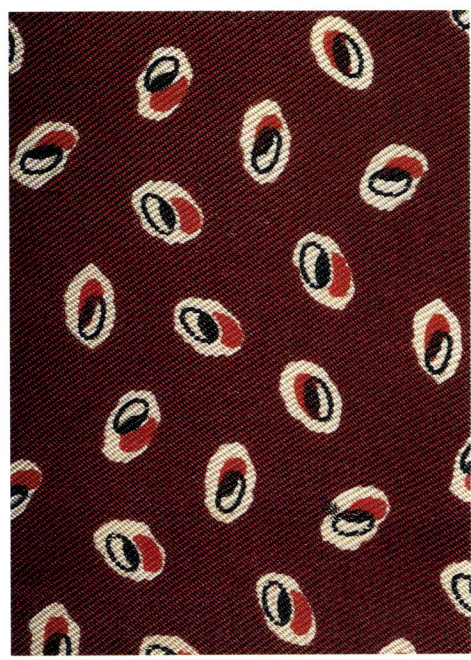

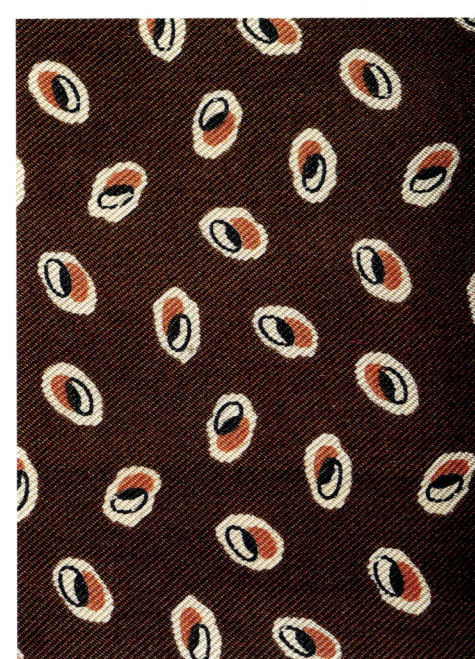

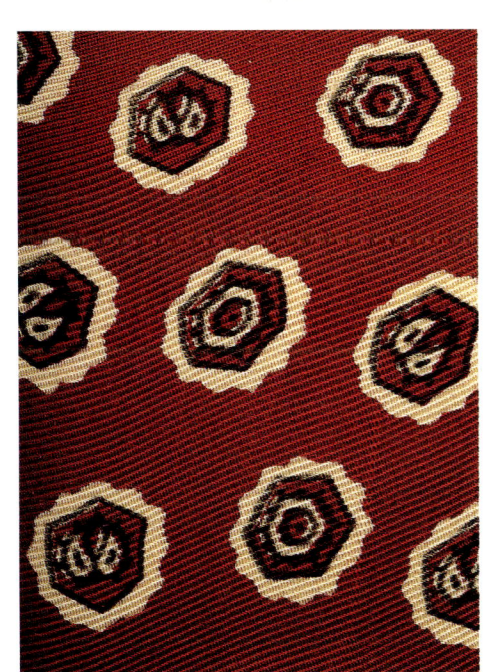

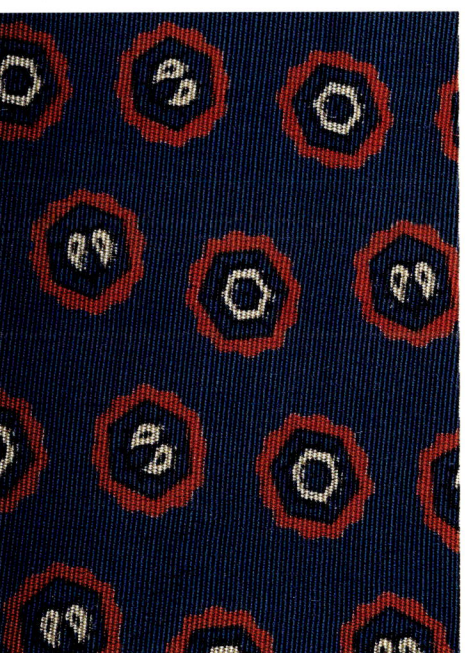

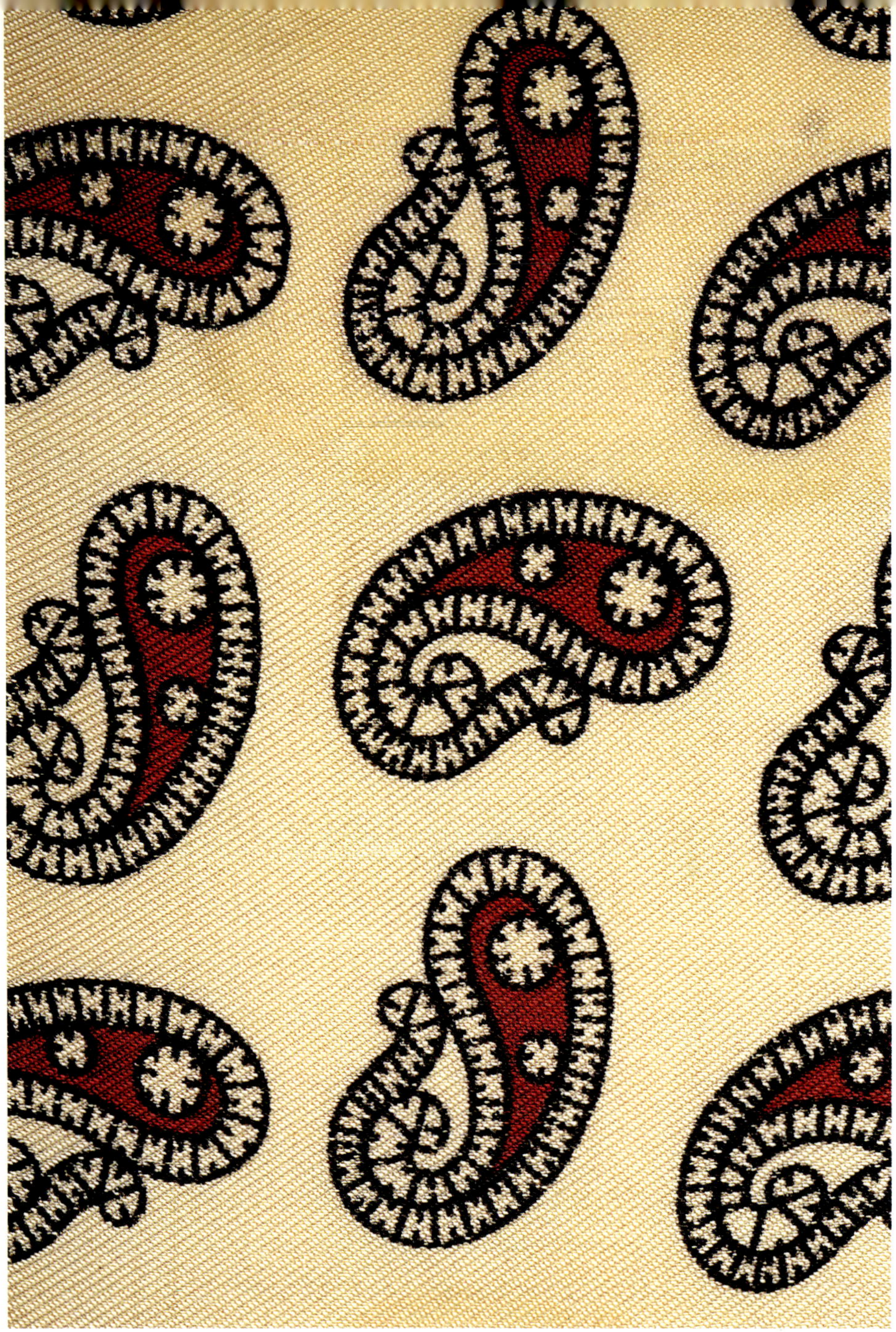

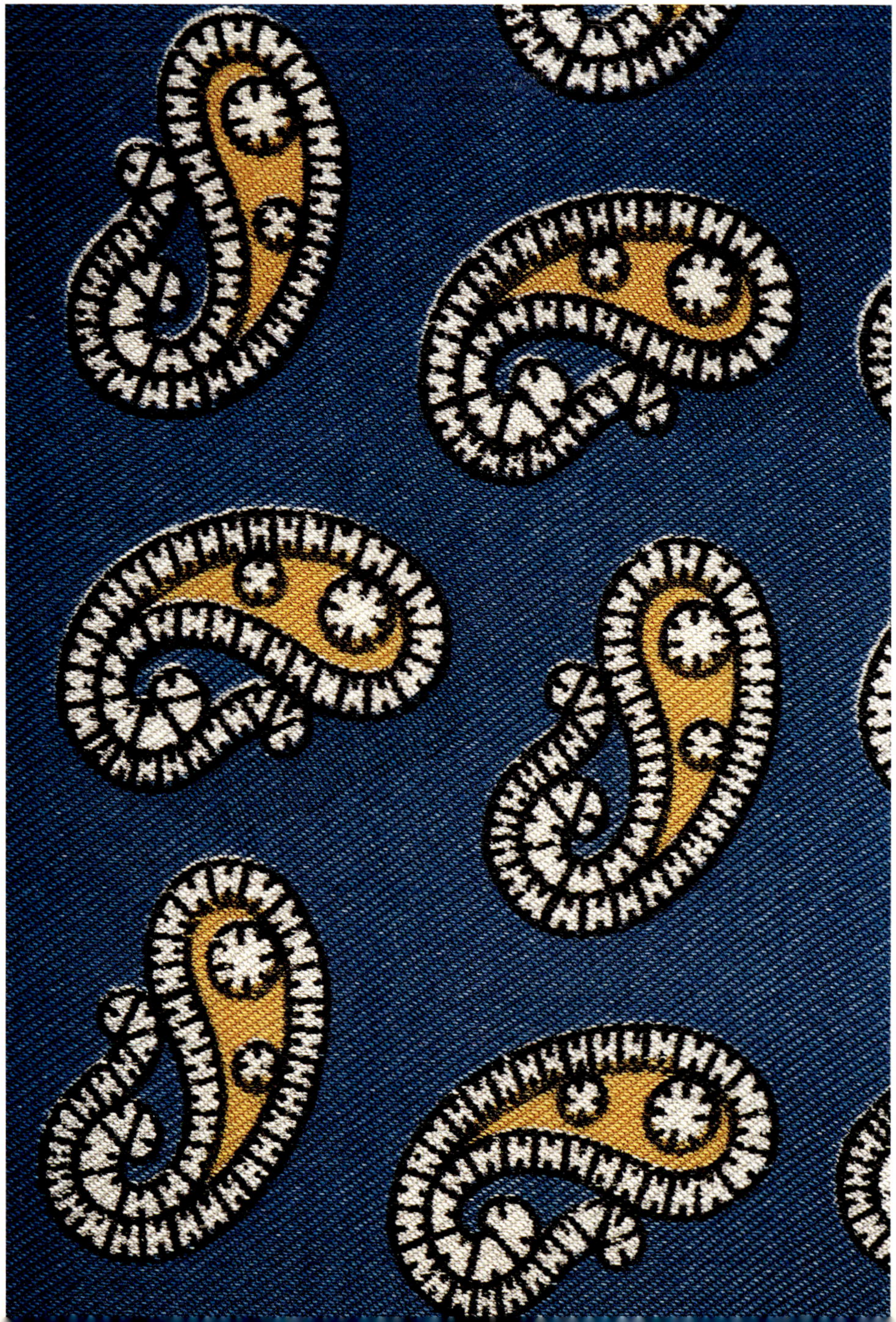

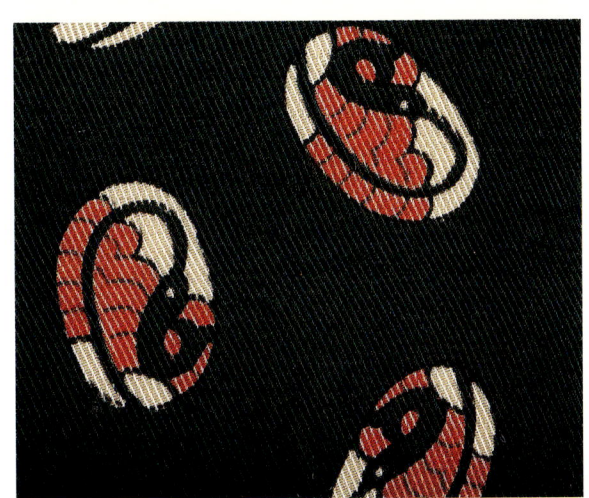

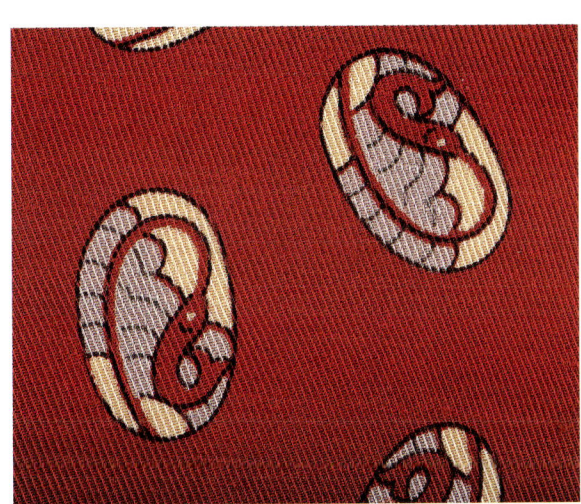

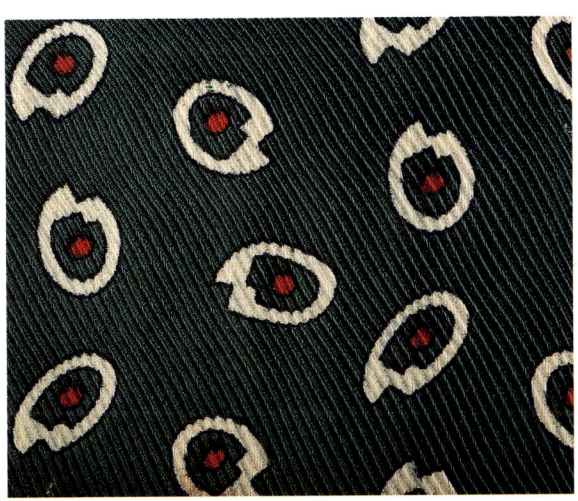

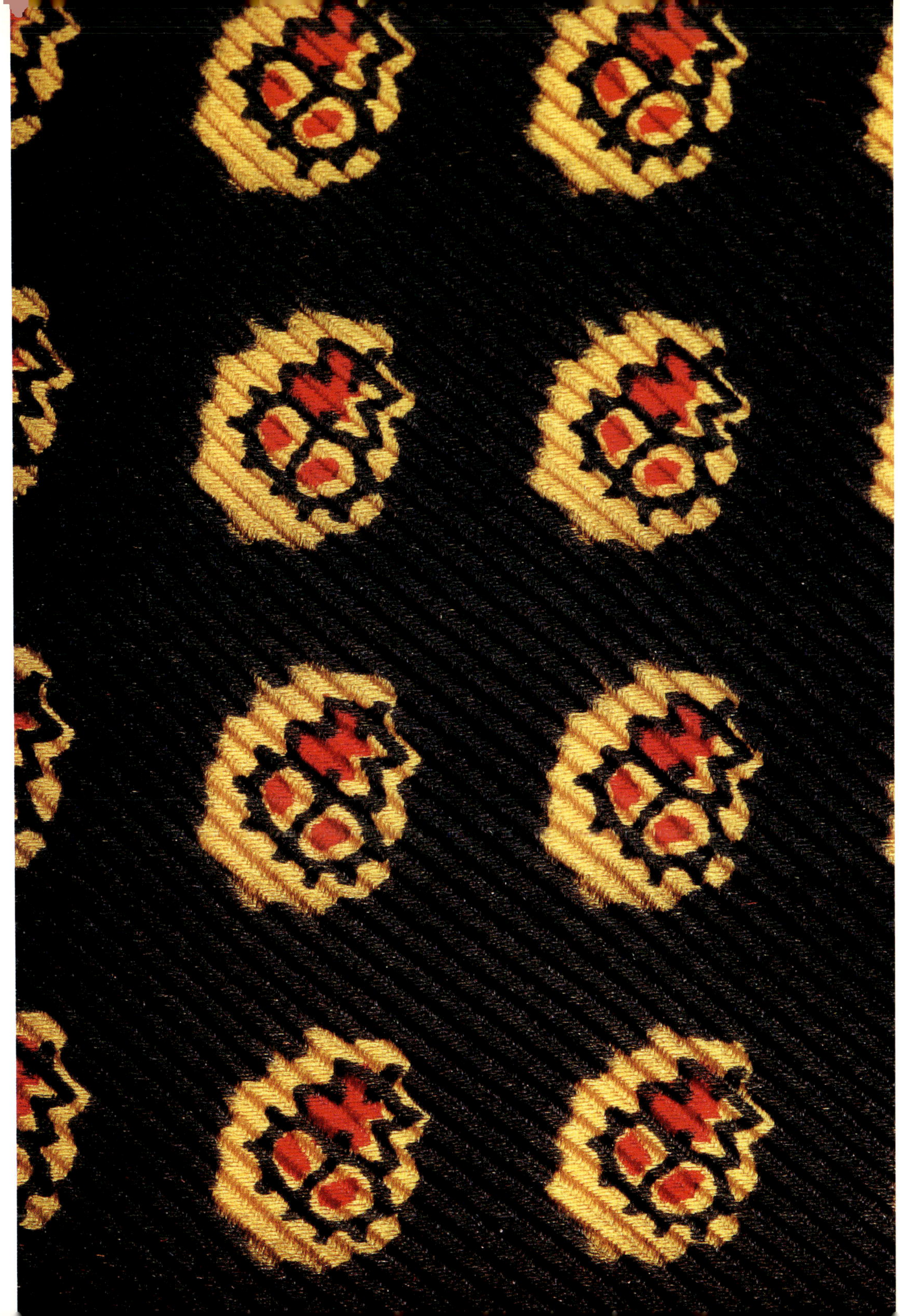

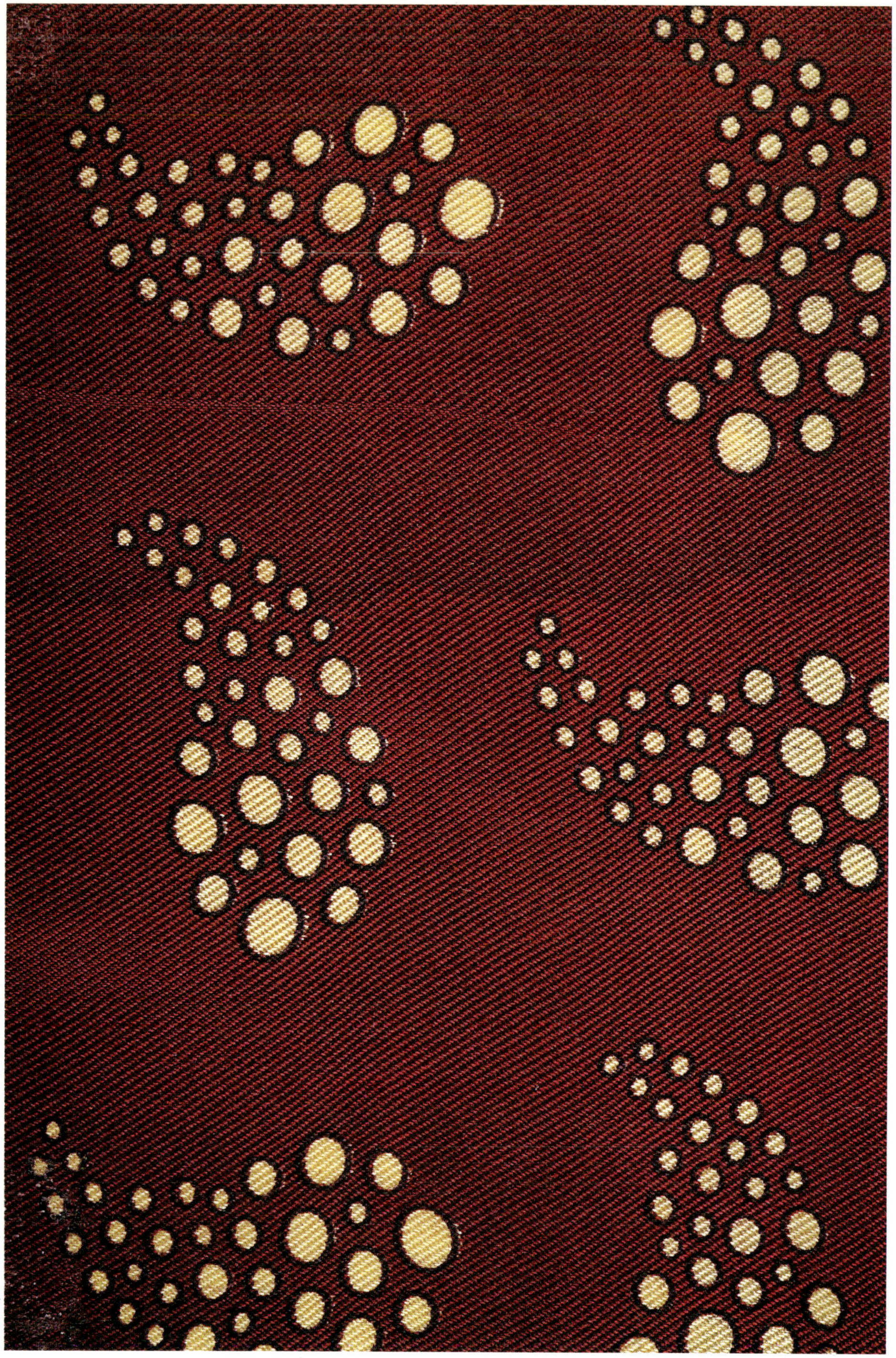

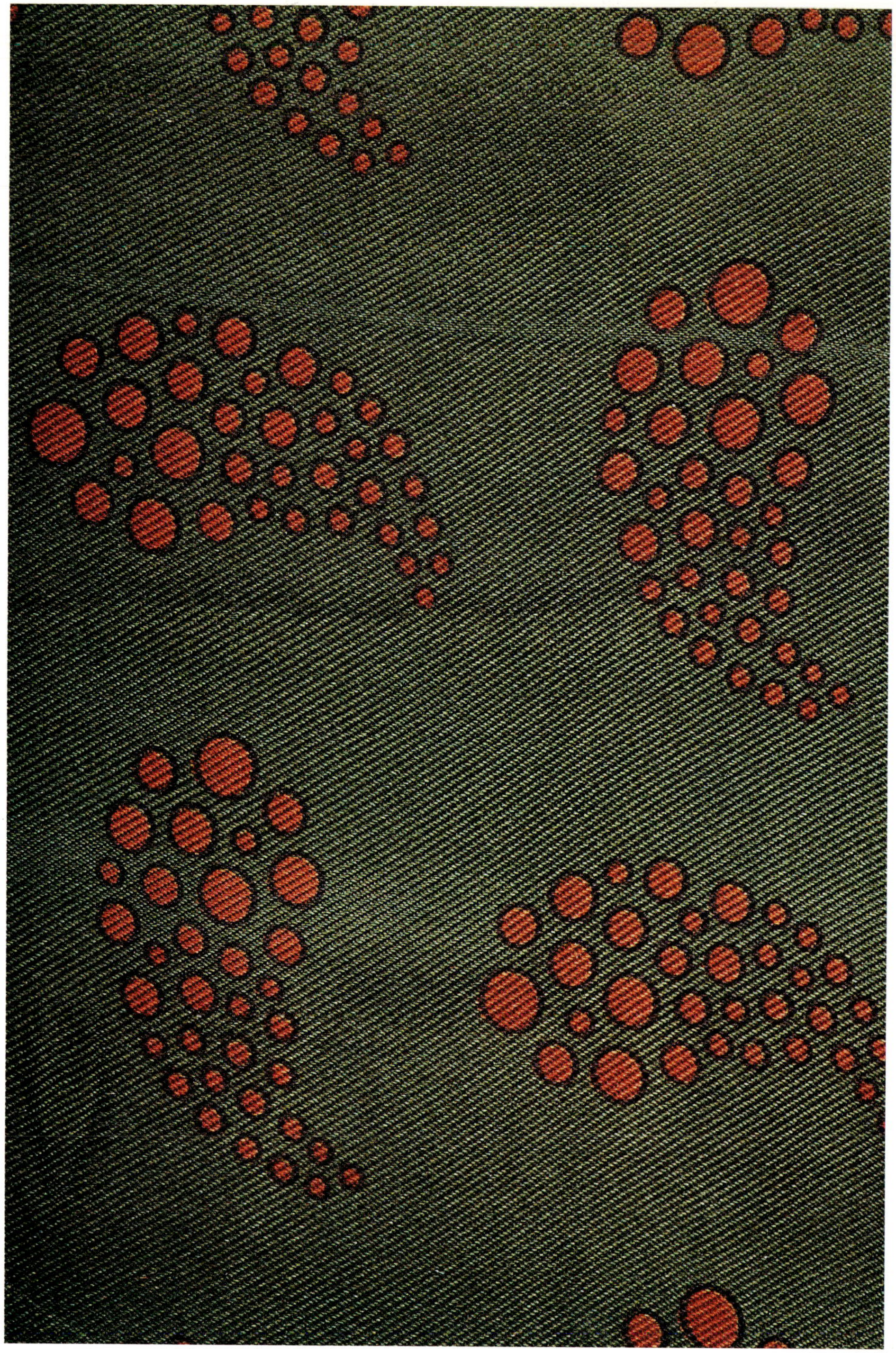

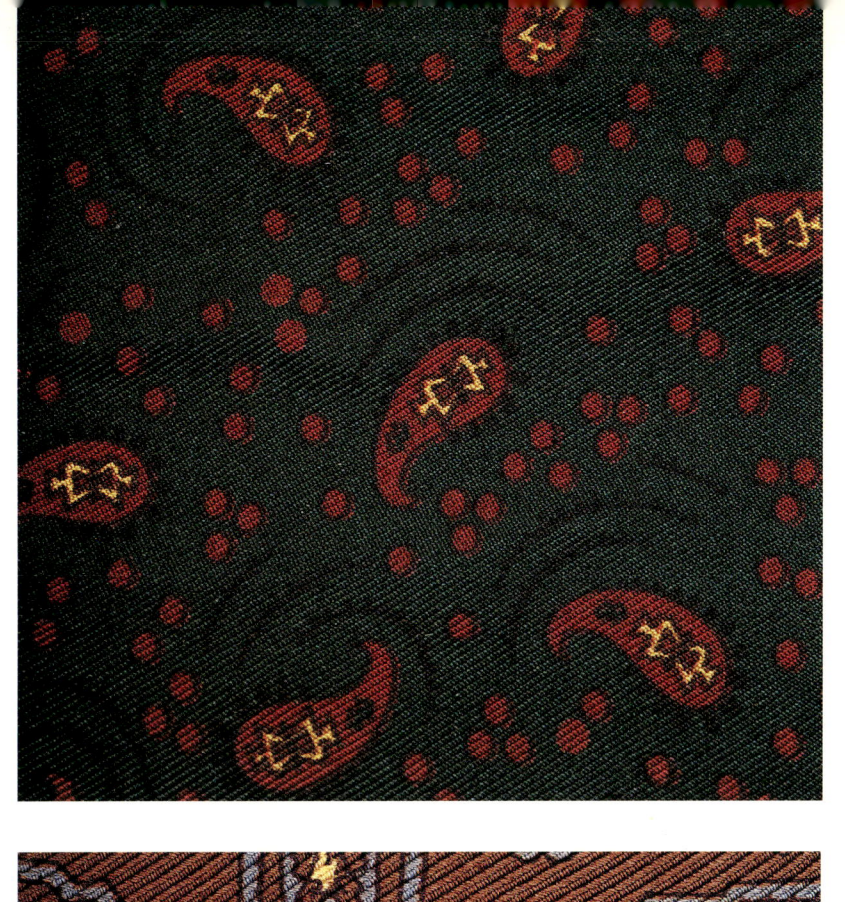
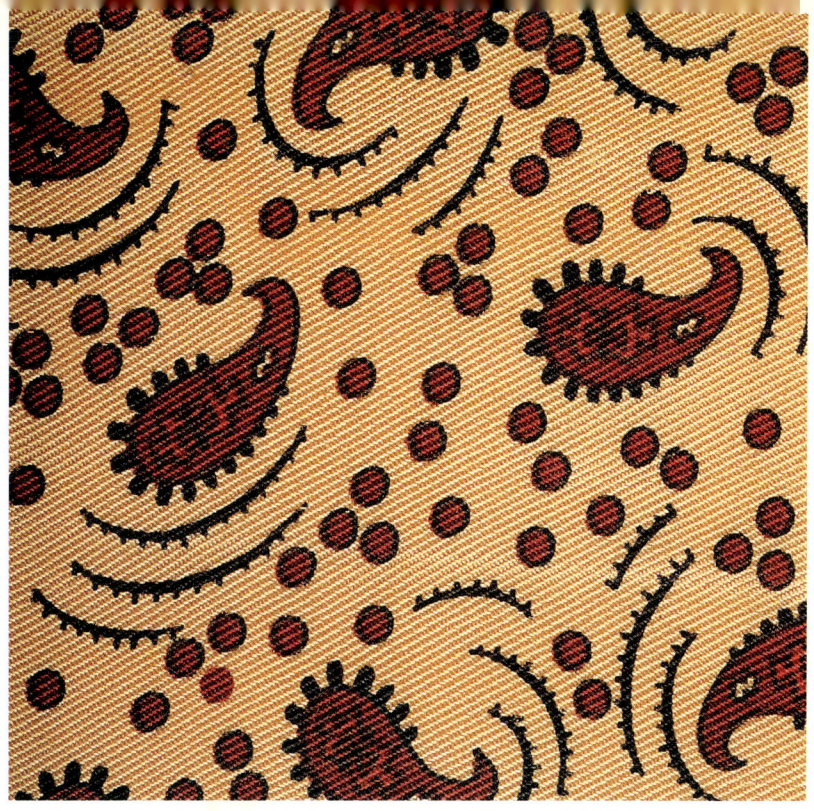
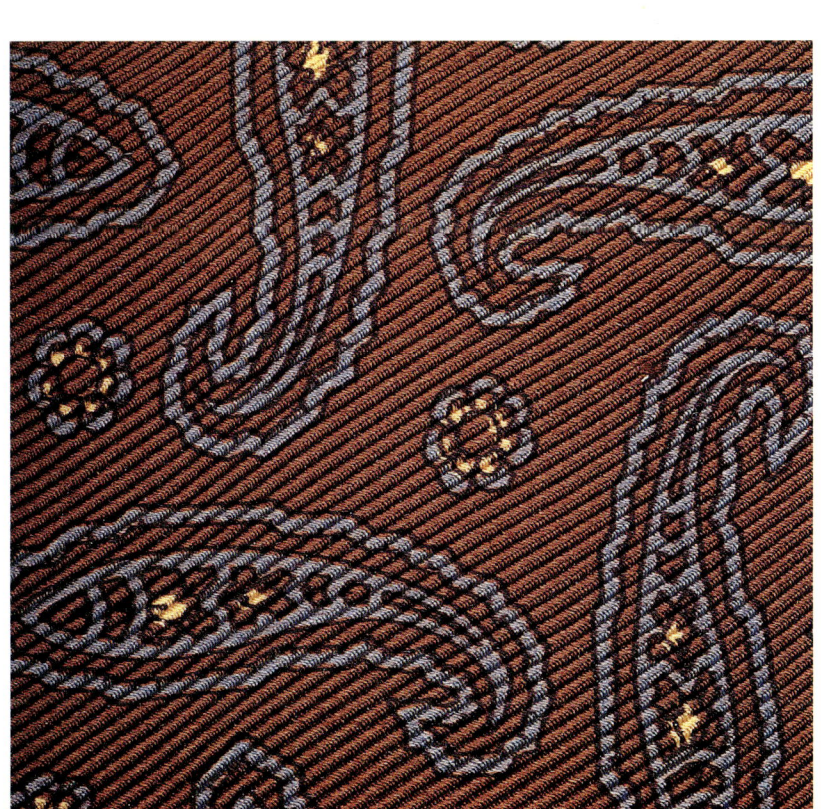
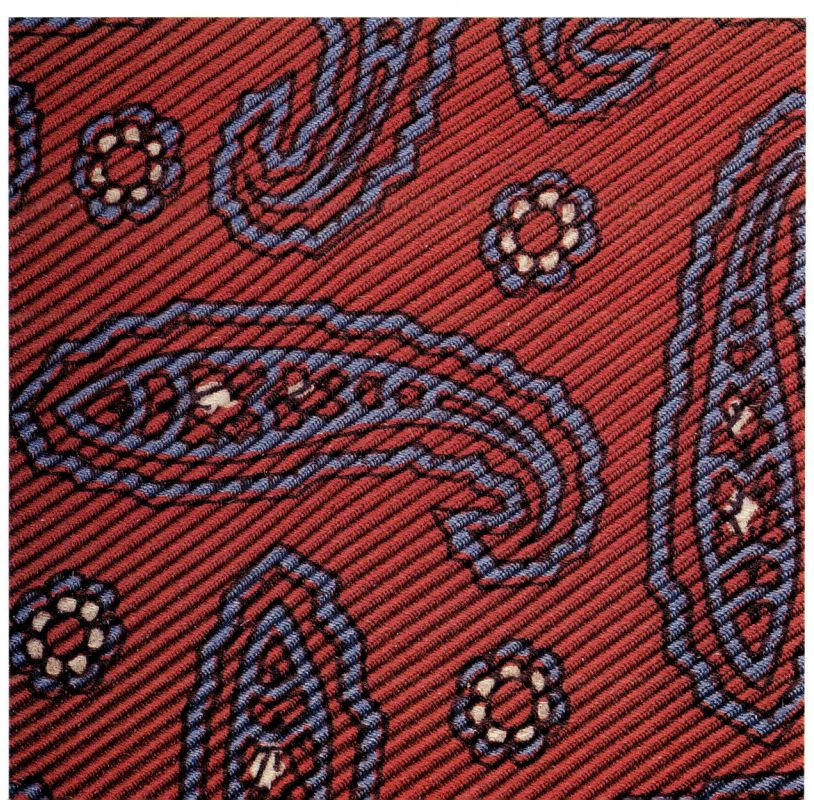

111

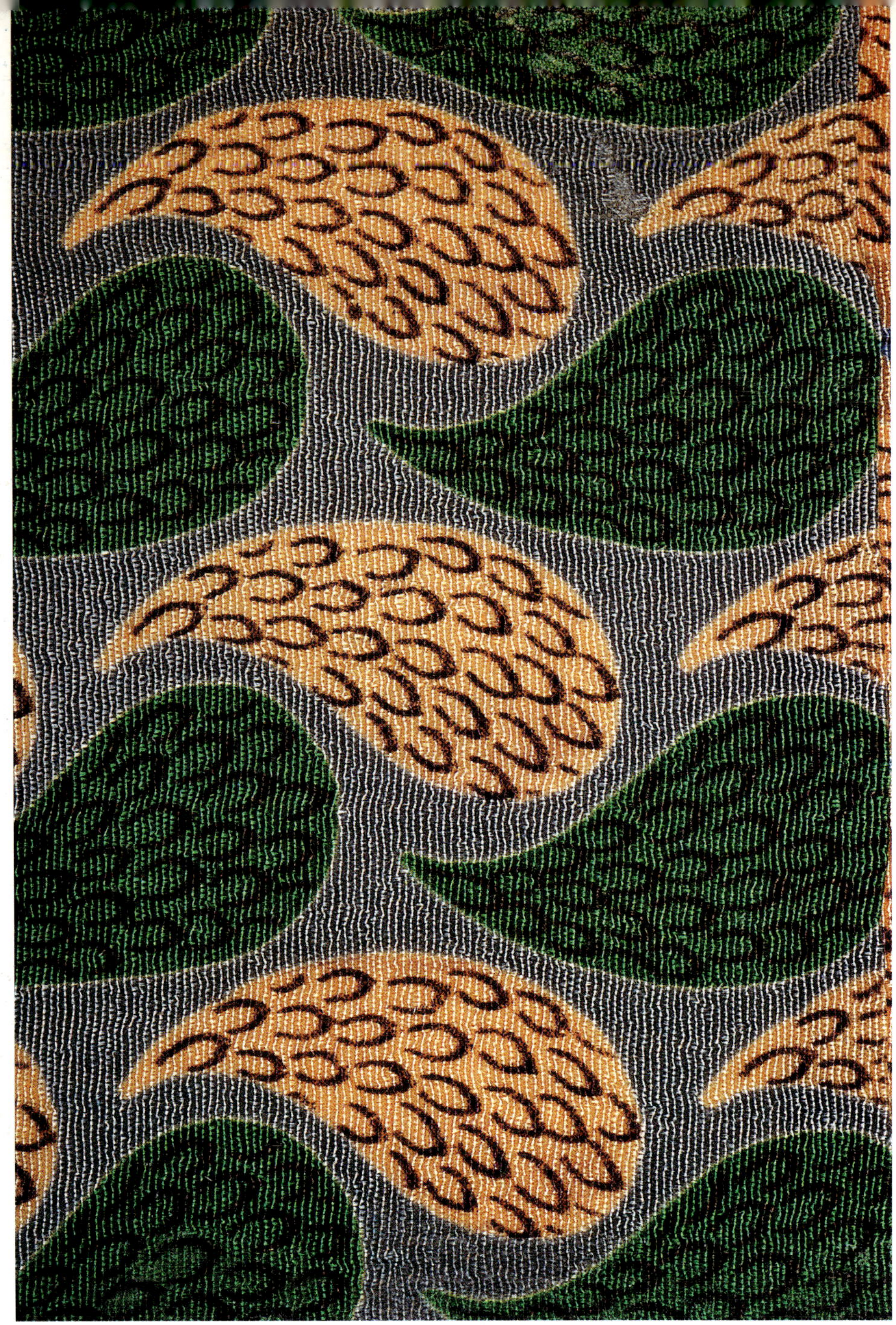